AF522096

ORGANIC CHEMISTRY

(AROMATIC, ALCOHOLS, ALDEHYDES & ACIDS)

ORGANIC CHEMISTRY

(AROMATIC, ALCOHOLS, ALDEHYDES & ACIDS)

By

Amit Arora

DISCOVERY PUBLISHING HOUSE

NEW DELHI-110002

First Published-2006
Reprint : 2012
ISBN 81-8356-189-6

Published by

DISCOVERY PUBLISHING HOUSE
4831/24, Ansari Road, Prahlad Street,
Darya Ganj, New Delhi-110002 (India)
Phone: 23279245 • Fax: 91-11-23253475
E-mail:dphtemp@indiatimes.com

Printed at:
Dynamic Printers

Preface

This book is written for B.Sc., B.Sc. (Hons.) and M.Sc. students of various universities of India. In this book my aim has been describe the fundamental principles of organic chemistry. Since I do not consider the chemistry of natural products to be fundamental chemistry but rather the application of fundamental principles. The subject matter described in this book covers much of the basic organic chemistry that is needed by a student who wish to study chemistry as a main subject at degree level. The arrangement of the subject-matter is based on homologous series and in general, descriptions of reactions are followed by discussion of their mechanisms and these includes an elementary account of the sort of evidence that led workers to suggest mechanisms that are acceptable at the present time.

We are sure you will appreciate the thought and care that has gone into making of this text. This book has everything you want to have in your advanced organic chemistry text.

The author wish to thank all those who helped in the preparation of the text and illustration, suggestion for improvement and criticism will be gratefully received and acknowledge.

Author

CONTENTS

1

DIAZONIUM SALTS AND THEIR RELATED COMPOUNDS

INTRODUCTION

When a primary aliphatic amine is treated with nitrous acid, the nature of the product depends on the amine used nitrogen is always evolved:

$$RNH_2 + HNO_2 \longrightarrow ROH + N_2 + H_2O.$$

When, however, a primary ·.omatic amine is treated with nitrous acid in a well-cooled solution the product is an unstable compound known as a diazonium salt:

$$ArNH_2 + HNO_2 + HCl \longrightarrow ArN_2{}^+Cl^- + H_2O.$$

Diazo-compounds may be obtained with primary aliphatic amines provided that the amino group is attached to a carbon atom which is adjacent to a – I group such as acyl, carbalkoxy, or cyano; *e.g.*, aminoacetic ester forms diazoacetic ester.

The formation of a diazonium compound by the interaction of sodium nitrite, an inorganic acid (usually) and a primary aromatic amine, in ice-cold solution, is known as *diazotization.* This reaction was discovered by Griess in 1858.

Diazonium salts have the structure $[Ar\overset{+}{N} \equiv N:]X^-$ (they are electrolytes in solution), and are salts of the strong base, diazonium hydroxide, $[Ar\overset{+}{N} \equiv N:]OH^-$, which has not yet been isolated, but is known in aqueous solution. Most diazonium salts of the inorganic acids are colourless solids, extremely soluble in water and many, particularly the nitrate, are explosive (in the solid state). They form complex salts with many metallic salts, of which one of the most important is zinc chloride, $(ArN_2)_2^{2+}\ ZnCl_4^{2-}$. These complex salts are stable in solution, and hence offer a means of stabilising a diazonium salt solution.

The stability of diazonium salts (as compared to RN_2^+) is attributed to resonance:

$$C_6H_5-N=\overset{+}{N}: \longleftrightarrow C_6H_5-\overset{+}{N}\equiv N: \longleftrightarrow \overset{+}{C_6H_5}=N=N^- \longleftrightarrow \cdots$$

The diazonium salts are very important synthetic reagents, being the starting-point in the preparation of various aromatic compounds, dyes and drugs. As pointed out previously, the preparation of a diazonium salt in the solid state is usually a hazardous process. Fortunately, however, the aqueous solutions of the diazonium salts undergo practically all the necessary reactions (cf. the Grignard reagents).

There are various methods of preparing a solution of a diazonium salt, but the usual procedure is to dissolve (or suspend) the amine in excess of dilute inorganic acid (usually hydrochloric or sulphuric) cooled in ice and to add slowly, with stirring, a cooled aqueous solution of sodium nitrite, the addition of which is completed when the reaction mixture produces a blue colour with potassium iodide-starch paper, thereby showing the presence of free nitrous acid.

The mechanism of diazotisation is still not settled. Several factors affect the rate law of diazotisation, one being the acidity of the solution. In weakly aqueous acid solutions (and also in more strongly acid solutions), the rate-determining step is nitrosation of the free amine (to a primary nitrosamine), and a mechanism consistent with the kinetic studies is:

$$HO-NO + H^+ \xrightleftharpoons{\text{fast}} H_2O^+-NO \xrightleftharpoons{\text{fast}} H_2O + NO^+$$

$$ArNH_2 + NO^+ \xrightarrow{\text{slow}} ArNH_2^+NO \xrightarrow[\text{fast}]{-H^+} ArNHNO$$

$$\xrightarrow[\text{(rearr.)}]{\text{fast}} Ar\ddot{N}=\ddot{N}-OH \xrightarrow[\text{fast}]{H^+} Ar-\ddot{N}=\ddot{N}-\overset{+}{O}H_2$$

$$\xrightarrow{\text{fast}} Ar-\overset{+}{N}\equiv N: + H_2O$$

In the presence of certain substituents, diazotisation may lead to anomalous results. Thus, diazonium salts derived from amino

(I): benzene ring bearing N_2^+ and, ortho to it, CO_2^-

(II): benzene ring bearing N_2^+ and, para to it, SO_3^-

carboxylic and aminosulphonic acids are dipolar ions, *e.g.*, diazotised anthranilic (I) an sulphanilic (II) acids.

Nomenclature

The name of the diazonium salt is obtained by adding diazonium chloride, diazonium sulphate etc., to the name of the parent hydrocarbon, *e.g.*, $C_6H_5N_2^+Cl^-$, benzenediazonium chloride p-$CH_3C_6H_4N_2^+HSO_4^-$ p-toluenediazonium sulphate.

Reactions of the Diazonium Salts

These may be divided into two groups; those which involve, the liberation of nitrogen gas and the displacement of the diazo-group, $—N_2^+$ by another univalent group, and those in which the two nitrogen atoms are retained.

REPLACEMENT REACTIONS

1. Replacement by hydroxyl. When a diazonium sulphate solution is boiled or steam distilled the diazo-group is replaced by hydroxyl, *e.g.*,

$$C_6H_5N_2^+HSO_4^- + H_2O \longrightarrow C_6H_5OH + N_2 + H_2SO_4 \text{ (g)}$$

This reaction is *nucleophilic* and occurs by the unimolecular mechanism involving an arene cation, and in the presence of halide ions (nucleophiles), aryl halides may also be formed.

$$Ar—N_2^+ \xrightarrow{\text{slow}} Ar—N_2 + Ar^+ \xrightarrow{\text{fast}} \begin{cases} \xrightarrow{H_2O} ArOH \\ \xrightarrow{X^-} ArX \end{cases}$$

This mechanism has much to support it, *e.g.*, in water, the decomposition of benzenediazonium chloride is a first-order reaction. There are, however, some experimental results which are in apparent conflict with this mechanism.

Electron-withdrawing groups, *e.g.*, CO_2H, in the p-position, retard the reaction. This is the anticipated result on the basis that heterolysis of the $Ar—N_2^+$ bond is the r/d step.

By similar argument, the presence of a p-electron-releasing group, *e.g.*, OMe, would be expected to accelerate the decomposition. In practice, the opposite result is often observed. This has been explained as follows.

$$\mathrm{Me\ddot{O}}-\mathrm{C_6H_4}-\overset{+}{\mathrm{N}}\equiv\mathrm{N} \longleftrightarrow \mathrm{Me}\overset{+}{\mathrm{O}}=\mathrm{C_6H_4}=\mathrm{N}-\overset{-}{\mathrm{N}} \longleftrightarrow \cdots$$

Resonance gives the Ar— N_2^+ bond some double bond character, with consequent increase in strength. Since the breaking of this bond is involved in the r/d step, the activation energy of this step is raised and so the reaction is retarded. This retardation is observed for groups which are electron-releasing by the + R effect (or by the hyperconjugative effect, *e.g.*, Me). On the other hand, electron-releasing groups in the *m*-position accelerate the reaction. In this case, the +R effect is absent; only the +I effect is operating.

It might be noted here that the arene cation cannot be stabilised by resonance, since a covalent pair of σ-electrons is lost, resulting in a vacant orbital perpendicular to the π-orbital (*i.e.*, π-orbitals) of the ring.

2. Replacement by hydrogen. Replacement of the diazo-group by hydrogen offers a means of preparing substituted aromatic compounds in which the substituents are in positions which they would not take up by direct substitution. According to Kornblum (1944, 1949), the most reliable method of replacing the diazo-group by hydrogen is by means of hypophosphorous acid. Also, this reaction is catalysed by cuprous salts.

A highly favoured mechanism for this reaction involves a free-radical chain:

Chain initiation

$$\mathrm{Ar\,N_2^+Cl^- + H_2P(O)PH \longrightarrow HCl + Ar{-}N = N{-}PH(O)OH}$$

$$\mathrm{\longrightarrow Ar + N_2 + PH(O)OH}$$

Chain propagation

$$\mathrm{Ar\cdot + H_2P(O)OH \longrightarrow ArH + \cdot PH(O)OH}$$

$$\mathrm{Ar\,N_2^+ + \cdot PH(O)OH \longrightarrow ArN_2\cdot + {}^+P.H(O)OH}$$

$$\downarrow \qquad\qquad \downarrow \mathrm{H_2O}$$

$$\mathrm{Ar\cdot + N_2 \qquad H_3PO_3 + H^+}$$

Another common method of replacing the diazo-group by hydrogen is to dissolve the amine in a mixture of ethanol and concentrated sulphuric acid, add sodium nitrite and then warm (sometimes in the presence of copper as catalyst). Two products are usually obtained, one being the phenolic ether and the other the replacement of the diazo group by hydrogen.

$$ArOEt \xleftarrow{EtOH} Ar\,N_2^+\,HSO_4^- \xrightarrow{EtOH} AtH + CH_3CHO$$

The relative amounts of these two products depend predominantly on the nature of the diazo-compound and of the alcohol; also the pH of the solution and whether water is present affect the proportions of the products. If a –I group is present in the nucleus, and ethanol is used, the replacement of the diazo-group by hydrogen predominates. On the other hand, if methanol is used, it predominant product is an aryl methyl ether, so much so that this offers a means of preparing this the of ether.

The mechanism of this reaction is uncertain, but there is evidence io suggest that ethers are formed via phenyl cations which react with the alcohol (cf. reaction I; replace H_2O by ROH). On the other hand, the deamination reaction is believed to proceed via a free-radical mechanism involving aryl free radicals.

Some examples of replacement by hydrogen are:

m-*Bromotoluene*

CH_3, NH_2, Br (arene) $\xrightarrow[H_2SO_4]{NaNO_2}$ CH_3, Br, $N_2^+HSO_4^-$ (arene) $\xrightarrow[Cu]{C_2H_5OH}$ CH_3, Br (arene) (54–59%)

m-*Nitrotoluene*

CH_3, NH_2, NO_2 (arene) $\xrightarrow[H_2SO_4]{NaNO_2}$ CH_3, NO_2, $N_2^+HSO_4^-$ (arene) $\xrightarrow[(warm)]{C_2H_5OH}$ CH_3, NO_2 (arene) (62–72%)

The diazo-group may be replaced by hydrogen in many other ways, *e.g.*, (i) decomposing the fluoroborate in ethanol in the presence of zinc dust (47-85%; Roe *et al.*, 1952); (ii) dissolving the hexafluorophosphate, $ArN_2^+\,PF_6^-$, in tetramethylurea (Rutherford *et al.*, 1963).

Apart from the factors described above that affect the nature of the deaminated product, there is also the fact that the diazo-group is a strong activator in nucleophilic substitution. This is illustrated by the following example, in which replacement by hydrogen occurs together with activated nucleophilic substitution:

$$\text{2,6-Br}_2\text{C}_6\text{H}_3\text{N}_2^+\text{Cl}^- \xrightarrow{\text{EtOH}} \text{1-Br-3-Cl-C}_6\text{H}_4 + N_2 + Br^-$$

3. Replacement by halogen. (i) *Sandmeyer reaction* (1884). When a diazonium salt solution is run into a solution of cuprous halide dissolved in the corresponding halogen acid, the diazo- group is replaced by a halogen atom, *e.g.*,

m-*Chloronitrobenzene*

$$m\text{-}NO_2C_6H_4NH_2 \xrightarrow[\text{HCl}]{\text{NaNO}_2} m\text{-}NO_2C_6H_4N_2^+Cl^- \xrightarrow{\text{CuCl/HCl}} m\text{-}NO_2C_6H_4Cl \quad (68\text{–}71\%)$$

p-*Bromotoluene*

$$p\text{-}CH_3C_6H_4NH_2 \xrightarrow[\text{H}_2\text{SO}_4]{\text{NaNO}_2} p\text{-}CH_3C_6H_4N_2^+HSO_4^- \xrightarrow{\text{CuBr/HBr}} p\text{-}CH_3C_6H_4Br \quad (70\text{–}73\%)$$

The diazotisation may be carried out with hydrobromic acid, but in practice sulphuric acid is used since it is cheaper and only slightly affects the vield of aryl bromide.

The mechanism of the reaction is uncertain. A possibility is' as follows. The 'cuprous chloride' is actually $CuCl_2^-$,(Cowdrey *et al.*, 1949).

$$CuCl + Cl^- \longrightarrow CuCl_2^-$$

$$ArN_2^+ + Cl\text{—}Cu\text{—}Cl^- \longrightarrow N_2 + Ar\cdot + Cl\text{—}Cu\text{—}Cl$$

$$\xrightarrow{\text{fast}} ArCl + CuCl$$

(ii) **Gattermann reaction (1890).** This reaction is carried out by warming the diazonium salt solution in the presence of copper powder, *e.g.*,

$$o\text{-}CH_3C_6H_4NH_2 \xrightarrow{\text{NaNO}_2/\text{HBr}} o\text{-}CH_3C_6H_4N_2^+Br^- \xrightarrow{\text{Cu}} o\text{-}CH_3C_6H_4Br \quad (43\text{–}47\%)$$

Iodo-compounds may be prepared by boiling the diazonium salt solution with aqueous potassium iodide:

$$C_6H_5N_2^+Cl^- + KI \longrightarrow C_6H_5I + KCl + N_2 \quad (74\text{–}76\%)$$

This is usually the best method for introducing iodine into the benzene ring.

The mechanism of this reaction is uncertain, but a possibility is that it occurs by a bimolecular mechanism, the displacement of the diazo-group being facilitated by the strong nucleophilic iodide ion. However, there is evidence that I_3^-, ions are also the attacking species. These are produced by oxidation of I^- ions to I_2, by the diazonium ion or nitrous acid, followed by

$$I_2 + I^- \rightleftharpoons I_3^-$$

Fluoro-compounds may be prepared by the **Schiemann reaction** (1927). When fluoroboric acid is added to a diazonium salt solution, the insoluble diazonium fluoroborate is precipitated. This, on gentle heating, gives the aryl fluoride:

$$PhN_2^+Cl^- + HBF_4 \longrightarrow PhN_2^+BF_4^-$$

$$\longrightarrow N_2 + Ph^+ + BF_4^-$$

$$\longrightarrow PhF + BF_3 \quad (51\text{–}57\%)$$

Fluoroboric acid may be prepared by adding hydrochloric acid to sodium fluoroborate, or by dissolving boric acid in 50 per cent hydrofluoric acid (contained in a lead vessel); the latter usually gives better yields. A better method is to prepare the diazonium hexafluorophosphate; this is less water-soluble than the fluoroborate:

$$ArN_2^+Cl^- + HPF_6 \longrightarrow HCl + ArN_2^+PF_6^-$$

$$\xrightarrow{\text{heat}} ArF + N_2 + PF_5$$

Suschitzky *et al.* (1969) have shown that when the arene contains electron-withdrawing substituents, *e.g.*, NO_2, CO_2H, decomposition of the arenediazonium hexafluoroantimonate gives the best yield of aryl fluoride.

$$ArN_2^+AsF_6^- \xrightarrow{\text{heat}} ArF + AsF_5 + N_2$$

4. Replacement by a cyano-group. This is a special case of the and meyer and Gattermann reactions, and is carried out by treating a diazonium salt solution with cuprous cyanide dissolved in aqueous potassium cyanide or with aqueous potassium cyanide in the presence of copper powder:

$$C_6H_5N_2^+Cl^- \xrightarrow{K_3Cu(CN)_4} C_6H_5CN + N_2 \quad (65\%)$$

$$C_6H_5N_2^+HSO_4^- \xrightarrow{Cu} C_6H_5CN + KHSO_4 + N_2 \ (60\%)$$

5. Replacement by a nitre-group. This may be carried out in several ways. A common method is to decompose the diazonium fluoroborate with aqueous sodium nitrite containing copper powder:

$$p\text{-}NO_2C_6H_4NH_2 \xrightarrow{NaNO_2/HBF_4} p\text{-}NO_2C_6H_4N_2^+BF_4^- \xrightarrow[Cu]{NaNO_2} p\text{-}C_6H_4(NO_2)_2 + NaBF_4 + N_2 \quad (67\text{–}82\%)$$

6. Replacement by an arsonk acid group, AsO_3H_2. One method is **Bart's reaction** (1910). This is carried out by decomposing a diazonium salt with sodium arsenite in the presence of a copper salt: *e.g.*, phenylarsonic acid:

$$C_6H_5N_2^+Cl^- + Na_3AsO_3 \xrightarrow{CuSO_4} C_6H_5AsO_3Na_2 + N_2 + NaCl$$

$$\xrightarrow{HCl} C_6H_5AsO_3H_2 \ (39\text{–}45\%)$$

The yield may be increased by buffering the solution with sodium carbonate (Bias, 1940).

7. Replacement by an aryl group. This may be carried out by treating a diazonium sulphate with ethanol and copper powder, *e.g.*, biphenyl from benzenediazonium sulphate:

$$2C_6H_5N_2^+HSO_4^- \xrightarrow[Cu]{C_2H_5OH} C_6H_5C_6H_5 + 2N_2 + 2H_2SO_4 \quad (25\%)$$

This is really a special case of the Gattermann reaction. Alternatively, a biaryl may be prepared by adding an aromatic compound to an *alkaline* solution of a diazonium salt, *e.g.*, *p*-bromo-biphenyl from *p*-bromobenzenediazonium chloride and benzene :

$$p\text{-}BrC_6H_4N_2^+Cl^- + C_6H_6 \xrightarrow{NaOH} p\text{-}BrC_6H_4\text{–}C_6H_5 + N_2 + NaCl + H_2O \quad (34\text{–}35\%)$$

This method of preparation is known as the **Gomberg reaction** (1924), and experiment has shown that whatever is the nature of a substituent in the second component, o- or p-substitution always occurs predominantly, but some w-product is also formed, *e.g.*, benzene-diazonium chloride forms p-nitrobiphenyl when treated with nitrobenzene:

$$C_6H_5N_2^+Cl^- + C_6H_5NO_2 \xrightarrow{NaOH} C_6H_5{-}C_6H_4NO_2$$

Thus, the reaction takes place by a free-radical mechanism.

$$C_6H_5N_2^+\ Cl^- \xrightarrow{OH^-} C_6H_5{-}N{=}N{-}OH \longrightarrow C_6H_5\cdot + N_2 + \cdot OH$$

$$C_6H_5\cdot + C_6H_5NO_2 + \cdot OH \longrightarrow p\text{-}C_6H_5{-}C_6H_4NO_2 + H_2O$$

Meerwein reaction (1939). This is the reaction between diazonium salts and compounds containing an activated ethylenic bond to give arylated products. The reaction is carried out in acetone solution in the presence of a cupric salt. *e.g.*, benzenediazonium chloride adds to vinyl cyanide with elimination of nitrogen. The mechanism is uncertain, but it is generally accepted to be homolytic, addition occurring at the β-position (this free radical is more stable than the one formed from α-addition).

$$C_6H_5\cdot + CH_2 = CHCN \longrightarrow C_6H_5{-}CH_2{-}\dot{C}HCN \xrightarrow{Cu^{2+}}$$

$$Cu^+ + C_6H_5{-}CH_2{-}\overset{+}{C}HCN \xrightarrow{Cl^-} C_6H_5CH_2CHClCN$$

When the alkene is an α, β-unsaturated acid, then arylation may occur with decarboxylation, *e.g.*, benzenediazonium chloride and cinnamic acid give stilbene:

$$C_6H_5N_2^+Cl^- + C_6H_5CH{=}CHCO_2H$$

$$\longrightarrow C_6H_5CH{=}CHC_6H_5 + N_2 + CO_2 + HCl$$

8. Replacement by a chloromercuri-group. This is an example of indirect mercuration, and may be carried out by heating the complex of a diazonium chloride with mercuric chloride in acetone or ethanol solution with copper powder:

$$ArN_2{\cdot}HgCl_3 + 2Cu \longrightarrow ArHgCl + N_2 + 2CuCl \quad (40\%)$$

The diazo-group may be replaced by many other groups, *e.g.*

(i) $ArN_2^+Cl^- \xrightarrow{CuSCN} ArSCN$

(ii) $ArN_2^+Cl \xrightarrow{KNCO/Cu} ArNCO$

(iii) $2\,ArN_2^+Cl \xrightarrow{(NH_4)_2S} Ar_2S$

(iv) $ArN_2^+HSO_4^- + SO_2 + Cu \longrightarrow ArSO_2H + N_2 + CuSO_4$

(v) $ArN_2^+Cl^- + S{=}C(SK)(OC_2H_5) \longrightarrow S{=}C(SAr)(OC_2H_5) \xrightarrow{H\cdot} ArSH$

An alternative route to the thiol is:

$$ArN_2^+Cl^- \xrightarrow{NaSH} ArSH$$

(vi) Beech (1954) has shown that formaldoxime reacts with diazonium salts to give products which, on hydrolysis, yield aldehydes. Other oximes, under similar conditions, give alkyl aryl ketones (R = H or R):

$$Ar\,N_2^+Cl^- + RCH{=}NOH \xrightarrow{NaOH} ArCR{=}NOH \xrightarrow{H^+} ArCOR$$

(vii) Cookson *et al.* (1962) have shown that the addition of acetic acid to diazonium fluoroborates in tetracarbonylnickel gives arenecarboxylic acids (25-75%).

In certain cases, diazotisation may lead to abnormal reactions; *e.g.*,

(i) Diazotisation of *o*-diamines produces triazoles.

(ii) Treatment of *m*-phenylenediamine with nitrous acid gives Bismarck Brown.

REACTIONS OF THE DIAZONIUM SALTS IN WHICH THE NITROGEN ATOMS ARE RETAINED

1. When bromine and hydrobromic acid are added to a diazonium salt solution, a crystalline precipitate of *diazonium perbromide*, $ArN_2^+Br_3^-$ is obtained. This, on heating, forms the aryl bromide. On the other hand, if the perbromide is treated with aqueous ammonia, the azide, ArN_3, is produced.

Phenyl azide may also be prepared by the action of hydrazoic acid on nitrosobenzene (Maffei el al., 1954):

$$C_6H_5NO + 2HN_3 \longrightarrow C_6H_5N_3 + H_2O + 2N_2 \qquad (92\%)$$

2. When reduced with stannous chloride and hydrochloric acid or with sodium sulphite, diazonium salts form phenylhydrazines. If vigorous reducing agents are used, *e.g.*, zinc and hydrochloric acid, the product is an aromatic amine:

$$ArN_2^+Cl^- \xrightarrow{Zn/HCl} [ArNHNH_2] \xrightarrow{Zn/HCl} ArNH_2 + NH_3$$

3. Diazonium salts readily undergo *coupling* reactions with phenols, naphthols, and aromatic amines, to form highly coloured azo-compounds, *e.g.*, benzenediazonium chloride couples with phenol in faintly alkaline solution to form p-hydroxyazobenzene, and with dimethylaniline in faintly acid solution to form p-dimethylaminoazobenzene.

$$C_6H_5{-}N{=}N{-}C_6H_4{-}OH \xleftarrow[OH^-]{C_6H_5OH} C_6H_5N_2^+Cl^- \xrightarrow[H^+]{C_6H_5NMe_2}$$

$$C_6H_5{-}N{=}N{-}C_6H_4{-}NMe_2$$

Coupling also occurs (in sodium acetate buffered solution) with compounds containing an active methylene group, *e.g.*, D.E.M,, E.A.A.; the product is usually the phenylhydrazone, formed by the tautomerisation of the azo-compound to the more thermodynamically stable hydrazone.

$$C_6H_5N_2^+ + {-}CH(CO_2Et)_2 \xrightarrow{AcONa} C_6H_5{-}N{=}N{-}CH(CO_2Et)_2$$

$$\rightleftharpoons C_6H_5NH{-}N{=}C(CO_2Et)_2$$

This coupling reaction is often used to detect the presence of enols, and is then referred to as the **Japp-Klingermann reaction** (1887).

A detailed study of the coupling reaction has shown that the mechanism is electrophilic substitution, the electrophile being the diazonium cation and the substrate being the free amine or the phenoxide ion:

$$\underset{(I)}{C_6H_5{-}\overset{+}{N}{\equiv}N:} \longleftrightarrow \underset{(II)}{C_6H_5{-}\ddot{N}{=}\ddot{N}^+} + C_6H_5{-}\ddot{N}(CH_3)_2 \longrightarrow$$

$$C_6H_5{-}\ddot{N}{=}\ddot{N}{-}C_6H_5(H){=}\overset{+}{N}(CH_3)_2 \xrightarrow{-H^+} C_6H_5{-}\ddot{N}{=}\ddot{N}{-}C_6H_4{-}\ddot{N}(CH_3)_2$$

The phenoxide ion also undergoes the resonance effect in a similar manner.

Since (II) is the 'reactive' contributing resonating structure, any factor that increases its contribution will increase the reactivity of that diazonium cation. It can be anticipated that electron-withdrawing groups would favour (II) and electron-releasing groups would favour (I). This has been found to be so in practice, *e.g.*, the -nitrobenzenediazonium cation is more than 10000 times as reactive as the p-methoxy ion under the same conditions (Conant *et al.*, 1930). The 2,4,6-trinitrobenzene-diazonium cation is so reactive that it couples with certain hydrocarbons, *e.g.*, mesitylene.

The mechanism described above is also in keeping with the following facts.

(i) Primary, secondary, and tertiary aromatic amines undergo coupling in weakly acid solutions but not in strongly acid solution; the rate of coupling decreases as the pH changes from 6 to 2. In the latter case, the substrate is the amine salt, and in this the ring is deactivated by the pH effect of the ammonium cation group. Since m-substitution never occurs, this position is not sufficiently active to react even with highly reactive diazonium salts. It is because of this salt formation that diazotisation of primary amines is carried out in strongly acid solution in order to avoid coupling.

(ii) Phenols couple very readily in weakly alkaline solution (the rate of coupling increases as the pH changes from 5 to 8), and this is consistent with the fact that the ring is far more activated in the phenoxide ion than in the neutral phenol molecule.

Coupling with benzene substrates occurs preferentially in the p-position to the hydroxyl or amino-group, but if this is blocked, then o-coupling occurs, *e.g.*, p-cresol gives the o-azo- compound:

$$ArN_2^+Cl^- + \text{Me—}C_6H_4\text{—OH} \xrightarrow{OH^-} \text{Me—}C_6H_3(\text{OH})\text{—N=NAr}$$

This is also exemplified by the fact that phenol forms bis-(2,4-) and tris-azo-compound (2,4,6-) with an excess of diazonium salts.

The presence of another activating group in the m-position of the phenol or amine accelerates coupling, *e.g.*, resorcinol forms the mono, bis-and tris-azo-compounds more readily than does phenol:

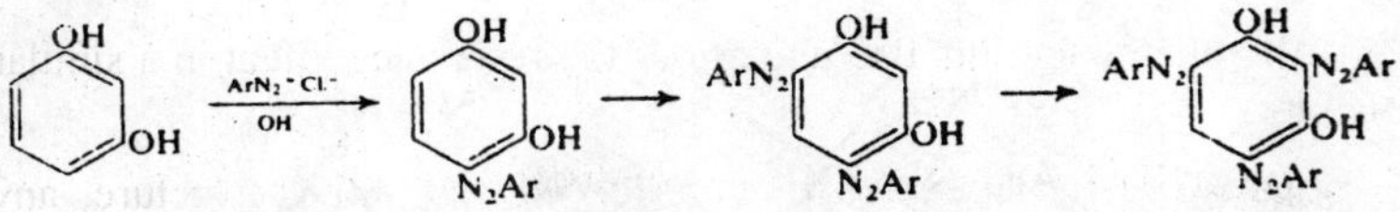

$$\text{resorcinol} \xrightarrow[\text{OH}]{ArN_2^+Cl^-} \text{4-}N_2Ar\text{-resorcinol} \longrightarrow \text{2,4-}(N_2Ar)_2 \longrightarrow \text{2,4,6-}(N_2Ar)_3$$

1- and 2-Naphthols in alkaline solution couple with diazonium salts respectively in the 4- and I-position. The formation of the azo-2-naphthol is used as a test for primary aromatic amine:

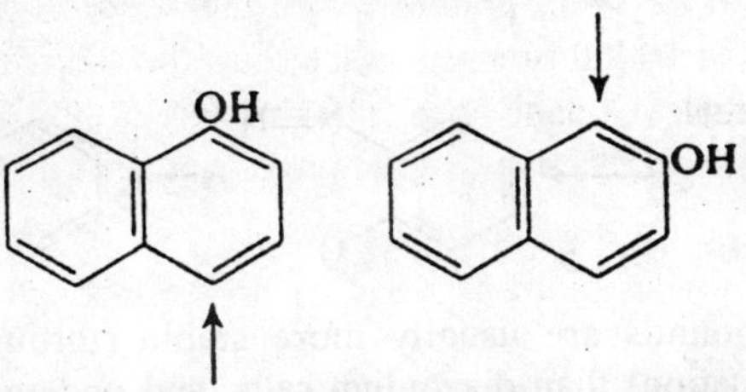

Diazo-compounds. Diazonium salts have the formula $[Ar—N≡N]^+X^-$, where X is Cl, HSO_4, NO_3, etc. These are the salts of diazonium hydroxide, $[Ar—N≡N]^+OH^-$, which has never been isolated. When a diazonium salt solution is made alkaline, the diazonium hydroxide liberated rapidly rearranges to the diazohydroxide, the salts of which are called the *diazoates* (*diazoates*). The mechanism of these changes may be as follows:

$$[Ar—N≡N:]+ Cl^- \xrightarrow{NaOH} [Ar—N≡N:]^+OH^-$$

$$Ar—\overset{+}{N}≡N: \longleftrightarrow Ar—\ddot{N} \overset{OH^-}{\rightleftharpoons} Ar—\ddot{N}=\ddot{N}—OH$$

$$\xrightarrow{NaOH} \ddot{N}=\ddot{N}—O\text{–}Na^+$$

These diazoates exist in two geometrical isomeric forms, the syn- (cis-) and anti- (irons-):

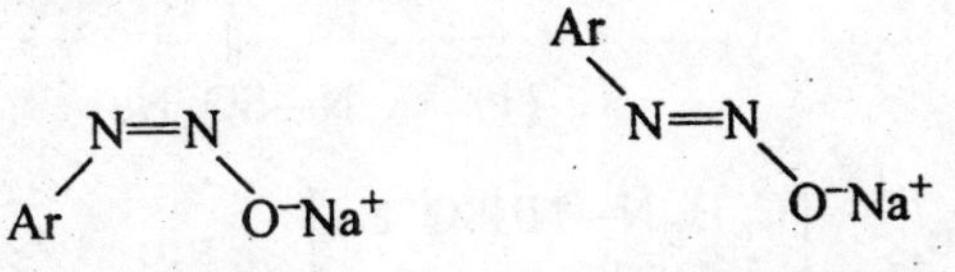

$$\begin{array}{cc} \text{N=N (Ar, }O^-Na^+\text{ on same side)} & \text{N=N (Ar, }O^-Na^+\text{ on opposite sides)} \end{array}$$

sodium n-diazoate
syn-form

sodium isodiazoate
anti-form

The syn-form, the n-diazoate, is produced in weakly alkaline solution and is unstable, slowly changing into the *anti*-form, the isodiazoate; this change is accelerated by making the solution strongly alkaline.

Other diazo-compounds that exhibit geometrical isomerism are the diazosulphonates (X = CN): and the *diazocyanides* (X = CN):

Syn-form anti-form

Diazo-oxides (diazo-phenols). When diazonium salts derived from o- and p-aminophenols are treated with alkali, the products are diazo-oxides, *e.g.*,

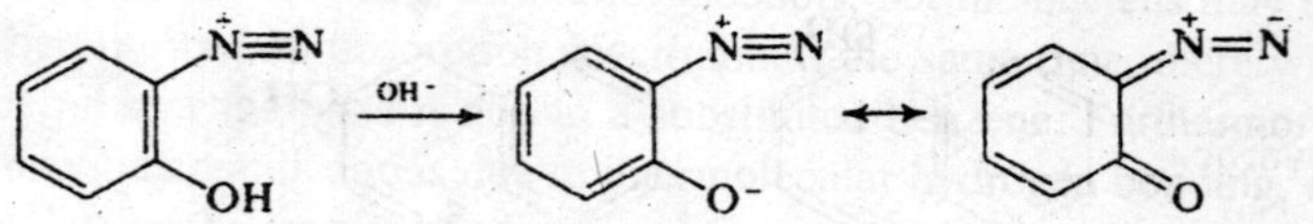

These compounds are usually more stable (through increased resonance stabilisation) than diazonium salts, and undergo the general reactions of diazonium salts.

Hydrazines. The most important substituted hydrazine is phenylhydrazine, b.p. 241° C. This may be prepared by the reduction of benzenediazonium chloride with stannous chloride and hydrochloric acid:

$$C_6H_5N_2^+Cl^- \xrightarrow{SnCl_2/HCl} C_6H_5NHNH_2{\cdot}HCl \ (90\%)$$

The reduction is carried out commercially with sodium sulphite.

The mechanism of the reduction may be as follows, sodium benzene-*anti*-diazosulphonate (I), and the sodium salt of phenylhydrazine sulphonate (II), being intermediate products:

$$[C_6H_5N{\equiv}N]^+Cl^- \xrightarrow{Na_2SO_3} C_6H_5{-}N{=}N{-}SO_3Na \quad (I)$$

$$\xrightarrow{H_2SO_4} \left[\begin{array}{c} C_6H_5N{-}NHSO_3Na \\ | \\ SO_3H \end{array}\right] \xrightarrow{H_2O}$$

$$\underset{(II)}{C_6H_5NHNHSO_3Na} + H_2SO_4 \xrightarrow{HCl} C_6H_5NHNH_2{\cdot}HCl$$

$$\xrightarrow{NaOH} C_6H_5NHNH_2 \quad (80\text{–}84\%)$$

Phenylhydrazine is readily oxidised when exposed to the air. It is strongly basic and forms well-defined salts, *e.g.*, phenylhydrazine hydrochloride, $C_6H_5NHN\overset{+}{H_3}Cl^-$; these salts are relatively stable to atmospheric oxidation. Phenylhydrazine is a powerful reducing agent, reducing

Fehling's solution in the cold:

$$C_6H_5NHNH_2 + 2CuO \longrightarrow Cu_2O + H_2O + C_6H_5N{=}NH$$

$$\longrightarrow C_6H_6 + N_2$$

When reduced with vigorous reducing agents, *e.g.*, zinc and hydrochloric acid, phenyl hydrazine gives aniline and ammonia:

$$C_6H_5NHNH_2 + 2CuO \xrightarrow{Zn/HCl} C_6H_5NH_2 + NH_3$$

The most important reaction of phenylhydrazine is with carbonyl compounds to *phenyl-hydrazones*, hydrazones, and with simple sugars to form *osazones*. These derivatives are usually well- defined crystalline solids, and so are useful for the identification of carbonyl compounds. Many phenylhydrazones are decomposed by strong inorganic acids to reform the carbonyl compound. When reduced with zinc and acetic acid, phenylhydrazones form primary amines:

$$C_6H_5NHN{=}CHR \xrightarrow{e;\,H^+} C_6H_5NH_2 + RCH_2NH_2$$

Various derivatives of phenylhydrazine are used instead of phenylhydrazine, since these produce compounds which crystallise more readily, *e.g.*, p-nitro-, 2,4-dinitro- and p-bromo- phenylhydrazine. Of these, 2,4-dinitrophenylhydrazine is most commonly used, and this is prepared by the action of hydrazine sulphate and potassium acetate on l-chloro-2,4- dinitrobenzene;

$$\text{(Cl, }NO_2\text{, }NO_2\text{-benzene)} + NH_2NH_2{\cdot}H_2SO_4 \xrightarrow{CH_3CO_2K} \text{(}NHNH_2\text{, }NO_2\text{, }NO_2\text{-benzene)} \quad (80\%)$$

It is a red crystalline solid, m.p. 194°C.

DIAZOAMINO- AND AMINOAZO-COMPOUNDS

Diazo-compounds have the structure Ar—N=N—Y, where Y is a group that is not attached to the nitrogen atom by carbon. Azo-compounds, on the other hand, have the structure Ar—N=N—Ar, each aryl group being attached to a nitrogen atom by carbon.

Coupling between diazonium salts and primary or secondary amines to form azo-compound has already been discussed. Under different conditions, however, particularly the pH of the solution, coupling occurs at the nitrogen atom to form diazoamino-compounds.

Diazoaminobenzene may be prepared by treating aniline hydrochloride solution with sodium nitrite just sufficient to diazotise *half* of the aniline, and then adding sodium acetate:

$$C_6H_5N_2^+Cl^- + C_6H_5NH_2 + CH_3CO_2Na \longrightarrow$$

$$C_6H_5N{=}NNHC_6H_5 + NaCl + CH_3CO_2H \quad (82\text{–}85\%)$$

Diazoaminobenzene is often formed during the diazotisation of aniline, especially when the solution is only weakly acid. It exists in two forms, one as golden-yellow prisms, m.p. 98°, and the other, yellow prisms, m.p. 80°C. It is feebly basic and does not form stable salts. When boiled with dilute sulphuric acid, diazoaminobenzene liberates nitrogen forming phenol are aniline:

$$C_6H_5NHN{=}NC_6H_5 + H_2SO_4 \longrightarrow C_6H_5NH_2 + C_6H_5N_2^+HSO_4^-$$

$$\xrightarrow{H_2O} C_6H_5OH + N_2 + H_2SO_4$$

When boiled with concentrated hydrobromic acid, diazoaminobenzene forms bromobenzene and aniline (the former being produced via benzenediazonium bromide), and when treated with sodium nitrite and hydrochloric acid, it forms benzenediazonium chloride.

A very important property of diazoaminobenzene is its tendency to rearrange to aminoazo-benzene.

Tautomerism of the diazoamino compounds. Diazoamino-compounds are tautomeric, and this has been demonstrated in diazoaminobenzene itself by the use of this compound labelled with ^{15}N as shown in (1).

$$\underset{\text{(I)}}{PhN{=}\dot{N}HPh} \rightleftharpoons \underset{\text{(II)}}{PhNHN{=}\dot{N}Ph}$$

When (I) was boiled with concentrated hydrochloric acid, aniline, chlorobenzene, and nitrogen were produced (Shemyakin *et al.*, 1957). If the structure were (1), then all the labelled nitrogen would be in the aniline (cf. the reaction above with HBr). In actual fact, the isotopic nitrogen was found to be distributed between the aniline and the nitrogen liberated. This is consistent with diazoaminobenzene being a tautomeric compound of forms (I) and (II).

This tautomeric system is known as the diazoamino triad system, the three nitrogen system, or the triazene system. Since such a system has never been separated into the two forms (as is the case when a mobile hydrogen atom is attached to a carbon atom). Hunter (1945) called this type of tautomerism mesohydric tautomerism.

p-Aminoazobenzene (orange-yellow solid, m.p. 126°C) is produced by the diazoamino-aminoazo rearrangement of diazoaminobenzene when this is warmed with hydrochloric acid.

$$C_6H_5N{=}N{-}NH{-}C_6H_5 \longrightarrow C_6H_5N{=}N{-}C_6H_4{-}NH_2 \quad (80\%)$$

If the p-position in the anilino-group is occupied, then the isomeric change occurs more slowly, the azo-group taking up the o-position to the amino-group:

$$C_6H_5N{=}N{-}NH{-}C_6H_4{-}CH_3 \longrightarrow C_6H_5N{=}N{-}C_6H_3(NH_2)(CH_3)$$

The rearrangement is better effected by aniline hydrochloride than by hydrochloric acid. This is in keeping with the following mechanism, the aniline hydrochloride serving both as a proton donor and as a source of aniline:

$$C_6H_5{-}N(H){-}N{=}N{-}C_6H_5 + C_6H_5\overset{+}{N}H_3 \rightleftharpoons$$

$$\left[C_6H_5{-}NH_2{-}\ddot{N}{=}N{-}C_6H_5\right]^+ + C_6H_5NH_2$$

$$\Downarrow$$

$$C_6H_5NH_2 + [C_6H_5{-}N{\equiv}N]^+ \longrightarrow C_6H_5N{-}N{=}N{-}C_6H_4{-}NH_2$$

Aminoazobenzene forms two series of salts with acids. When one equivalent of hydrochloric acid is used, the salt formed is yellow and unstable, and probably has the structure (I) (since this has the same type of spectrum as azobenzene).

$$C_6H_5N{=}NC_6H_4\overset{+}{N}H_3Cl^- \qquad C_6H_5NH{-}N{=}C_6H_4{=}\overset{+}{N}H_2Cl^-$$

(I) II

When, however, a large excess of hydrochloric acid is used, the salt produced is dark violet and stable, and its structure is probably (II), *i.e.*, the salt of the quinone-iminohydrazone; the spectrum of this salt is different from that of azobenzene.

On vigorous reduction, aminoazobenzene is converted into aniline and P-phenylenediamine:

$$C_6H_5{-}N{=}N{-}C_6H_4{-}NH_2 \xrightarrow{Sn/HCl} C_6H_5NH_2 + NH_2C_6H_4NH_2$$

If titanous chloride is used as the reducing agent, the reaction is quantitative, and so may be used for the volumetric determination of aminoazo-dyes.

Aminoazobenzene is oxidised by manganese dioxide and sulphuric acid to p-benzoquinone.

Secondary amines behave similarly to aniline in their reaction towards diazonium salts; *e.g.*, N-methylaniline couples with benzenediazonium chloride to form methyldiazoaminobenzene:

$$C_6H_5N_2^+Cl + C_6H_5NHCH_3 \rightarrow C_6H_5N{=}N{-}(CH_3)C_6H_5 + HCl$$

At the same time, however, C-azo-coupling takes place, some methylaminoazobenzene being formed:

$$C_6H_5N_2^+Cl^- + C_6H_5NHCH_3 \longrightarrow C_6H_5N{=}NC_6H_4NHCH_3 + HCl$$

This is also formed by the rearrangement of methyldiazoamino-benzene.

With tertiary amines, the formation of a diazoamino-compound is impossible; the amino compound is always formed by direct coupling in the p-position *e.g.*, dimethylaminoazobenzene from dimethylaniline and benzenediazonium chloride:

$$C_6H_5N_2^+Cl^- + C_6H_5N(CH_3)_2 \longrightarrow C_6H_5N{=}N{-}C_6H_4{-}N(CH_3)_2 + HCl$$

Reduction of azo-compounds with sodium hyposulphite or stannous chloride and hydrochloric acid offers a relatively simple method of preparing diamines (or aminophenols) in a pure state: *e.g.*, 2.5-diamino-toluene from m-toluidine:

$$C_6H_5N_2^+Cl^- + CH_3C_6H_4NH_2 \xrightarrow{NaOH} C_6H_5{-}N{=}N{-}C_6H_3(CH_3){-}NH_2 \xrightarrow{Na_2S_2O_4} C_6H_5NH_2 + NH_2C_6H_3(CH_3)NH_2$$

It should be noted that the azo-compound. 4-amino-2 methyl-azobenzene, is formed by direct coupling with the nucleus. Generally, the diazoamino-compound (N-coupling) is formed by reaction between diazonium salts and primary aromatic amines; m-toluidine and the naphthyl-amines always form the C-azo-compound.

Azoxybenzene may be prepared by reducing nitrobenzene with various reducing agents, *e.g.*, methanolic sodium hydroxide, alkaline sodium arsenite, or an alkaline solution of glucose. It is generally accepted that azoxybenzene is formed by interaction of the intermediates nitroso-benzene and phenylhydroxylamine.

$$C_6H_5NO + C_6H_5NHOH \xrightarrow{-H_2O} C_6H_5N{=}N{+}(O^-)C_6H_5$$

On the other hand, oxidation of suitable substrates with peroxyacetic acid gives azoxybenzene, *e.g.*, azobenzene. This method is useful for preparing unsymmetrical azoxybenzenes from unsymmetrical azobenzenes.

Azoxybenzene (yellow crystalline solid, m.p. 36°C), when warmed with iron filings, is reduced to azobenzene. When reduced with ammonium sulphide, azoxybenzene gives hydrazobenzene, and with metal and acid it gives aniline.

Azoxybenzene exists in two forms which are geometrical isomers: the *trans*-isomer is ordinary azoxybenzene.

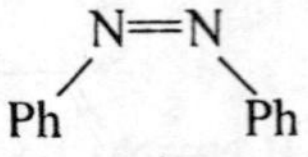

N=N (Ph, Ph) — cis. m.p. 71.4° C

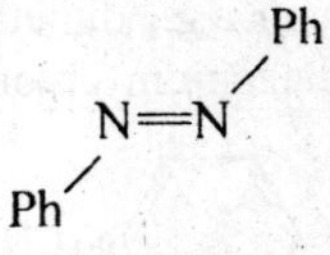

Ph–N=N–Ph — trans, m.p. 68°C

Wallach rearrangement (1880). When warmed with concentrated sulphuric acid, azoxy-benzenes rearrange to give predominantly p-hydroxyazobenzenes and a small amount of the o-isomer. The mechanism is still not settled.

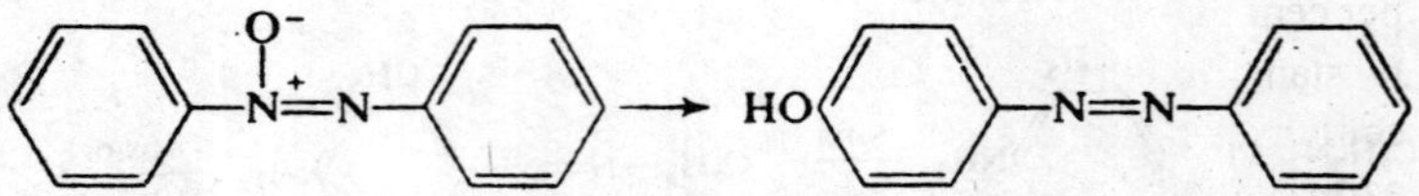

$$C_6H_5-\overset{O^-}{\overset{|}{N^+}}=N-C_6H_5 \longrightarrow HO-C_6H_4-N=N-C_6H_5$$

If, however, the rearrangement is effected by exposure to ultraviolet light, the product is; exclusively the o-isomer (Badger *et al.*, 1954). These authors have proposed an intramolecular mechanism for this o-rearrangement:

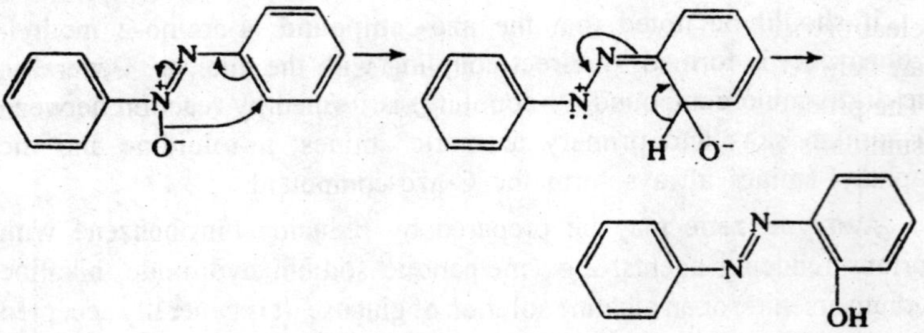

Azobenzene may be conveniently prepared by reducing nitrobenzene with iron filings, sodium amalgam, alkaline sodium stannite, or best with zinc dust and methanolic sodium hydroxide:

$$2C_6H_5NO_2 \xrightarrow{Zn/NaOH} C_6H_5N = NC_6H_5 + 4H_2O \ (84\text{–}86\%)$$

Lithium aluminium hydride also reduces nitrobenzene to azobenzene. There is good evidence to suggest that these reductions occur via the intermediate formation of azoxybenzene.

General methods of preparing azo-compounds are the oxidation of hydrazo-compounds with, *e.g.*, mercuric oxide, and by the condensation between nitroso-compounds and anilines.

$$Ar^1NO + Ar^2NH_2 \longrightarrow Ar^1N = NAr^2 + H_2O$$

Azobenzene (orange-red plates) exists are as geometrical isomers, the *trans*-form being 'ordinary' azobenzene:

$$\underset{\text{cis. m.p. } 71.4^\circ C}{Ph-N=N-Ph} \qquad \underset{\text{trans, m.p. } 68^\circ C}{Ph-N=N-Ph}$$

Azobenzene is reduced by zinc dust and aqueous sodium hydroxide, or by sodium borohydride in water or methanol in the presence of 10 per cent palladised charcoal (Wood *et al.*, 1962), to hydrazobenzene, and by stannous chloride or titanous chloride in acid solution, and by alkaline sodium hyposulphite, to aniline. The most outstanding property of the azo-compounds is their colour, and this is used to great advantage in the manufacture of azo-dyes.

Hydrazobenzene (s-*diphenylhydrazine*) may be prepared by reducing nitrobenzene or azo-benzene with zinc dust and aqueous sodium hydroxide. It is also formed when azoxybenzene is reduced electrolytically. All of these reductions starting with oxygenated substrates are believed to proceed via the formation of azobenzene as an intermediate. The products by reduction of nitrobenzene under various conditions are summarised in the chart.

$$\xrightarrow[\text{NaOH}]{\text{MeOH}} [PhNO + PhNHOH] \longrightarrow PhN{=}\overset{+}{N}(O^-)Ph$$

$$\xrightarrow[\text{or LAH}]{\text{Zn/MeOH/NaOH}} PhN{=}NPh \quad (PhN{=}\overset{+}{N}(O^-)Ph \xrightarrow{\text{Fe}} PhN{=}NPh;\ PhN{=}NPh \xrightarrow{\text{MeCOO}_2\text{H}} PhN{=}\overset{+}{N}(O^-)Ph)$$

$$PhN{=}NPh \xrightarrow{\text{Zn/NaOH}} PhNHNHPh; \quad PhNHNHPh \xrightarrow{\text{NaIO}_3} PhN{=}NPh$$

$$\xrightarrow{\text{Zn/NaOH}} PhNHNHPh$$

$$\xrightarrow[\text{HCl}]{\text{SnCl}_2} 2\,PhNH_2$$

Hydrazobenzene, m.p. 126°C, is slowly oxidised by atmospheric oxygen to azobenzene; this oxidation is rapid with sodium hypobromite or periodate. Hydrazobenzene is reduced by stannous chloride and hydrochloric acid to aniline, and when heated, forms azobenzene and aniline.

Benzidine rearrangement. This term refers to the acid-catalysed conversions of hydrazoarenes into *diaminobiaryls* and *aminodiaryamines*. These rearrangements are usually carried out in aqueous or ethanolic solutions of hydrochloric or sulphuric acid. The types of products that may be produced are illustrated in the chart:

$H_2N-C_6H_4-C_6H_4-NH_2$ (benzidine)

hydrazobenzene $C_6H_5-NH-NH-C_6H_5 \xrightarrow{HCl}$ diphenyline; o-semidine; p-semidine; o-benzidine

In addition to these rearrangements, products formed by *disproportionation* may also be obtained:

$$2\,C_6H_5-NH-NH-C_6H_5 \longrightarrow C_6H_5NH_2 + H_2N-C_6H_4-C_6H_4-NH_2 + C_6H_5-N{=}N-C_6H_5$$

The actual amount of each product depends on the structure of the hydrazoarene under consideration, *e.g.*, hydrazobenzene gives benzidine (~70%), diphenyline (~30%), and other minor products; 2,2′-hydrazonaphthalene gives almost I (X) per cent 2, 2′-diamino-l,1′-binaphthyl (o-benzidine type).

When the p-position of one nucleus is occupied, the main product of the rearrangement depends the nature of the substituent, *e.g.*, for OMe, OEt (strongly + R groups), an o-semidine is the main product together with a small amount of the p-semidine.:

$$MeO-C_6H_4-NHNH-C_6H_5 \xrightarrow{HCl} (NH_2)(MeO)C_6H_3-NH-C_6H_5 + MeO-C_6H_4-NH-C_6H_4-NH_2$$

When the p-substituent is halogen (fairly strong –I groups), both an diphenyline and an o-semidine are formed:

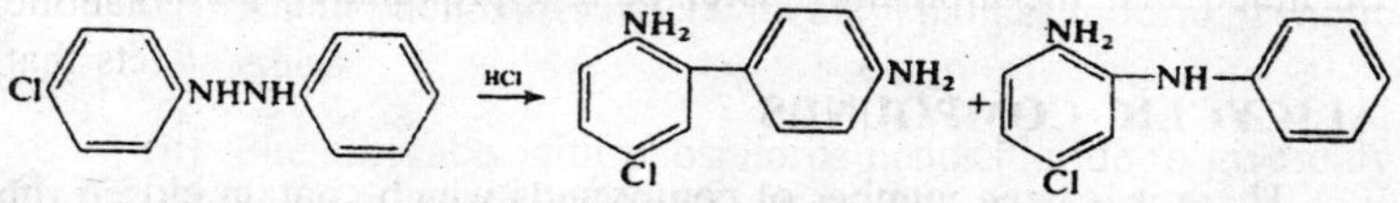

Cl–C₆H₄–NHNH–C₆H₅ $\xrightarrow{HCl}$ diphenyline (NH₂, Cl / NH₂) + o-semidine (NH₂, NH–C₆H₅, Cl)

When both p-positions are occupied, the main product is an o-semidine:

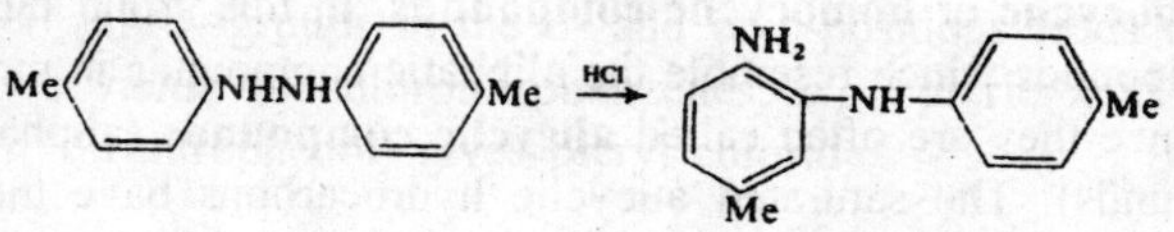

Me–C₆H₄–NHNH–C₆H₄–Me $\xrightarrow{HCl}$ o-semidine (NH₂, Me; NH–C₆H₄–Me)

Some groups, particularly carboxyl and sulpho, may be eliminated during the rearrangement, *e.g.* hydrazobenzene-4-carboxylic acid gives benzidine in very high yield.

The mechanism of the benzidine rearrangement is still not settled. Many mechanisms have been proposed, but common to all is that the rearrangement is intramolecular. Hammond *et al.* (1950) showed that for hydrazobenzene itself, the rearrangement has third-order kinetics; first-order in hydrazobenzene and second-order in hydrogen ion, *i.e.*,

$$\text{rate} = k[\text{PhNHNHPh}][\text{H}^+]^2$$

This led to the suggestion that the rearrangement proceeds via a diprotonated species. Ingold et at. (1962, 1964), however, have shown that there is a whole range of orders (1 to 2) in hydrogen ion, *e.g.*, the rearrangement of l.l'-hydrazonaphthalene has second-order kinetic' first-order in both substrate and hydrogen ion.

This has led to the proposal that two independent mechanisms can operate in the benzidine rearrangement, a *one-proton* and a *two-proton* mechanism. Only the latter is described here: this applies to hydrazobenzene.

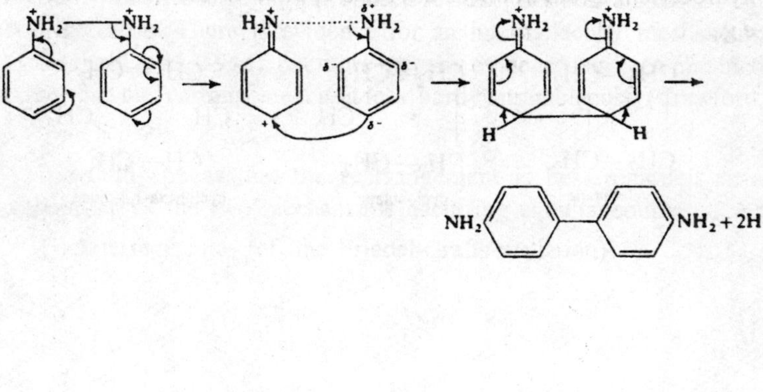

$\overset{+}{N}H_2$–$\overset{+}{N}H_2$ → $H_2\overset{+}{N}$·········$\overset{\delta+}{N}H_2$ (δ+, δ−) → $\overset{+}{N}H_2$, $\overset{+}{N}H_2$ (H, H) →

NH_2–C₆H₄–C₆H₄–$NH_2 + 2H^+$

This mechanism leads to C—C *linking*, and can give rise to the 4,4′-product (benzidine) and/or to the 2,4′-product (diphenyline). It should be noted that the aromatic nuclei lie in approximately parallel planes.

ALICYCLIC COMPOUNDS

There is a large number of compounds which contain closed rings comprised of carbon atoms only. These compounds are known collectively as **carbocyclic** or **homocyclic compounds.** In this group falls a class of compounds which resemble the aliphatic compounds in many ways, and hence they are often called **alicyclic compounds** (aliphatic *cyclic* compounds). The saturated alicyclic hydrocarbons have the general formula C_nH_{2n} (the same as that of the alkenes); they do not contain a double bond but possess a ring structure.

When the molecular formula of a *saturated* hydrocarbon corresponds to the general formula C_nH_{2n-2}, then the compound contains two rings; if to C_nH_{2n-4}, three rings, etc.

Nomenclature

Since the saturated alicyclic hydrocarbons contain a number of methylene groups joined together to form a ring, they are known as the **polymethylenes,** the number of carbon atoms in the ring being indicated by a Greek or Latin prefix, *e.g.*,

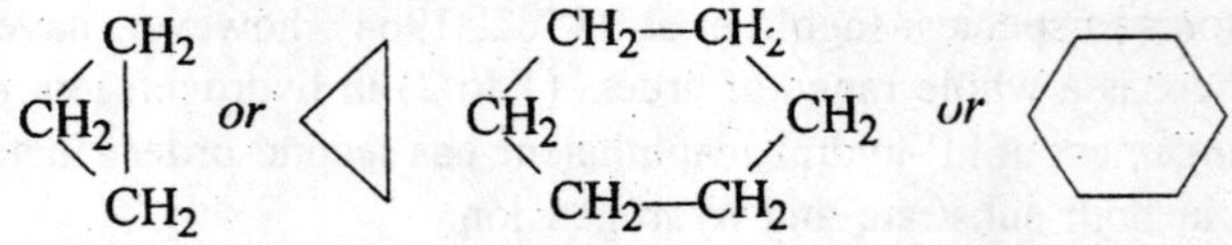

CH_2
CH_2 (ring) *or* (triangle)
CH_2

CH_2—CH_2
CH_2 CH_2 *or* (hexagon)
CH_2—CH_2

According to the I.U.P.A.C. system, the saturated monocyclic hydrocarbons take the names of the corresponding open-chain saturated hydrocarbons, preceded by the prefix cyclo- and they are known collectively as the **cycloparaffins** or **cycloalkanes,** and if the alicyclic hydrocarbon is unsaturated, the rules applied to the alkenes are used, *e.g.*,

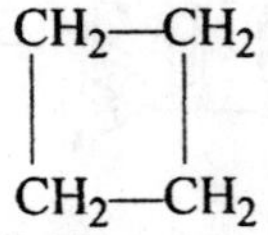

CH_2—CH_2
| |
CH_2—CH_2
cyclobutane

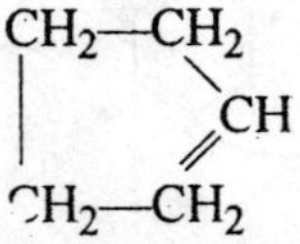

CH_2—CH_2
| CH
CH_2—CH_2
cyclobutane

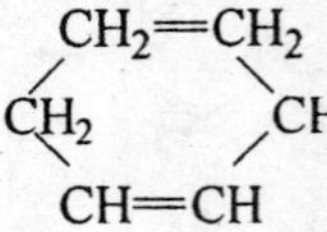

CH_2=CH_2
CH_2 CH
CH=CH
cyclohexa-1,4-diene

In addition to the simple monocyclic compounds, there are more complicated compounds with bridges linked across the ring, *e.g.*,

```
        CH                    CH—CH2
      / | \                  /  |   |
  H2C   |   CH2           CH2   |  CH2
   |   CH2 |                 \  |   |
  H2C   |   CH2               CH—CH2
      \ | /
        CH                bicyclo[3,1,0]hexane

bicyclo[2,2,1]heptane
```

According to the I.U.P.A.C. system, cycloalkanes consisting of two rings only and having two or more atoms in common, take the prefix *bicycle* followed by the name of the alkane *containing the same total number of carbon atoms.* The number of carbon atoms in each of the three bridges connecting the two tertiary carbon atoms is indicated in brackets in descending order. Numbering begins with one of the bridgeheads and proceeds by the longest possible path to the second bridgehead; numbering is then continued from this atom by the longer unnumbered path back to the first bridgehead and is completed by the shortest path, *e.g.*,

```
               CH3
          7     |1      2
         CH2 —— C ———— CHC2H5
          |     |8       |
          |   CHCH3    3CH2
         6|     |5       |4
        CHCl —— CH ———— CH2
```

6-chloro-2-ethyl-1,8-dimethylbicyclo[3,2,1]octane

N.B. A bridged system is considered to have a number of rings equal to the number of scissions required to convert the system into an acyclic compound.

CYCLOALKANES AND CYCLOALKENES

Five- and six-membered cycloalkanes occur in petroleum (the *naphthenes*); three-, four- and five-membered rings occur in terpenes, the most important class of alicyclic compounds. Many cyclic acids also occur in petroleum; these are known as the *naphthenic acids* and are mainly cyclopentane derivatives.

Some cyclopentene derivatives of the fatty acids occur naturally, and are important in medicine.

General Methods of Preparation of Alicyclic Compounds

It is interesting to note that up to about 1880, organic chemists thought that there were only two classes of compounds, aliphatic and aromatic; *e.g.*, V. Meyer (1876) believed that rings smaller than six carbon atoms were never likely to be obtained. Since 1882, however, many methods have been introduced to prepare various-sized rings. The following methods are typical.

1. When α,ω-dihalogen derivatives of the alkanes are treated with sodium or zinc, the corresponding cycloalkane is formed (Freund, 1882), *e.g.*, 1,3-dibromopropane forms cyclopropane.

$$CH_2(CH_2Br)_2 + Zn \longrightarrow \text{cyclo-}(CH_2)_3 + ZnBr_2$$

α,ω-Dihalogen derivatives of the alkanes in which the two halogen atoms are further apart than the 1,6 positions, do not form ring compounds but undergo the Wurtz reaction, *e.g.*, $Br(CH_2)_nBr$ ($n > 6$) gives, with sodium, $Br(CH_2)_{2n}Br$, $Br(CH_2)_{3n}Br$, etc.

2. When the calcium or barium salt of a dicarboxylic acid is distilled, a cyclic ketone is formed (Wislicenus, 1893), *e.g.*, barium adipate gives cyclopentanone. The mechanism is uncertain, but a possibility is

$$CH_2(CH_2CH_2CO\text{–}O)_2Ba \longrightarrow \text{cyclic intermediate: } CH\text{–}CH_2\text{–}CH_2\text{–}CH_2\text{–}C(OH)(O\text{–})\text{, with } CH\text{–}CO\text{–}O\text{–}Ba\text{–} \xrightarrow{-BaO}$$

$$\text{CH(CO}_2\text{H)–CH}_2\text{–CH}_2\text{–CH}_2\text{–CO (ring)} \xrightarrow{-BaO} \text{cyclo-}(CH_2)_4CO$$

This method is useful for the preparation of only five-, six-, and seven-ring ketones.

Cyclic ketones may readily be converted into the corresponding cycloalkanes by means of the Clemmensen reduction, *e.g.*,

$$\text{cyclohexanone} \xrightarrow[\text{HCl}]{\text{Zn/Hg}} \text{cyclohexane} + H_2O\ (g.)$$

Alternatively, the conversion may be effected by either of the following two methods.

$$\text{cyclopentanone} \xrightarrow[\text{EtOH}]{\text{Na}} \text{cyclopentanol (–OH)} \xrightarrow[0^\circ C]{\text{HI}} \text{iodocyclopentane (–I)} \xrightarrow[\text{(Zn/HCl)}]{e;\ H^+} \text{cyclopentane}$$

$$\text{cyclohexanone} \xrightarrow[\text{EtOH}]{\text{Na}} \text{cyclohexanol (–OH)} \xrightarrow[(-H_2O)]{H_3O^+} \text{cyclohexene} \xrightarrow[\text{Ni}]{H_2} \text{cyclohexane}$$

3. Six-membered alicyclic compounds may very conveniently be prepared by the reduction of benzene and its derivatives. Catalytic reduction under pressure using nickel is the most satisfactory, *e.g.*, phenol is almost quantitatively converted into cyclohexanol.

$$C_6H_5OH + 3H_2 \xrightarrow[200^\circ C]{\text{Ni}} C_6H_{11}OH$$

4. Various alicyclic compounds may be prepared by the condensation between certain dihalogen derivatives of the alkanes and sodium malonic ester or sodium acetoacetic ester (Perkin junior, 1883), *e.g.*,

(i) α, ω-Dibromides, $Br(CH_2)_nBr(n = 2 - 5)$, condense with *one* molecule of malonic ester in the presence of *two* molecules of sodium ethoxide to form cycloalkane- 1,1 -dicarboxylic esters. These may be converted into the monocarboxylic acid by the usual procedure used in malonic ester syntheses.

$$\begin{matrix} CH_2Br \\ | \\ CH_2Br \end{matrix} \xrightarrow{Na^{+-}CH(CO_2C_2H_5)_2} \begin{matrix} CH_2CH(CO_2C_2H_5)_2 \\ | \\ CH_2Br \end{matrix} \xrightarrow{\text{EtONa}}$$

$$\left[\begin{matrix} CH_2\bar{C}(CO_2C_2H_5)_2 \\ | \\ CH_2 — Br \end{matrix}\right] \longrightarrow \begin{matrix} CH_2 \\ | \quad \diagdown \\ | \quad C(CO_2C_2H_2)_2 \\ | \quad \diagup \\ CH_2 \end{matrix} \xrightarrow[\text{(ii) HCl}]{\text{(i) KOH}}$$

$$\begin{matrix} CH_2 \\ | \quad \diagdown \\ | \quad C(CO_2H)_2 \\ | \quad \diagup \\ CH_2 \end{matrix} \xrightarrow{\text{heat}} \begin{matrix} CH_2 \\ | \quad \diagdown \\ | \quad CHCO_2H + CO_2 \\ | \quad \diagup \\ CH_2 \end{matrix}$$

(ii) The above α, ω-dibromides also condense with two molecules of sodium malonic ester, *e.g.*, butane-1,2,4-tetracarboxylic ester from ethylene dibromide.

$$\begin{array}{l} CH_2Br \\ | \\ CH_2Br \end{array} + 2CHNa(CO_2C_2H_5)_2 \longrightarrow \begin{array}{l} CH_2CH(CO_2C_2H_5)_2 \\ | \\ CH_2CH(CO_2C_2H_5)_2 \end{array} + 2NaBr$$

On treatment with excess of sodium ethoxide, this tetracarboxylic ester forms the disodium derivative which, when treated with iodine, is converted into a cyclobutane derivative. If methylene di-iodide is used instead of iodine, the cyclopentane derivative is obtained.

$$\begin{array}{l} CH_2CH(CO_2C_2H_5)_2 \\ | \\ CH_2CH(CO_2C_2H_5)_2 \end{array} \xrightarrow{2NaOC_2H_5} \begin{array}{l} CH_2\bar{C}(CO_2C_2H_5)_2 \\ | \\ CH_2\bar{C}H(CO_2C_2H_5)_2 \end{array} \xrightarrow{I_2}$$

$$\begin{array}{l} CH_2C(CO_2C_2H_5)_2 \\ |\quad\ | \\ CH_2C(CO_2C_2H_5)_2 \end{array} \xrightarrow[\text{(ii) HCl}]{\text{(i) KOH}} \begin{array}{l} CH_2-C(CO_2H)_2 \\ |\qquad\ | \\ CH_2-C(CO_2H)_2 \end{array} \xrightarrow{heat}$$

$$\begin{array}{l} CH_2-CHCO_2H \\ |\qquad\ | \\ CH_2-CHCO_2H \end{array} + 2CO_2$$

Thus, by using the appropriate dihalogen derivatives of the alkanes under suitable conditions, it is possible to prepare rings containing three to seven carbon atoms (the yield being highest for the five-and lowest for the seven-membered ring).

Acetoacetic ester may also be used to prepare ring compounds, *e.g.*,

$$CH_2\begin{array}{l} \diagup CH_2-CH_2Br \\ \diagdown CH_2-CH_2Br \end{array} + CH_2\begin{array}{l} \diagup COCH_3 \\ \diagdown COC_2C_2H_5 \end{array} \xrightarrow[\text{(2 steps)}]{2NaOC_2H_5}$$

$$CH_2\begin{array}{l} \diagup CH_2-CH_2 \diagdown \\ \diagdown CH_2-CH_2 \diagup \end{array} C \begin{array}{l} \diagup COCH_3 \\ \diagdown CO_2C_2H_2 \end{array}$$

1-acetyl-cyclohexane-1-carboxylic ester

Hydrolysis and decarboxylation of this compound produce cyclohexyl methyl ketone.

By using acetoacetic ester, it is possible to prepare rings containing three, five six and seven of a dihydropyran derivative.

$$(CH_2)_3Br_2 + Na^+\,CH(CO_2C_2H_5)(COCH_3) \longrightarrow \text{[}H_2C(CH_3)\text{–}CCO_2C_2H_5{=}CCH_3\text{–}O^-\text{, }H_2C\text{–}H_2C\text{–}Br\text{]} \longrightarrow \text{dihydropyran: } CH_2\text{, }H_2C\text{, }H_2C\text{, }CCO_2C_2H_5{=}CCH_3\text{, }O$$

5. Certain cyclic ketones can be obtained by the **Dieckmann reaction** (1901). This reaction, which is really an intramolecular Clasien condensation, is carried out by treating the esters of adipic, pimelic or suberic acids with sodium, or better, with sodium ethoxide (Mayer *et al.*, 1959), whereupon five-, six -or seven membered rings, respectively, are obtained, *e.g.*, adipic ester forms cyclopentanone.

$$\begin{array}{l} CH_2CH_2CO_2Et \\ | \\ CH_2CH_2CO_2Et \end{array} \xrightarrow{Na} \begin{array}{l} CH_2\bar{C}HCO_2Et \\ | \\ CH_2CH_2COEt \\ \quad\quad\quad\; \| \\ \quad\quad\quad\; O \end{array} \longrightarrow$$

$$\begin{array}{l} CH_2\text{—}CH\text{—}CO_2Et \\ |\qquad\quad C\text{—}O^- \\ CH_2\text{—}CH\quad OEt \end{array} \xrightarrow{-OEt^-} \begin{array}{l} CH_2\text{—}CHCO_2Et \\ |\qquad\quad C{=}O \\ CH_2\text{—}CH_2 \end{array}$$

$$\xrightarrow{OEt^-} \begin{array}{l} CH_2\text{—}CCO_2Et \\ |\qquad\quad C\text{—}O^- \\ CH_2\text{—}CH_2 \end{array} \xrightarrow{H^+} \begin{array}{l} CH_2\text{—}CHCO_2Et \\ |\qquad\quad CO \\ CH_2\text{—}CH_2 \end{array}$$

$$\xrightarrow[-CO_2]{heat} \begin{array}{l} CH_2\text{—}CH_2 \\ |\qquad\quad CO \quad (74\text{–}86\%) \\ CH_2\text{—}CH_2 \end{array}$$

Esters lower than adipic ester may form products by *intermolecular* condensation and cyclisation *e.g.*, in the presence of sodium or sodium ethoxide, ethyl succinate forms *succino-succinic ester* (cyclohexane-2,5-dione- 1,4-dicarboxylic ester).

$CO_2C_2H_5$
HC⁻
H_2C
O=C
OEt
OC_2H_2
C=O
CH_2
⁻CH
$CO_2C_2H_2$

$\xrightarrow{-2EtO^-}$

$CO_2C_2H_5$
CH
H_2C CO
OC CH_2
CH
$CH_2C_2H_5$
(80%)

On the other hand, five membered ring compounds may be prepared by the intermolecular condensation between oxalic and glutaric esters.

O
C_2H_5O—C
C_2H_5O—C
O
⁻$CHCO_2C_2H_5$
CH_2
⁻$CHCO_2C_2H_5$

$\xrightarrow{-2EtO^-}$

CO—$CHCO_2C_2H_5$
CH_2
CO—$CHCO_2C_2H_5$
(80%)

Certain ketones may also be used with ethyl oxalate to give five-membered rings, *e.g.*, acetone gives cyclopentane1,2,3-trione.

6. Various types of bifunctional compounds undergo intramolecular reactions under suitable conditions to form cyclic compounds, particularly when the product is a five- or six-membered ring, *e.g.*,

(i) CH_2 (CH_2COMe, CH_2COMe) $\xrightarrow[\text{(ii) } H_3O^+]{\text{(i) Mg/Hg}}$ CH_2 (CH_2—C(OH)Me, CH_2—C(OH)Me)

CH_2COMe
CH_2
CH_2COMe

CH_2—C(OH)Me
CH_2
CH_2—C(OH)Me

(ii) $(CH_2)_3$ (COMe, CH_2Br) $\xrightarrow{\text{Mg}}$ [$(CH_2)_3$ (COMe, CH_2MgBr)] $\longrightarrow$

COMe
$(CH_2)_3$
CH_2Br

[COMe
$(CH_2)_3$
CH_2MgBr]

CH_2—CH_2 Me
C
CH_2—CH_2 OH

(iii) 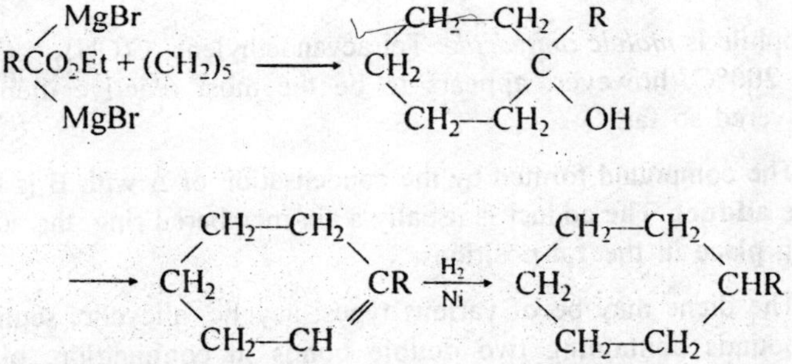 $RCO_2Et + (CH_2)_5(MgBr)_2 \longrightarrow$ cyclic tertiary alcohol: ring CH_2—CH_2—CH_2—CH_2—CH_2—C(R)(OH)

$\longrightarrow$ cyclohexene CH_2—CH_2—CH_2—CH_2—CH=CR $\xrightarrow[Ni]{H_2}$ cyclohexane CH_2—CH_2—CH_2—CH_2—CH_2—CHR

7. Diels-Alder reaction or **diene synthesis** (1928). This is the most important type of the class of reactions known as *cyclo-additions*. These are addition reactions in which ring systems are formed without elimination of any compounds. In general terms, the Diels-Alder reaction may be written.

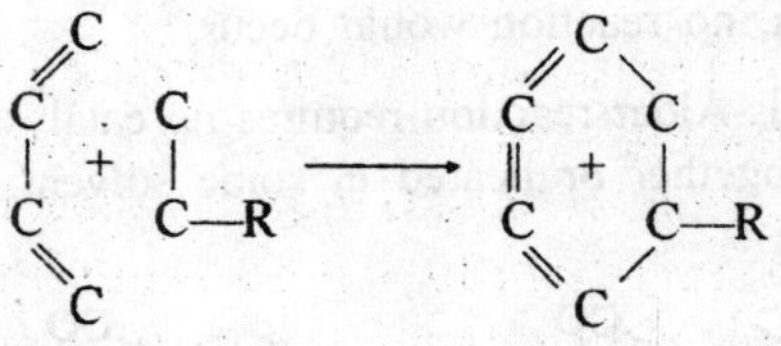

C=C—C=C + C=C—R $\longrightarrow$ six-membered ring C—C=C—C—C(—R)—C

and

C=C—C=C + C≡C—R $\longrightarrow$ six-membered ring C—C=C—C—C(—R)=C

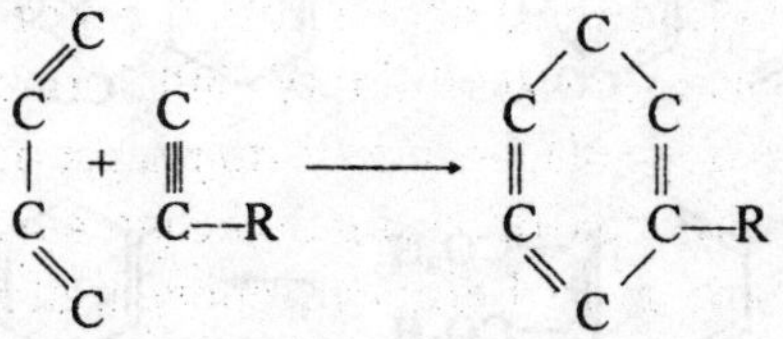

A B

Compound A is usually referred to as the **diene** (whether it be a conjugated diene, polyene, enyne or diyne), and compound B is usually referred to as the **dienophile.** R is usually a group which contains a carbonyl group attached to one of the ethylenic or acetylenic carbon atoms, *e.g.*, the dienophile is usually an α, β-unsaturated carbonyl compound, *e.g.*, α,β-unsaturated acids, acid anhydrides, esters, aldehydes or ketones; the dienophile may also be a quinone. The presence of a carbonyl group (a – I group) in the dienophile is not essential: compounds which contain other – I groups such as nitro- or cyano-, can also behave as dienophiles. In certain cases, the dienophile may even be an unsaturated hydrocarbon. Nevertheless, the diene synthesis takes place most readily when the dienophile contains a carbonyl group, and the most useful dienophile is *maleic anhydride.* Tetracyanoethylene, $C(CN)_2 = C(CN)_2$ m.p. 200°C, however, appears to be the most reactive dienophile discovered so far.

The compound formed by the condensation of A with B is known as the **adduct.** The adduct is usually a six-membered ring, the addition taking place in the 1,4-positions.

The diene may be of various types: acyclic, alicyclic, semicyclic compounds containing two double bonds in conjugation, bicyclic compounds, aromatic hydrocarbons containing at least three *linear* benzene rings *(e.g.,* anthracene), and certain heterocyclic compounds *(eg.,* furan. It is important to note that only a cisoid and *not* a transoid diene enters into the Diels-Alder reaction, *e.g.,* in example (v), the conformation given is cisoid; if this could be 'frozen' in the transoid conformation, no reaction would occur.

The Diels-Alder reaction requires no catalyst; the two compounds are heated together or heated in some solvent, *e.g.,* benzene. Some examples are :

(i) + CO, O, CO → CO, O, CO

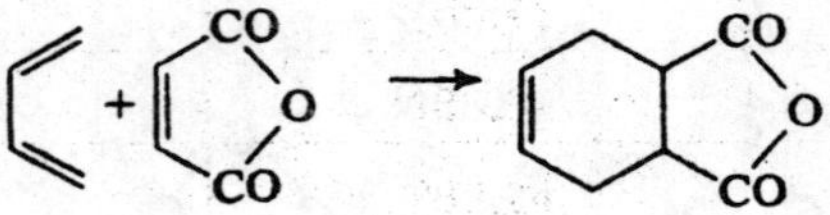

(ii) + C—CO_2H, C—CO_2H → CO_2H, CO_2H

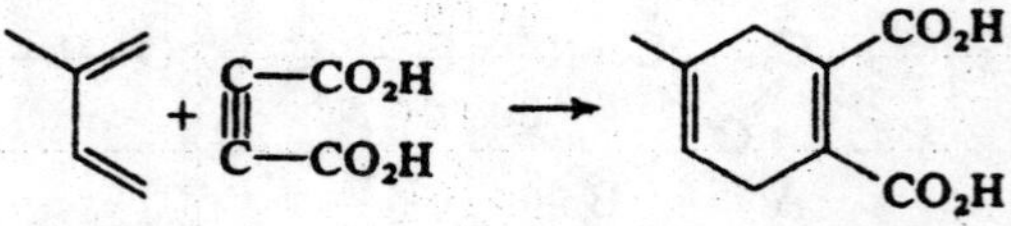

(iii) + CHO → CHO

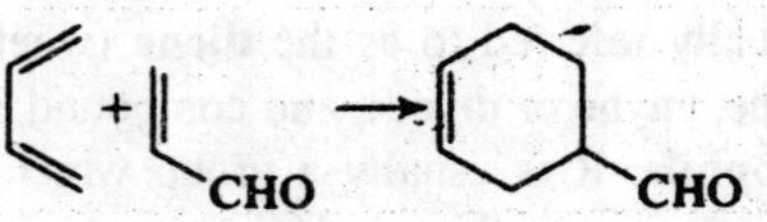

(iv) + CO, O, CO → CO, O, CO

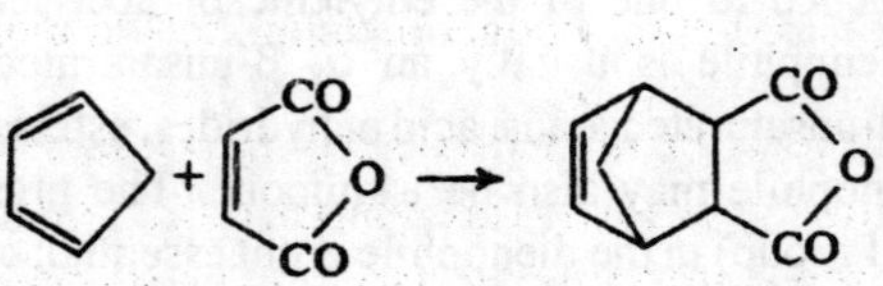

(v) + CO, O, CO → CO, CO—O

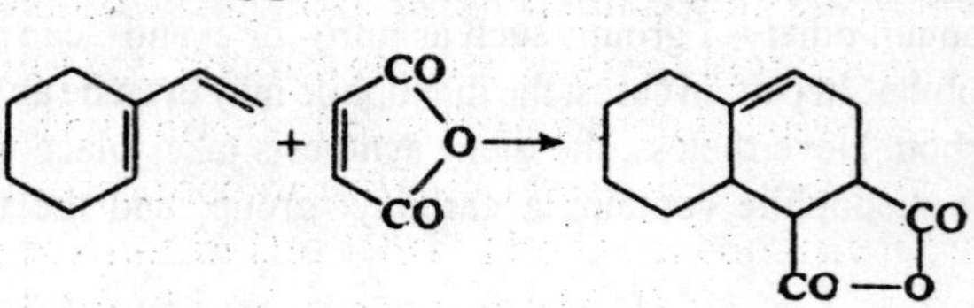

The reason why only polynuclear aromatics containing at least three linear benzene rings undergo the Diels-Alder reaction is uncertain. One possible explanation is as follows. The reaction is exothermic (ΔH = 50.2–75.3 kJ mol^{-1}), and in formation of the adduct, resonance energy of the diene is lost. If ΔH – R.E. (loss) is still positive, the reaction is thermodynamically favourable, and so may proceed. In the examples described here (benzene, naphthalene, and anthracene), the R.E. of the adduct is given by 150.6 nkJ mol^{-1}; where *n* is the number of benzene rings remaining intact (the R.E. of benzene is 150.6 kJ mol^{-1}), $R.E_{obs}$ is the experimental value of each aromatic compound.

R.E. (obs.) kJ mol^{-1}	150.6	255.2	321.5
R.E. (adduct)	0	150.6	301.3
R.E. loss	150.6	104.6	50.2

Only for anthracene is ΔH – R.E. (loss) positive.

In a number of cases, addition in the Diels- Alder reaction is 1,2, to give a *four-membered ring,* and this is particularly the case when tetrafluorethylene is the dienophile, *e.g.*,

(i) CH_2=CHMe + CF_2=CF_2 —225°C→ (Me, F_2, F_2 cyclobutane) (ii) butadiene + CF_2=CF_2 —125°C→ (F_2, F_2 cyclobutane)–CH=CH_2

(iii) cyclopentadiene + CF_2=CF_2 —200°C→ (F_2, F_2 bicyclic adduct) + (F_2, F_2 fused cyclobutane)

Stereochemistry of the addition. Addition is always stereospecifically *cis with respect to the dienophile, .e.,* butadiene reacts with maleic acid to give *cis-* 1, 2, 3, tetrahydrophthalic acid, whereas fumaric acid gives the *trans* derivative.

(H, CO_2H / H, CO_2H maleic acid) ← CH_2=CH–CH=CH_2 → (H, CO_2H / HO_2C, H fumaric acid)

cis- (H, H; HO_2C, CO_2H)   *trans-* (H, HO_2C; HO_2C, H)

When the diene is cyclic, two cis- additions are possible, one giving the endo-compound and the other the exo-compound, *e.g.*,

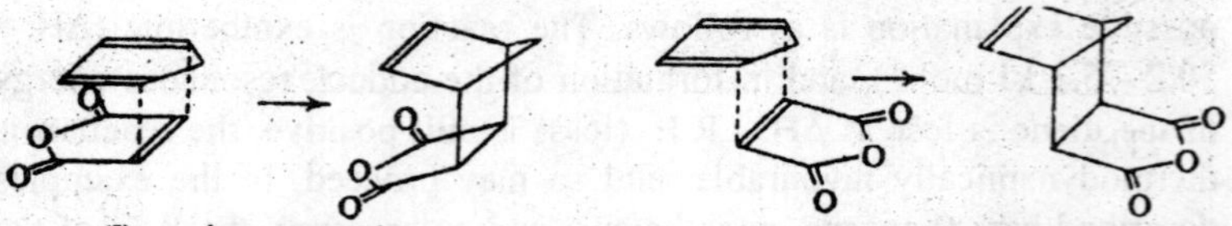

(I) ; *endo*-compound (II) ; *exo*-compound

It has been found experimentally that the *endo-product* is usually obtained exclusively. It has been shown, however, that the diene synthesis is reversible and exothermic, and therefore raising the temperature shifts the equilibrium to the reactants (Wassermann, 1938). It has also been shown that the endo-compound is the kinetically controlled product and the exo-compound the thermodynamically controlled product. Thus raising the temperature of the reaction favours the formation of the exo-compound.

Steric effects may operate in the Diels-Alder reaction when there are large groups at positions 2 and 3 in the diene, *e.g.*, 2, 3-di-t-butylbutadiene does not react.

Mechanism of the Diels-Alder reaction. Although a great deal of work has been done, the mechanism of this reaction is still uncertain. Wassermann (1950) proposed that ring formation occurs by *both* new bonds (of the addition) being formed *simultaneously*. This is imagined to proceed through a *non-planar* transition state, and is produced by bonds being formed simultaneously between diene and dienophile, which are in parallel planes, by overlap of the p-orbitals in an *endwise* fashion instead of the usual sideways overlap (see (I) and (II), the broken lines represent the overlapping orbitals). This mechanism accounts for stereospecific addition and also for steric effects (see above), but it does not account for polar effects, *e.g.*, acraldehyde reacts faster with isoprene than with butadiene. The presence of the methyl group (+ I effect) would not be expected to affect the rate with the above mechanism. It has been argued, however, that the less symmetrical the diene and dienophile are, the less symmetrical are the charge distributions in each, and consequently the formation of the transition state is facilitated. At the same time, this polarisation would account for the formation of the predominant product when two orientations are possible, *e.g.*, isoprene and acraldehyde react to give predominantly the 4-methyl adduct (4-methyl-1,2,5,6-tetrahydrobenzaldehyde).

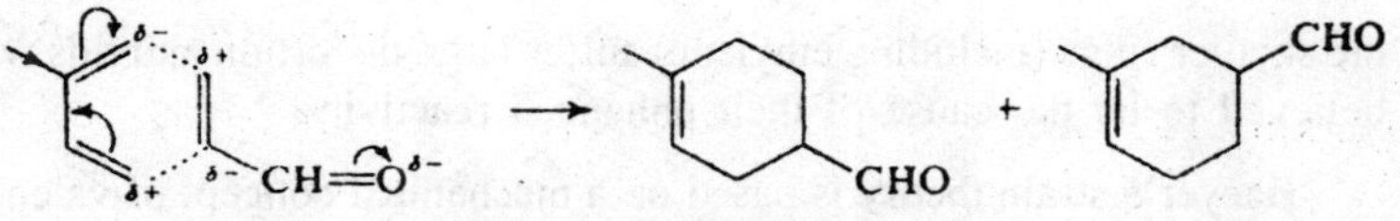

CH=O → (major product) CHO + CHO

These mechanisms are alsodiscussed.

CLASSIFICATION OF MONOCYCLIC SYSTEMS

Monocyclic systems have been classified according to the number of carbon atoms in the ring: **small rings,** 3 – 4; **common rings,** 5 – 7; **medium rings,** 8 – 11; **large rings,** 12 –. As we shall see, many chemical properties depend on the class of the cycloalkane, and these differences in behaviour have been explained largely in terms of steric strain.

Baeyer's strain theory. Baeyer (1885) was the first to point out that the angle subtended by the corners and centre of a regular tetrahedron –109° 28′ – lies between the values of the angles in a regular pentagon (108°) and a regular hexagon (120°). On this observation was based the **Baeyer strain theory.** According to the strain theory, the valency angle can be altered from this normal value (109° 28′), but when altered, a strain is set up in the molecule, and the greater the deviation from the normal angle, the greater is the strain.

Thus, according to Baeyer, five- and six-membered rings form most readily, and are the most stable because they involve the least strain (or distortion) from the normal valency angle. If all the rings are *assumed* to be planar, the distortions for each ring-size can be readily calculated, *e.g.,* in cyclopropane the three carbon atoms each occupy a corner of an equilateral triangle. Since the angles of an equilateral triangle are 60°, the distortion in cyclopropane will be 1/2 (109° 28′ – 60°) = + 24° 44′. It should be noted that the distortion of the bond angle has been assumed to be *equally* shared between the two bonds. Table 19.1 gives a list of distortions, and it can be seen that these decrease rapidly from ethylene (counting this as a 'two-membered' ring) to cyclopentane and then increase more slowly. Thus,

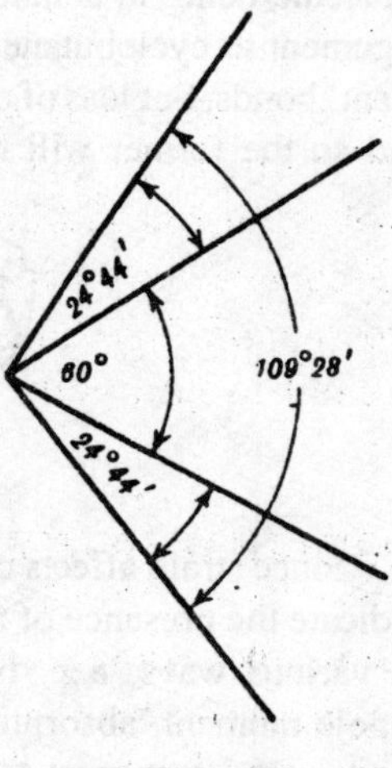

24°44′ 60° 109°28′ 24°44′

Fig. 1.1

the smaller rings (including ethylene) suffer large distortion and this was believed to be the cause of their enhanced reactivity.

Baeyer's strain theory is based on a mechanical concept of valency and on the assumption that *all* the rings are planar. This has led to conclusions which have now been shown to be wrong. In the first place, physical methods have shown that all rings, except cyclopropane, are not planar (cyclopropane is a three-point system and consequently *must* be planar). In the second place, quantum mechanical calculations do not permit very large distortions of bond angles. If cyclopropane were an equilateral Fig. 1.1 triangle with bonds directed *along* the lines joining the carbon atoms, then the ring valency angle of each carbon atom would be 60°. This value is impossible, since the carbon valency angle can never be less than 90° (when they are pure p-orbitals).

Furthermore, mixture *of p* with s-orbitals *opens* the valency angle. According to Coulson *et al.* (1949), calculation has shown that the smallest carbon valency angle that one can reasonably expect to have is 104°. Coulson has therefore suggested that in cyclopropane, the carbon hybridised orbitals are not pointing towards one another in the same straight line, and consequently there is a loss of overlap (Fig. 1.2a). It is this loss of overlap that gives rise to instability, the cyclopropane molecule being in a state of 'strain' due to 'bent' bonds. Applying this argument to cyclobutane (Fig. 1*.2b),* we see that this molecule also has 'bent' bonds, but loss of overlap is less in this case than for cyclopropane, and so the former will be more stable than the latter.

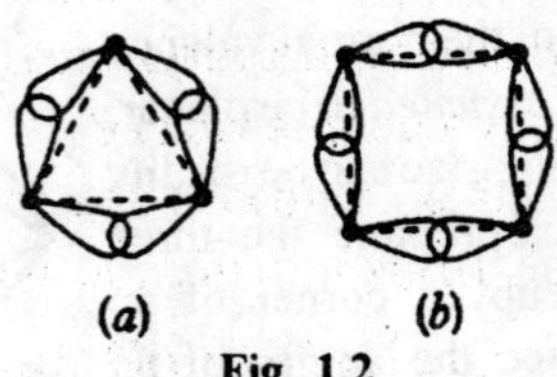

Fig. 1.2

Since strain affects chemical stability, measurement of the latter will indicate the presence of the former. Chemical stability may be measured in various ways, *e.g.,* by the heat of formation, heat of combustion, dipole moment, absorption spectra, etc. One of the most convenient to work with, in respect to hydrocarbons is the heat of combustion; this gives a measure **of thermochemical stability** and is the *total* strain, *i.e,,* the sum of the *Baeyer strain* (bond angle deformation only) and other sources of strain (steric repulsions, etc.). If the strain in a ring changes with the size of the ring, then this should be observed by changes in the heat of combustion.

The *total strain* in the molecule may be calculated as follows.

(number of C atoms in ring) × (observed heat of combustion/CH_2 – observed heat of combustion CH_2 for n-alkanes)

According to the results shown in Table 1.1, stability increases up to the six-membered ring, then decreases from seven to eleven-membered rings, and from the twelve-membered onwards attains the stability of the six-membered ring. These conclusions have been reached from heats of combustion obtained experimentally, and included in Table 1.1 are large rings which, according to Baeyer's strain theory are difficult, if at all possible, to prepare.

Table 1.1

Number of carbon atoms in the ring	*Angle between valency bonds*	*Distortion*	*Heat of combustion in kJ/CH_2*	*Total strain (kJ)*
2	0°	54° 44′	711	108
3	60°	24° 44′	697	120
4	90°	9° 44′	685	112
5	108°	0°44′	664	35
6	120°	–5° 16′	659	12
7	128° 34′	–9° 33′	662	35
8–11	135°-147° 16′	–12° 46′–18° 54′	661–665	32–88
12–	150°–	–20° 16′	657–661	0–48
n–alkanes	109° 28′	0°	657	0

General Properties of the Cycloalkanes

In many chemical respects the cycloalkanes are similar to the alkanes, but in others they are different in that the lower members form addition products with *ring fission.* Chlorine and bromine, in the presence of a catalyst, form halogenocycloalkanes, but with cyclopropane, the product is a mixture of 1,1-, 1,2-, and 1,3-dihalogenopropanes. Concentrated hydrogen iodide has no action on cycloalkanes except cyclopropane (at room temperature), which gives w-propyl iodide, and cyclobutane (on heating) which gives n-butyl iodide. Hydrogen in the presence of nickel at elevated temperatures does not attack cycloalkanes except cyclopropane (at 80°C to give propane) and cyclobutane (at 200°C to give «-butane).

Depending on the nature of the compound and on the reagents used, various rings may be opened or changed in size.

The infra-red absorption region of the CH_2 group (C – H stretch) in cycloalkanes is the same as that for alkanes except in the case of cyclopropane where the region is 3 080 – 3040cm^{-1} (v).

On the other hand, the absorption region of the carbonyl group (C = O stretch) in cycloalkanones depends on the size of the ring: 4-ring, 1780–1760 cm^{-1} (s), 5–ring, 1750 – 1740 cm^{-1} (s), 6- and 7 -rings, 1720–1700 cm^{-1} (s). Thus the first two rings can be distinguished from acyclic ketones, but not the last two. However, the molecular formula will help here.

Small-and common-ring compounds

Cyclopropane b.p. — 34°C, may be prepared by the action of zinc on trimethylene dibromide

Cyclopropane is one of the best anaesthetics known. When heated to a high temperature, preferably in the presence of a catalyst, *e.g.*, platinum, cyclopropane is converted into propene.

Cyclopropanes are readily prepared by addition reactions involving alkenes and methylenes.

(i) $CH_3CH_2CH = CH_2 + :CCl_2 \xrightarrow[Bu^tOK]{CHCl_3} CH_3CH_2CH\text{—}CH_2$ (ring closed by CCl_2)

(ii) $Me_2C = CMe_2 + :CHCl \xrightarrow[BuLi]{CH_2Cl_2} MeC\text{—}CMe_2$ (ring closed by CHCl)

Cyclopropanes may also be produced by reaction between an alkene, methylene di-iodide and a zinc-copper couple. Since this method does not result in insertion reaction (which can occur with alkenes), the reaction is believed to proceed via iodomethylzinc iodide (Hoberg, 1962).

$$>C=C< + ICH_2ZnI \longrightarrow -\overset{|}{\underset{|}{C}}-ZnI,\ -\overset{|}{\underset{|}{C}}-CH_2I \longrightarrow >C(CH_2)C< + ZnI_2$$

Cyclopropanes may also be prepared by addition of diazomethane to activated double bonds, and then decomposing the pyrazoline by heat in the presence of copper powder or by photochemical decomposition.

$$\begin{matrix} CHCO_2Et \\ \| \\ CHCO_2Et \end{matrix} + CH_2N_2 \longrightarrow \begin{matrix} CH_2 - CHCO_2Et \\ | \quad\quad | \\ HN \quad CCO_2Et \\ \diagdown \;\; /\!\!/ \\ N \end{matrix} \xrightarrow[\text{or } h\nu]{\text{Ch; heat}}$$

$$CH_2\begin{matrix} \diagup CHCO_2Et \\ \\ \diagdown CHCO_2Et \end{matrix} + N_2 \longrightarrow CH_2\begin{matrix} \diagup CHCO_2Et \\ | \\ \diagdown CHCO_2Et \end{matrix}$$

Many cyclopropanecarboxylic acids are known, some exhibit geometrical and optical isomerism, *e.g.*, cyclopropane-1,2--dicarboxylic acid exists in the is-from (optically inactive, it has a plane of symmetry) and the trans-from which is optically active(it has neither a plane nor a centre of symmetry).

cis-acid m.p. 139°C (ring: CH_2 apex with H, H; CO_2H and H on C-1; CO_2H and H on C-2, both CO_2H on same side)

trans-acid m.p. 175°C (ring: CH_2 apex with H, H; CO_2H and H on C-1; H and CO_2H on C-2, CO_2H groups on opposite sides)

Cyclopropane, b.p. – 36°C, has been prepared from cyclopropyl-amine via the Hofmann exhaustive methylation reaction.

Cyclopropylamine, b.p. 50°C, on treatment with nitrous acid forms alkyl alcohol instead of the expected cyclopropanol.

$$\begin{matrix} CH_2 \\ | \diagdown \\ | \diagup \\ CH_2 \end{matrix} CHNH_2 + HNO_2 \longrightarrow CH_2{=}CHCH_2OH + N_2 + H_2O$$

Cyclopropanol has not yet been prepared in a pure form, it readily rearranges to propionaldehyde. It has been prepared in a crude form by several methods (Roberts *et al.*, 1951) *e.g.*,

$$\begin{matrix} CH_2 \\ | \diagdown \\ | \diagup \\ CH_2 \end{matrix} CHCl \xrightarrow{Mg} \begin{matrix} CH_2 \\ | \diagdown \\ | \diagup \\ CH_2 \end{matrix} CHMgCl \xrightarrow[\text{(ii) } H_2O]{\text{(i) } O_2} \begin{matrix} CH_2 \\ | \diagdown \\ | \diagup \\ CH_2 \end{matrix} CHOH$$

Cyclobutane, b.p. – 15°C. Cyclobutane derivatives may be prepared by means of the malonic ester synthesis. The yield of cyclobutane by ring closure of 1,4-dihalogen derivatives of n-butane is very poor. Many derivatives of cyclobutane were known long before the parent hydrocarbon was obtained. This was first prepared by Willstatter (1907), using the *Hofmann exhaustive methylation reaction.*

$$\begin{matrix} CH_2Br \\ | \\ CH_2CH_2Br \end{matrix} + CH_2(CO_2C_2H_5)_2 \xrightarrow{2NaOC_2H_5} \begin{matrix} CH_2-C(CO_2C_2H_5)_2 \\ | \quad\quad | \\ CH_2-CH_2 \end{matrix}$$

$$\xrightarrow{\text{hydrolysis}} \begin{matrix} CH_2-C(CO_2H)_2 \\ | \quad\quad | \\ CH_2-CH_2 \end{matrix} \xrightarrow{\text{heat}} \begin{matrix} CH_2-CHCO_2H \\ | \quad\quad | \\ CH_2-CH_2 \end{matrix}$$

$$\xrightarrow[\text{(ii) } NH_3]{\text{(i) } PCl_5} \begin{matrix} CH_2-CHCONH \\ | \quad\quad | \\ CH_2-CH_2 \end{matrix} \xrightarrow{Br_2/KOH}$$

$$\begin{matrix} CH_2-CHNH_2 \\ | \quad\quad | \\ CH_2-CH_2 \end{matrix} \xrightarrow[\text{(ii) AgOH}]{\text{(i) } CH_3I} \begin{matrix} CH_2-CH\overset{+}{N}(CH_3)_3OH^- \\ | \quad\quad | \\ CH_2-CH_2 \end{matrix} \xrightarrow{\text{heat}}$$

$$\underset{\text{cyclobutene}}{\begin{matrix} CH_2-CH \\ | \quad\quad \| \\ CH_2-CH \end{matrix}} \xrightarrow[100^\circ C]{H_2/Ni} \begin{matrix} CH_2-CH_2 \\ | \quad\quad | \\ CH_2-CH_2 \end{matrix}$$

It appears that attempts to decarboxylate cyclobutanecarboxylic acid (by heating) result in opening of the ring, hence the necessity for the steps such as given in the equation (see also below).

Cyclobutanecarboxylic acid may also be converted into cyclobutane by the following routes.

$$\square\text{-}CO_2H \xrightarrow[Br_2]{\text{Ag salt}} \square\text{-}Br \xrightarrow[\text{(ii) } H_3O^+]{\text{(i) Mg}} \square$$

$$\downarrow \text{LAH}$$

$$\square\text{-}CO_2H \xrightarrow[154^\circ C]{H_2/Ni} \underset{(75\%)}{\square} + \underset{(21\%)}{CH_3CH_2CH_2CH_2}$$

Many cyclobutane derivatives may be prepared by means of the Diels-Alder reaction.

$$\begin{array}{ccc} EtO_2C & & NEt_2 \\ & \square & \\ Et_2N & & CO_2Et \end{array}$$

Cyclobutylamine. b.p. 2°C, shows the ordinary olefinic reactions. **Cyclobutadiene** has not yet been isolated, but the tetra-substituted derivative shown has been prepared (Gompper *et al.*, 1968,

Cyclobutylamine b.p. 82°C, when treated with nitrous acid, gives a mixture of cyclobutanol and cyclopropylcarbinol, the latter being produced by ring contraction.

$$\begin{array}{l} CH_2\text{—}CHNH_2 \\ | \qquad\quad | \\ CH_2\text{—}CH_2 \end{array} \xrightarrow{HNO_2} \begin{array}{l} CH_2\text{—}CHOH \\ | \qquad\quad | \\ CH_2\text{—}CH_2 \end{array} + \begin{array}{l} CH_2 \\ | \diagdown \\ | \quad CH_2\text{—}C{=}CH_2 \\ | \diagup \\ CH_2 \end{array}$$

This rearrangement of cyclic amines is known as the **Demjanov rearrangement** (1903).

If cyclobutylmethylamine is treated with nitrous acid, ring expansion by the Demjanov rearrangement takes place, four produces are obtained, cyclobutycarbinol, methylenecyclo butane, cyclopentanol and cyclopentene,

$$\begin{array}{l} CH_2\text{—}CHCH_2NH_2 \\ | \qquad\quad | \\ CH_2\text{—}CH_2 \end{array} \xrightarrow{HNO_2} \begin{array}{l} CH_2\text{—}CHCH_2OH \\ | \qquad\quad | \\ CH_2\text{—}CH_2 \end{array}$$

$$+ \begin{array}{l} CH_2\text{—}C{=}CH_2 \\ | \qquad\quad | \\ CH_2\text{—}CH_2 \end{array} + \begin{array}{l} CH_2\text{—}CH_2 \\ | \qquad\quad \diagdown \\ | \qquad\qquad CHOH \\ | \qquad\quad \diagup \\ CH_2\text{—}CH_2 \end{array} + \begin{array}{l} CH_2\text{—}CH_2 \\ | \qquad\quad \diagdown \\ | \qquad\qquad CH \\ | \qquad\quad /\!/ \\ CH_2\text{—}CH \end{array}$$

All cycloalkylmethylamines from three - to eight- membered rings behave in this manner, and the yield of expanded products depends largely on the strain in the ring compound produced.

Thus, highest yields are obtained when the products are five, six- or seven-membered rings.

The mechanism of the Demjanov rearrangement (a 1,2-shift) is still uncertain. Whatever the mechanism is, it must explain why, *e.g.*, both cyclopropylmethylamine and cyclobutylamine, on treatment with nitrous acid, give the same mixture of products.

(48%) (47%) (5%)

These results indicate that the reaction proceeds through *common* intermediates, and can be explained on the basis that the ion formed form each substrate equilibrates rapidly with other structures through which the same mixture of products is formed, *e.g.*,

(I) (II) (III)

From the analysis of the products, it would appear that (I) and (III) have about the same stability, both being much more stable then (II).

Cyclopentane, b.p. 50°C, many be prepared by cyclising 1,5-dibromopentane with zinc or, better, by reducing cyclopentanone by the Clemmensen method.

Cyclopentanone, b.p. 130°C, when treated with diazomethane, undergoes ring expansion to form cyclohexamone. Since this is more reactive than cyclopentanone, as soon as it is formed most of it reacts with diazomethane to form cycloheptanone. The mechanism is uncertain, but a strong possibility is the 1,2-shift (cf, the Arndt- Eistert synthesis,

(small amount) (small amount)

Cyclopentene, b.p., 45°C, may be prepared by dehydrating cyclopentanol or by heating cyclopentyl bromide with ethanolic potassium hydroxide.

Some cyclopentene derivatives of the fatty acids are very important, since they are useful in the treatment of leprosy and tuberculosis. Two acids, which occur in chaulmoogra oil, have been used form early times in the treatment of leprosy, both acids are optically active, the naturally occurring form being the (+),

```
CH=CH                                  CH=CH
|     \                                |     \
|      CH(CH2)10CO2H                   |      CH(CH2)12CO2H
|     /                                |     /
CH2—CH2                                CH2—CH2
```

hydnocarpic acid, m.p. 60°C chaulmoogric acid, m.p. 68°C

Cyclopentadiene, b.p. 41°C, is found in the crude benzene that is obtained from coal tar. The hydrogen atoms of the methylene group are very reactive, *e.g.*, cyclopentadiene reacts with Grignard reagents to form a hydrocarbon and cyclopentadienylmagnesium halide.

```
CH=CH                              CH=CH
|     \                            |     \
|      CH2 + RMgX  ———→            |      CHMgX + RH
|     /                            |     /
CH=CH                              CH=CH
```

Borg *et al* (1958) have treated cyclopentadienylsodium with chloroform and obtained chlorobenzene. The reaction is believed to proceed via the formation of dichloromethylene.

$$\text{C}_5\text{H}_5^- + CCl_2 \longrightarrow \text{C}_5\text{H}_5\text{CCl}_2^- \xrightarrow{-Cl^-} \text{C}_6\text{H}_5\text{Cl}$$

Owing to the presence of the reactive methylene group, cyclopentadiene condenses with aldehydes or ketones in the presence of sodium ethoxide to form *fulvenes.*

$$\text{C}_5\text{H}_6 + R_2CO \xrightarrow{EtONa} \text{C}_5\text{H}_4{=}CR_2 \longleftrightarrow \text{C}_5\text{H}_4^{-}{-}\overset{+}{C}R_2$$

The parent compound, **fulvene** (R = H), is unstable. The fulvenes are yellow or red compounds, and dipole moment measurements indicate a large contribution of the dipolar structure. Chemically, fulvenes behave predominantly as olefinic compounds.

Hydrogen of the methylene group in cyclopentadiene is acidic, *e.g.*, treatment with sodium produces sodium cyclopentadienide.

Since the carbanion is a resonance hybrid, it is this which stabilises the conjugate base with respect to its acid, i,e., resonance (with the resulting delocalisation of the negative charge) is the driving force in the reaction. This cyclopentadienyl anion exhibits aromatic properties.

On standing, cyclopentadiene dimerises (it undergoes the Diels-Alder reaction) to form dicyclopentadiene, which regenerates the monomer on heating.

Cyclopentadiene complexes. Pauson *et al.*, (1951) treated cyclo-penta-dienylmagnesium bromide with ferric chloride and isolated dicyclo-pentadienyliron.

$$2C_5H_5MgBr \xrightarrow{FeCl_3} (C_5H_5)_2Fe$$

This iron (Fe II) complex was named **ferrocene** by Woodward *et al.*, (1952) to indicate its aromatic character.

Many dicyclopentadienyls have now been prepared, the methods being by the use of the Grignard reaction (as shown above), by direct reaction between cyclopentadiene vapour and a heated metal or metal carbonyl, or by reaction between the sodium salt of cyclopentadiene and a metal halide (in tetrahydrofuran or liquid ammonia solution); *e.g.*, $(C_5H_5)_2Cr$, $(C_5H_5)_2Mn$, $(C_5H_5)_2Co$, etc.

One of the best methods of preparing ferrocene is by reaction between ferrous chloride and cyclopentadiene in the presence of anhydrous sodium ethoxide (yield: 90%).

```
CH=CH                CH=CH
|     \             /    |
|      CH—Fe—CH         |
|     /             \    |
CH=CH                CH=CH
```

(I)

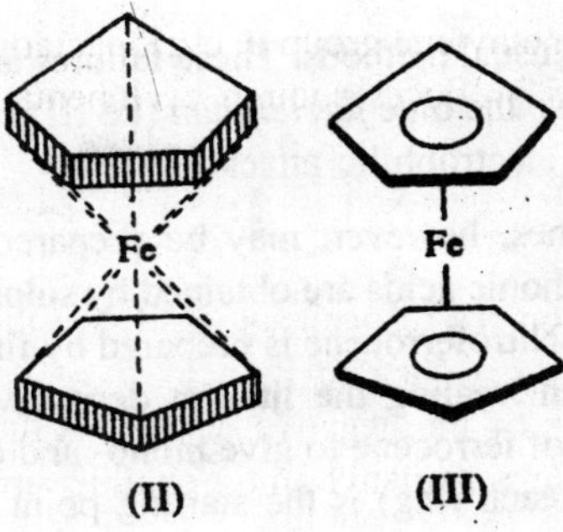

Fig. 1.3

Originally, structure (I) was assigned to ferrocene, but the high stability of the molecule showed that this was unlikely. Ferrocene has a zero dipole moment, and the molecule is therefore symmetrical. Furthermore, the infra-red spectrum showed that all of the C – H bonds are equivalent, and so, on this evidence, structure (II) (this is a 'sandwich' structure, and is the anti-prism one) was proposed and has been confirmed by X-ray analysis, *i.e.,* that the two five-membered rings lie in parallel planes with the iron atom placed symmetrically between the two. In ferrocene and other cyclopentadienyls of the transition metals (Cr, Mn, Co, etc.), the entire ring is bonded uniformly to the metal atom, the bonding occurs by overlap of the sextet of π-electrons of the ring with the d-orbitals of the metal, thereby giving a *delocalised* covalent bond between the metal atom and the cyclopentadienyl ring as a whole. This, in ferrocene, is represented by (III). Structurally related compounds are bisbenzenechromium, etc.

Derivatives of sodium, potassium, etc., are examples of ionic bonding; there is essentially an electron transfer from metal to ring, and compounds of this type are known as *cyclopentadienides.* On the other hand, there are cases where the key atom, *e.g.*, mercury, and *one* carbon atom of the ring each contribute one electron to form essentially a σ-bond.

Ferrocene is an orange solid, m.p. 173°C, and its reactions are aromatic, *e.g.,* it gives a mono-and diacetyl derivative by means of the Friedel – Crafts reaction under suitable. Conditions. Ferrocene forms an aldehyde by the N-methylformanilide method, and this undergoes the Cannizzaro reaction, the azlactone synthesis, etc. Ferrocene undergoes mercuration and metalation- but cannot be directly halogenated, nitrated, or sulphonated by the usual methods. These failures are due to oxidation of the iron atom to give the blue *ferricinium* (Fe III) *cation,* $(C_5H_5)_2Fe^+$, which is resistant to electrophilic attack.

Halogenoferrocenes, however, may be prepared via mercuration, and mono- and disulphonic acids are obtained by sulphonation- in acetic anhydride as solvent. Nitroferrocene is prepared by first metalating with butyl-lithium and then treating the lithium derivative with dinitrogen tetroxide. Metalation of ferrocene to give mono- and dilithiumferrocene (one lithium atom in each ring) is the starting point of a large variety of ferrocene derivatives. The aromatic character of ferrocene is further illustrated by the fact that it does *not* condense with maleic anhydride, and hence the conjugated system is absent in the rings.

COMe
Fe
Ac_2O
H_3PO_4
Fe
AcCl
$AlCl_3$
COMe
Fe
COMe

(major product)

Fig. 1.4

Cyclohexane, b.p. 81°C, occurs in petroleum. When reduced with hydrogen iodide at 250°C, benzene is converted into a mixture of cyclohexane and methylcyclopentane. Pure cyclohexane may be conveniently prepared by the Clemmensen reduction of cyclohexanone, and is prepared industrially by the hydrogenation of benzene in the presence of nickel at 200°C. Another industrial method is the hydrogenation of phenol, dehydrating the product cyclohexanol to cyclohexene, which is then catalytically hydrogenated to cyclohexane; the process is carried out without isolating the intermediate compounds.

OH
H_2
Ni
OH
Al_2O_3
350°C
H_2
Ni

Hot concentrated nitric acid oxidises cyclohexane to adipic acid, and fuming sulphuric acid converts it into benzenesulphonic acid.

$$\text{C}_6\text{H}_{12} \xrightarrow{HNO_3} HO_2C(CH_2)_4CO_2H$$

$$\text{C}_6\text{H}_{12} \xrightarrow{H_2SO_4} C_6H_5SO_3H$$

For a consideration of the spatial arrangement of the carbon atoms in cyclohexane,

Cyclohexane can be catalytically (Pt or Pd) dehydrogenated to benzene.

Provided the ring contains at least one double bond, then dehydrogenation can be readily effected with sulphur of selenium. In general, this dehydrogenation is confined to six-membered rings.

Cyclohexanol, m.p. 24°C, is prepared industrially by the catalytic hydrogenation (Ni) of phenol. It undergoes the general reactions of an aliphatic secondary alcohol.

It is converted by gentle oxidation (dilute nitric acid) into cyclohexanone; vigorous oxidation (concentrated nitric acid) produces adipic acid.

Cyclohexanone, b.p. 157°C, may be prepared by the oxidation of cyclohexanol, it is manufactured by the oxidation of cyclohexane with oxygen under pressure in the presence of a cobalt catalyst (cyclohexanol is also obtained). α-Halogenocyclohexanones undergo the **Favorsky rearrangement** (1894) when treated with alkali, *e.g.*, the six-membered ring changes to a five-membered ring carboxylic acid.

$$\begin{array}{ccc} CH_2\text{—}CH_2 & & CH_2\text{—}CH_2 \\ CH_2 \qquad\qquad CO & \xrightarrow[\text{(ii) HCl}]{\text{(i) NaOH}} & \qquad\quad CHCO_2H \\ CH_2\text{—}CH_2 & & CH_2\text{—}CH_2 \end{array}$$

The mechanism of the Favorsky rearrangement is uncertain, but a highly favoured one is via a cyclo-propanone intermediate; *e.g.*, in the presence of sodium methoxide.

Treatment of acyclic α-halogenoketones with sodium methoxide also results in the Favorsky rearrangement (α-methoxyketones have also been isolated. Turro *et al.*, 1969), *e.g.*,

$Me_2C(Br)COCH_3 \longrightarrow MeOH + Br^- + Me_2$(cyclopropanone) $\xrightarrow[MeOH]{MeO^-}$

$Me_3CCO_2Me + Me_2C(OMe)COMe$

Cyclohexanehexol, inositol, *hexahydroxycyclohexane,* can exist in eight geometrical isomeric forms, of which only one is optically active (see formula). The (+)- and (–)- forms of this isomer (m.p. 248°C) occur in plants as the hexa-phosphoric ester, which is known as *phytin.* Some of the optically inactive forms also occur as their hexaphosphoric esters. One of the optically inactive forms is present in the vitamin B complex; it appears to have growth-promoting properties in chicks, and is an anti-alopecia factor (anti-baldness) in mice.

Hexachlorocyclohexane, $C_6H_6Cl_6$, exists in a number of .geometrical isomeric forms, one of which is a powerful insecticide.

Cyclohexene *(tetrahydrobenzene),* b.p. 83°C, may conveniently be prepared by dehydrating cyclo-hexanol with sulphuric acid. It has the usual properties of an alkene.

Many cyclohexene derivatives may be prepared by means of the Diels- Alder reaction.

Cyckohexadienes *(dihydrobenzenes).* There are two isomeric cyclohexadienes, the 1,3- (b.p. 81°C) and the 1,4- (b.p. 86°C). They readily polymerise and undergo the usual reactions of an alkadiene. Since cyclohexa-l,3-diene contains a conjugated system of double bonds, it can undergo both the 1,2- and 1,4-addition reactions.

Cyclohexatriene is benzene, and differs enormously in its chemical properties from the cycloalkenes and cycloalkadienes. Benzene has *'aromatic properties'* (see benzene).

5,5-Dimethylcyclohexane-l,3-dione, dimedone, m.p. 148°C, is a very sensitive reagent for formaldehyde, which may also be estimated gravimetrically with this reagent.

$$2(CH_3)_2C\langle{}^{CH_2-CO}_{CH_2-CO}\rangle CH_2 + HCH \longrightarrow$$

$$(CH_3)_2C\langle{}^{CH_2-CO}_{CH_2-CO}\rangle CH-CH_2-CH\langle{}^{CH-CO_2}_{CH-CO_2}\rangle C(CH_3)_2 + H_2O$$

Cycloheptane, b.p. 118°C, occurs in petroleum, it may be prepared by the Clemmensen reduction of cycloheptanone.

Cycloheptariene (tropilidene), b.p. 115.5°C, may be prepared by the photolysis of diazomethane in the presence of benzene (note the ring expansion).

$$C_6H_6 + CH_2 \longrightarrow C_7H_8$$

Two derivatives of cycloheptatriene, **tropone** *(cycloheptatrienone)* and **tropolone** *(cycloheptatrienolone),* are particularly intersting because they posses aromatic character, *e.g.*

OMe (anisole) $\xrightarrow[\text{EtOH}]{\text{Na}-\text{NH}_3}$ OMe (cyclohexadiene) $\xrightarrow[(\text{CHBr}_3-\text{EtONa})]{:\text{CBr}_2}$ MeO, Br, Br (bicyclic adduct) $\xrightarrow[\text{H}_2\text{O}]{\text{AgNO}_3}$ O (tropone)

Although, their classical structures contain a keto group, both compounds show a lack of ketonic properties for a further discussion). Tropone may be prepared from anisole by the following sequence of reactions (Birch *et al.*, 1962)

Tropolone may be prepared from (i) cycloheptatriene, and (ii) cyclopentadiene.

(i) $\xrightarrow[\text{NaOH}]{\text{KMnO}_4}$ $\xrightarrow{\text{enolises}}$ OH

(ii) + $F_2C{=}CF_2$ → F_2 F_2 $\xrightarrow[\text{5 mm}]{700°C}$ F_2 F_2

+ F_2 F_2 $\xrightarrow[\text{AcOH}]{\text{KOAc}}$ OH

Fig. 1.5

Conformational analysis. This has already been discussed with respect to *acyclic* compounds. Here we shall deal with cyclic systems and, in particular, cyclohexane derivatives. The principles are the same but because of the 'rigidity' of cyclic systems, additional problems are involved. As we have seen, cyclohexane is a stable molecule (Table 1.1), and this was explained by Sachse (1890), who assumed that this ring was *puckered,* the *normal* valency being retained and thereby producing *strainless* rings; and according to him, cyclohexane exists in two forms, both of which are strainless (Fig. 1.6).

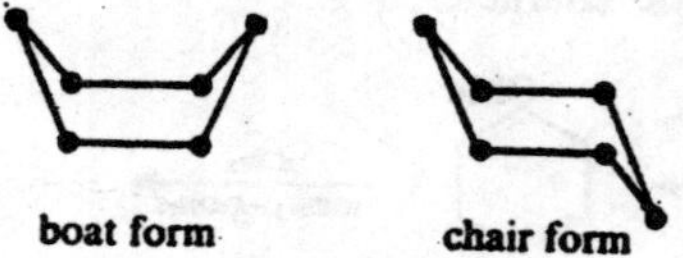

Fig. 1.6

These two forms are called the chair and boat conformations, and it is now known that neither is completely strainless. Both are free from angle strain, but because of differences in steric strain and bond opposition strain, the two forms differ in energy content. In the boat form chair form chair form (Fig. 1.7a) all the C — H bonds on adjacent carbon atoms are in the skew position (see Fig. 1.7e). In the boat '9' form (Fig. 1.7c), however, four of the C — H bonds are skew (1,2; 3,4; 4,5 and 6,1), and two are eclipsed (2,3 and 5,6). At the same time, there will also be some bond opposition strain for these two pairs of eclipsed bonds and also steric repulsion between the hydrogens pointing towards each other at 1 and 4 (Fig. 1.7c).

Hence, the total strain in the boat conformation is larger than that in the chair conformation, and consequently the former is less stable than the latter. The boat form, however, is flexible and can be readily distorted into many shapes, and in these the hydrogen eclipsings are reduced.

The chair form is 'rigid' (in the sense that it resists distortion), and when it is changed into the boat form, some angular deformation is necessary. The energy barrier for this transformation is about 37.7 – 46.0 kJ mol^{-1} (determined from N.M.R. spectral data). Calculations have also shown that the twist-boat contains 6-7 kJ mol^{-1} less energy than the classical boat conformation, and that the chair conformation contains 22.2 kJ mol^{-1} less energy than the twist-boat. These data are shown in Fig. 19.5. The energy barriers are not large enough to prevent interconversion of the chair and boat conformations (at room temperature), but are large enough to permit each conformation to retain its identity.

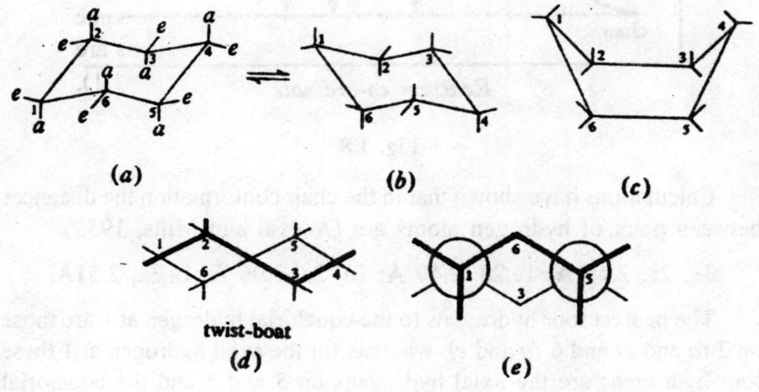

Fig. 1.7

Furthermore, because the energy barrier for the chair ⟶ twist-boat is about double that of the reverse change, the chair will be by far the predominant form in the equilibrium mixture. This has been shown to be the case by Hassel *et al.*, (1943) by means of electron-diffraction studies on cyclohexane (at room temperature), and thermo-dynamic calculations have shown that the population of the twist-boat is about one in a thousand.

Since the boat conformation occurs in relatively few cases, we shall confine our attention to the chair form. Consideration of the C — H bonds in this form shows that there are two sets of six. In one set the six C — H bonds are parallel to the axis of the ring (Fig. 1.7a), these are the axial *(a)* bonds. In the other set the six C — H bonds make an angle of 109° 28′ with the axis (or ± 19° 28′ with the horizontal plane of the ring; Fig. 1.7a); these are the **equatorial** (e) **bonds.** Each carbon atom has one axial and one equatorial bond, and because of the mobility of the chair conformation, one form (Fig. 1.7a) can readily change into the other (Fig. 1.7b), and when this occurs all hydrogens originally axial now become equatorial, and *vice versa*. The two forms are identical.

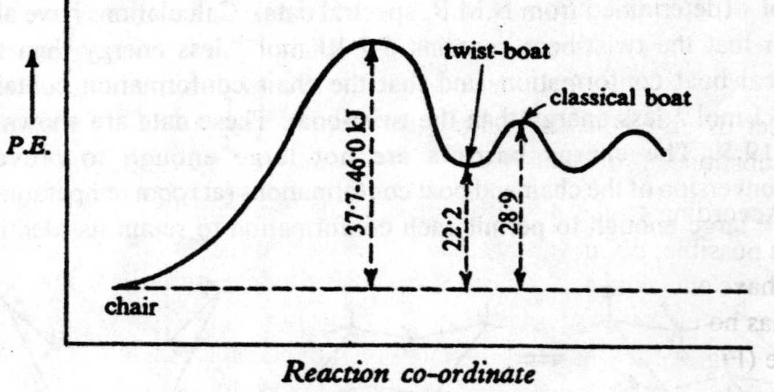

Fig. 1.8

Calculations have shown that in the chair conformation the distances between pairs of hydrogen atoms are (Angyal and Mills, 1952).

1e, 2e, 2.49 Å; 1e,2a, 2.49 Å; 1a, 2a, 3.06 Å, 1a,3a, 2.51Å.

The nearest four hydrogens to the equatorial hydrogen at 1 are those on 2 *(a* and *e)* and 6 *(a* and *e),* whereas for the axial hydrogen at 1 these four hydrogens are the axial hydrogens on 3 and 5 and the equatorial on 2 and 6 (see Fig. 1.7a). Thus, for hydrogen atoms, the interactions 1e,2e, 1e,2a and 1a,3a are about the same.

Now let us consider the problem when one hydrogen atom in cyclohexane is replaced by some substituent which, of necessity, must be larger than a hydrogen atom. A study of accurate scale models has shown that in monosubstituted cyclohexanes, an axial substituent at 1 is closer to the two axial hydrogens at 3 and 5 than an equatorial substituent at 1 is to the four hydrogens at 2 and 6. Thus, la,3a interactions will be greater than 1e, 2e or *1e,2a* interactions, and so, in general, a monosubstituted cyclohexane will assume the conformation in which the substituent occupies an equatorial position. This has been confirmed experimentally, *e.g.*, Hassel (1950) has shown from electron-diffraction studies that the methyl group in methylcyclohexane and the chlorine in chlorocyclohexane are equatorial (predominantly).

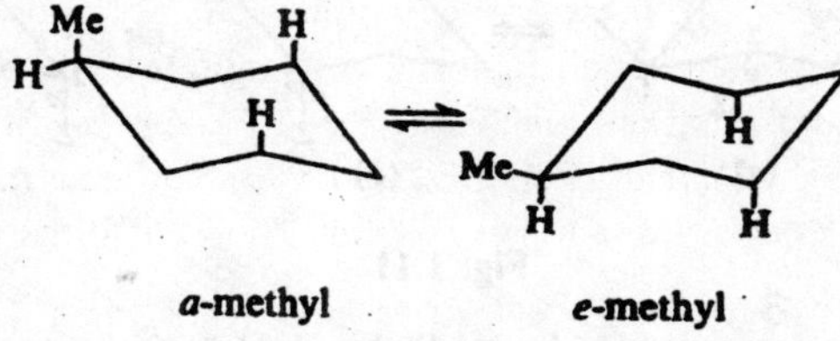

Fig. 1.9

Let us now consider 1,2-disubstituted cyclohexanes in which the two substituents are identical.

According to the classical ideas of stereochemistry, there are two forms possible, *cis* and *trans.* The conformation of the *cis* configuration *must* have one axial and one equatorial substituent (Fig. 1.10a), and since this has no elements of symmetry, it is not superimposable on its mirror image (Fig. 1.10c). When Fig. 1.10 (a) flips to give Fig. 1.10 (b), axial hydrogens ⟶ equatorial and *vice-versa,* and therefore Fig. 1.10 (c), which is equivalent of Fig. 19.6 (6), is the enantiomer of Fig. 1.10 (a). Furthermore, since strain is identical in both forms, their populations are equal, and consequently this 1,2-disubstituted cyclohexane is optically inactive by *external* compensation.

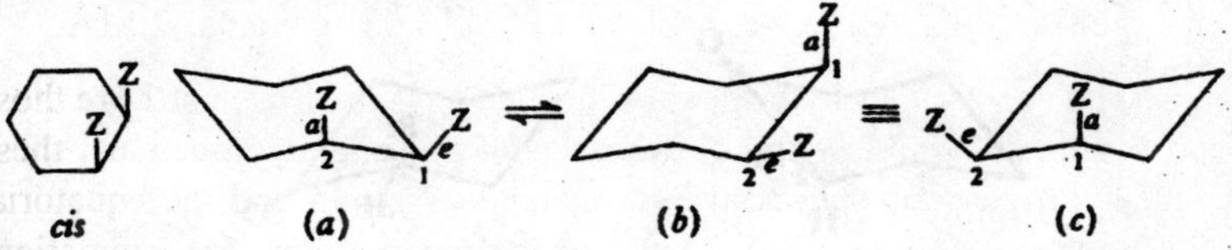

Fig. 1.10

However, this type of compound (*i.e.*, Z = Z) has never yet been resolved, and this is due to the fact that the two enantiomers are readily interconvertible. In effect, each enantiomer undergoes very rapid auto-racemisation.

The *trans* configuration can exist in two different conformations, 1a, 2a and le, 2e (Fig. 1.11), but from what has been said above, the trans-le,2e form will be more stable than the trans-1a,2a, and this trans form will be more stable than the *cis*. Also, neither conformation has any element of symmetry, and neither conformation can be converted into its mirror image by flipping. Hence, *trans*-1,2-disubstituted cyclohexanes exist as pairs of enantiomers.

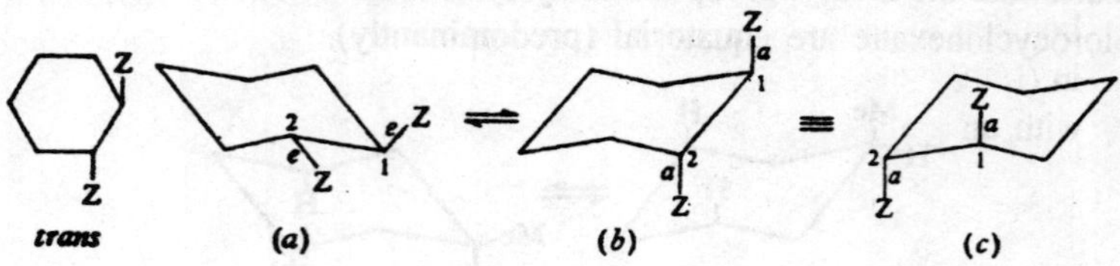

Fig. 1.11

The above arguments can be applied to the 1,3- and 1,4-disubstituted cyclohexanes, and to the polysubstituted derivatives, and it will be found that the chair conformation with the maximum number of equatorial substituents will be the preferred form. This generalisation, however, is only true when other forces due to, *e.g.*, dipole interactions, hydrogen bonding, are absent. When these 'disturbing' factors are present, they may be large enough to outweigh the 1,3-interactions, *e.g.*, infra-red spectra studies have shown that the bromine in 2-bromocyclohexanone is predominantly *axial* (Fig. 1.12). The C — Br and C = O bonds are both strongly polar, and when the bromine is equatorial the dipolar repulsion is a maximum, and is a minimum when the bromine is axial. Since the axial form predominates, the equatorial dipolar repulsion must be much larger than the 1,3-interactions. In cw-l,3-dihydroxycyclohexane, the diaxial form predominates, since this form is stabilised by hydrogen bonding.

Fig. 1.12

It is also instructive to consider one example of a 1,2-disubstituted cyclohexane in which the two substituents are different, *e.g.*, cis-2-

methylcyclohexanol. Since 1,3-interactions will be most powerful when the larger group is axial, the preferred form will therefore be the one in which the larger group is equatorial. Thus, since the methyl group is larger than hydroxyl, the preferred form of cis-2-methylcyclohexanol is 1a-OH, 2e-CH_3.

Effect of conformation on the course and rate of reactions. Because the environments of axial and equatorial groups are different, it may be expected that the reactivity of a given group will therefore depend on its orientation. Since S_N2 reactions always occur with inversion, it can be seen that the approach of the attacking group (Z) along the bonding line remote from the group to be expelled (Y) is less hindered in (I, eY) than in (II, eY). Thus, S_N2. reactions take place more readily with an axial substituent than with an equatorial.

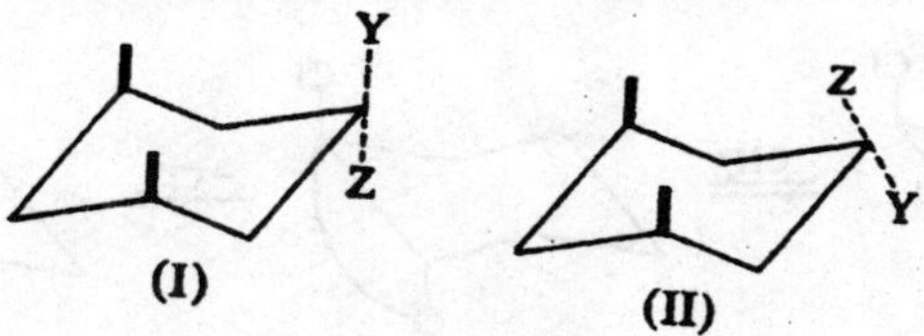

Fig. 1.13

It can be expected that an SN1 reaction will be sterically accelerated for an axial substituent, since the formation of the carbonium ion will relieve the steric strain due to the 1,3-interactions. As these interactions are absent for an equatorial substituent, in this case there will be no steric acceleration.

In elimination reactions, for the E2 mechanism to occur readily, the two groups concerned must be in the trans-position. In cyclohexane derivatives, this requirement is found only in trans-1,2-diaxial compounds, and these thus readily undergo elimination reactions.

E1 reactions are less easy to study because of the difficulty of ascertaining the geometry of the intermediates involved.

Let us consider the case of 4-t-butylcyclohexyl p-toluenesulphonate (Winstein *et al.*, 1955). Two forms, *cis* and *trans*, are possible, but because of the large bulk of the t-butyl group, this group is *always* equatorial, *i.e.*, the conformations of the two stereoisomers are made rigid. Under the-same conditions (sodium ethoxide in ethanol at 70°C), the *cis* form readily undergoes bimolecular elimination, but the *trans* does not.

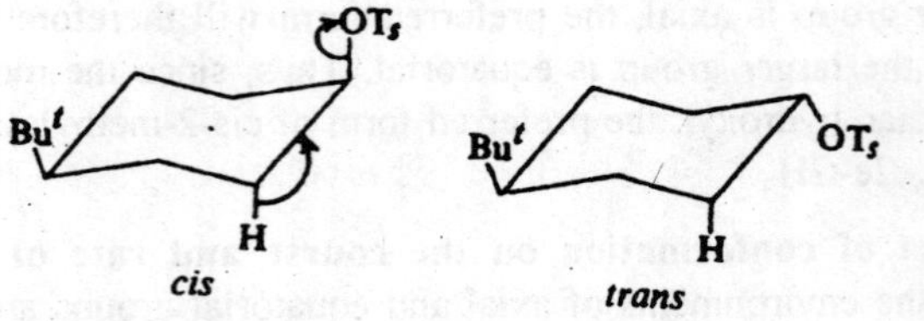

Fig. 1.14

The following example is of particular interest in that it illustrates the phenomenon of anchimeric assistance in neighbouring group participation. Bartlett (1935) has shown that alkali converts trans-2-chlorocyclohexanol into cyclohexene oxide, and proposed the following mechanism.

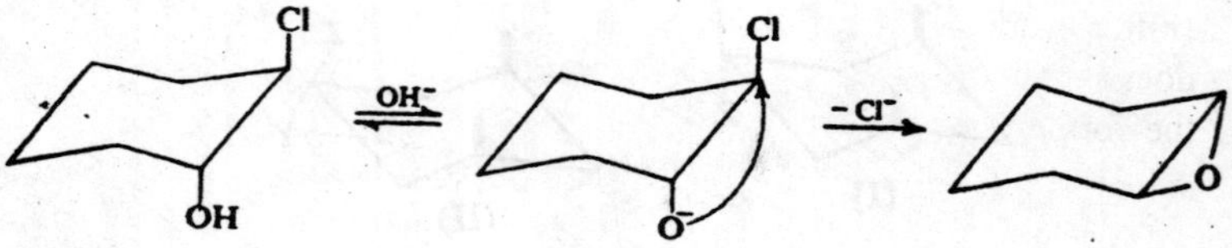

Fig. 1.15

Bergkvist (1947) showed that this reaction proceeds more than 100 times as fast as that when the *cis*-compound is used. This may be explained as follows. The *cis*-compound probably forms the alkoxide ion just as readily as does the trans/w-compound, but the cis-ion cannot attack at the rear of the adjacent carbon atom (where the chlorine is being ejected) without great distortion of the molecule.

So far, we have examined reactions in which the bond linking an atom to the ring is broken. Now let us consider reactions in which it does not occur, *e.g.*, esterification and ester hydrolysis by the $A_{AC}2$ mechanism. In the esterification of cyclohexanol, all stages involve groups larger than hydrogen becoming attached to the oxygen of the OH group. This will therefore increase non-bonded interactions, more so far an axial than for an equatorial OH. Hence, the former should esterify more slowly than the latter.

Hydrolysis by the $A_{AC}2$ mechanism is the reverse of esterification, but if we now consider the point that some steps involve change of sp^2 hybridisation (flat) of the carbon atom of CO_2R to sp^3 (tetrahedral), non-bonded interactions increase. Consequently, the axial ester should undergo hydrolysis more slowly than the equatorial one. These arguments hold good for hydrolysis by the $B_{AC}2$ mechanism, and also for esterification and hydrolysis of cyclohexanecarboxylic acid. These predictions are borne out in practice, *i.e.*, axial cyclohexanols (and esters) and cycloalkanecarboxylic acids (and esters) are more slowly esterified (and hydrolysed) than the corresponding equatorial isomers.

Rearrangements involving cyclohexanes. Many cyclohexane derivatives can undergo 1,2-shifts, but the course of the reaction depends on a number of factors. If the leaving group Z is *axial* and there is a group Y attached to an *adjacent* carbon atom which is also axial (*i.e.*, Z and Y are *trans-a,a*) and which is capable of migrating (*e.g.*, methyl, hydrogen), then the spatial requirements are satisfied and so a 1,2-shift can occur with formation of a *cyclohexane derivative* (see III). If, however, the leaving group Z is equatorial and Y is axial (*i.e.*, Z and Y are cis), then these two groups no longer satisfy the spatial requirements of a 1,2-shift. On the other hand, *an adjacent carbon atom of the ring system* does satisfy the spatial requirements of the 1,2-shift, and this occurs with the formation of a *cyclopentane derivative* (see IV).

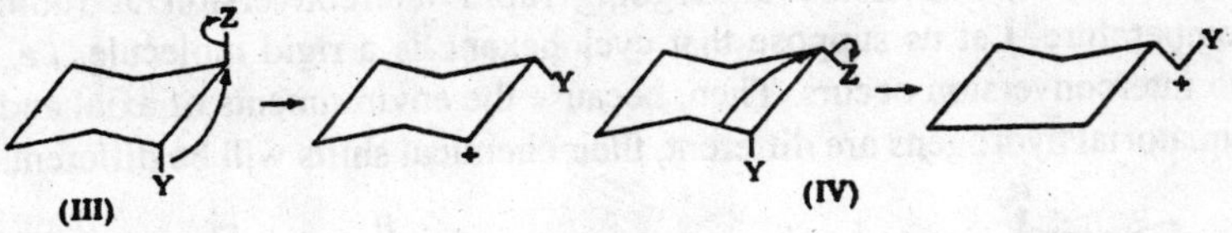

(III) (IV)

Fig. 1.16

These principles may be illustrated by the action of acid on *cis*-1,2-dimethylcyclohexane-1,2-diol to give 2,2-dimethylcyclohexanone and 1-acetyl-1-methylcyclopentane. Both products are formed by the pinacol-pinacone rearrangement.

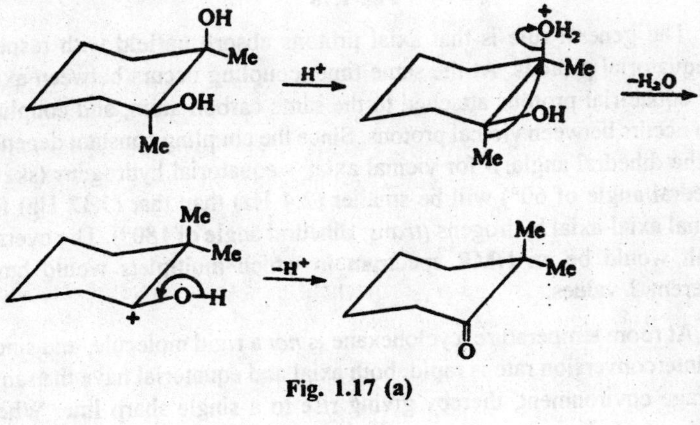

Fig. 1.17 (a)

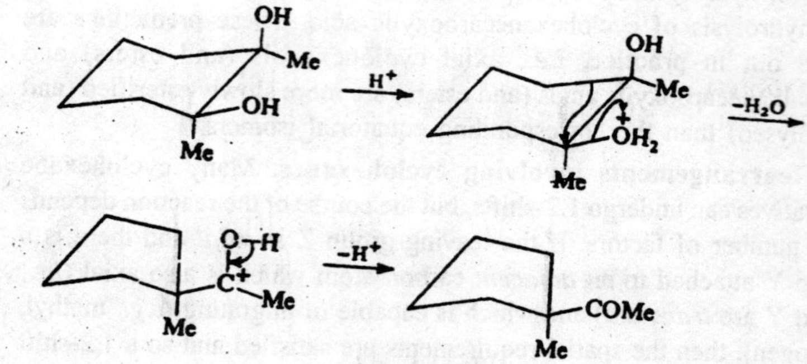

Fig. 1.17 (b)

NMR spectroscopy of cyclohexanes. The NMR spectra of unsubstituted cycloalkanes, except cyclopropane, show a t-value around 8.5 p.p.m. due to the methylene protons (*cf.* acyclic methylene value, 8.7). As we have seen, cyclohexane exists in two stable chair conformations, (V) and (VI) which are undergoing rapid interconversion at room temperature. Let us suppose that cyclohexane is a rigid molecule, *i.e.*, no interconversion occurs. Then, because the environments of axial and equatorial hydrogens are different, their chemical shifts will be different.

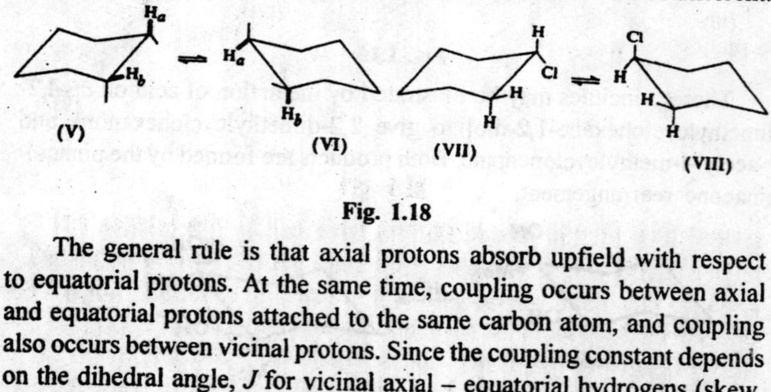

Fig. 1.18

The general rule is that axial protons absorb upfield with respect to equatorial protons. At the same time, coupling occurs between axial and equatorial protons attached to the same carbon atom, and coupling also occurs between vicinal protons. Since the coupling constant depends on the dihedral angle, *J* for vicinal axial – equatorial hydrogens (skew, dihedral angle of 60°) will be smaller (2.4 Hz) than that (5.12 Hz) for vicinal axial-axial hydrogens (*trans;* dihedral angle of 180°). The overall result would be an NMR spectrum in which multiplets would have different J values.

At room temperature, cyclohexane is *not* a rigid molecule, and since the interconversion rate is rapid, both axial and equatorial have the same average environment, thereby giving rise to a single sharp line. When the spectrum of cyclohexane is measured at about — 100°C, two signals are observed, one due to axial and the other to equatorial hydrogens. Under these conditions, the interconversion rate is sufficiently slowed down for each type of proton to show its own chemical shift.

Now let us consider a monosubstituted cyclohexane, *e.g.*, chlorocyclohexane (VII and VIII). Since the equatorial chlorine conformation (VII) is favoured, the populations of (VII) and (VIII) are different, and consequently the time spent by a given hydrogen atom (axial or equatorial) will be longer in conformation (VII) than in (VIII). Even so, because of rapid interconversion at room temperature, only one *broad* proton signal is observed. When the temperature is cooled below –55°C, two broad peaks are observed.

In the case of fused systems, *e.g.*, the decalins (see below), *cis*-decalin shows one proton peak, whereas *trans*-decalin shows two. The former isomer is capable of interconversion, but the latter is not. The latter is a rigid system, and so equatorial and axial protons are held in different environments, the result being two different chemical shifts.

Cycloalkenes. The presence of a double bond in cyclohexene causes the ring to assume the half-chair conformation (atoms 1,2,3, and 6 lie in one plane).

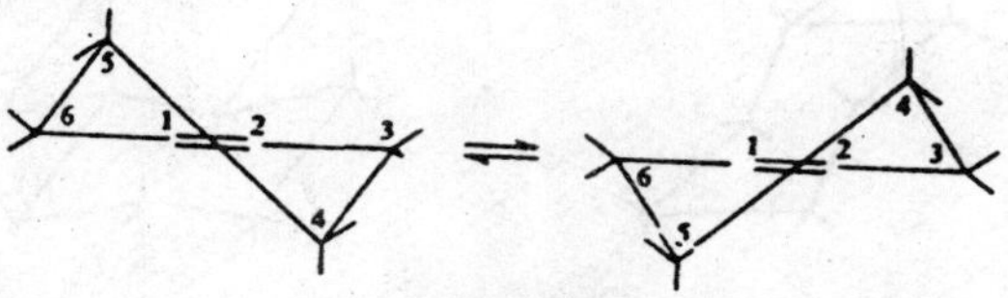

cyclohexene

Fig. 1.19

Fused systems. Since the boat and chair forms of cyclohexane are readily interconvertible, neither form can be isolated, Mohr (1918), however, elaborated Sachse's theory and predicted that the fusion of two cyclohexane rings, *e.g.*, in decalin, should produce *cis* and *trans* forms which should be stable enough to retain their identities. Both forms have now been prepared. Several conventions have been introduced to represent these isomers. One uses *full* lines to represent groups *above* the plane of the molecule, and *broken* lines to represent those *below*. Another convention uses a black dot to represent a hydrogen atom above the plane, and the ordinary lines of the formula to indicate a hydrogen atom below the plane. Thus, the decalins may be drawn.

H
H
cis
H
H
trans

Fig. 1.20

The configurations of decalin are complicated by the fact that a number of strainless modifications are possible which differ in the type of Mocking', *i.e.,* whether axial or equatorial bonds are used to fuse the rings. According to Hassel *el at.* (1946), *cis-* and *trans*-decalin are as shown in Fig. 1.21.

In both cases the cyclohexane rings are *all chair forms.* The *cis* form is produced by joining one axial and one equatorial bond of each ring, whereas the *trans* form is produced by joining the two rings by equatorial bonds only. Calculation has shown that the *cis* isomer has 10.0 kJ energy content more than the *trans.* It is also of interest to note that if the decalins are regarded as 1,2-disubstituted cyclohexanes, then the *trans* form (1e, 2e) would be expected to be more stable than the *cis* (1e, 2a).

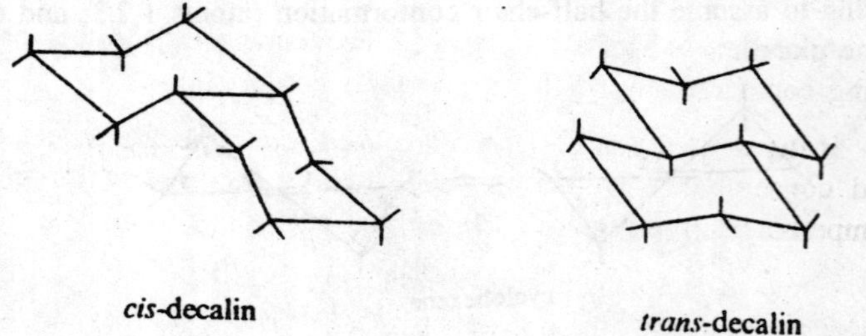

Fig. 1.21

Medium-ring Compounds

Cyclo-octane, m.p. 11-8°C, may be obtained by the catalytic reduction of cyclo-octene, cyclo-octadiene or cyclo-octatetraene (Willstatter 1911).

(1,3,5,7)- Cyclo-octatetraene, b.p. 142–143°C, was first obtained by Willstatter (1911, 1913), who prepared it form the alkaloid *pseudo*-pelletierine (Ψ-pelletierine) by means of the Hofmann exhaustive methylation method.

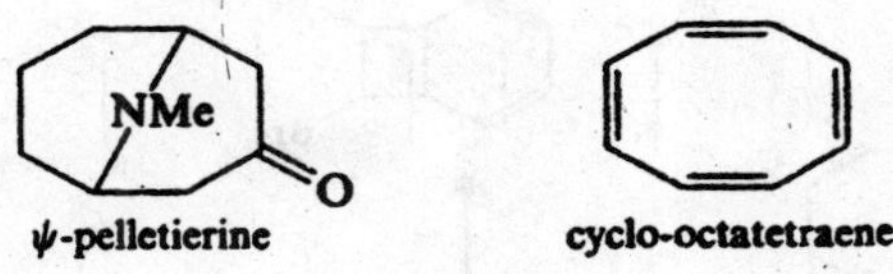

Fig. 1.22

Reppe (1940) prepared cyclo-octatetraene in large quantities by the polymerisation of acetylene under pressure in the presence of a nickel compound, *e.g.*, nickel cyanide, as catalyst, in tetrahydrofuran solution.

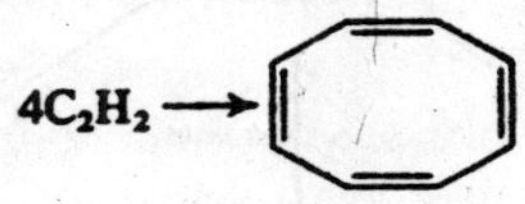

Fig. 1.23

Infra-red and Raman spectra studies have shown that cyclo-octatetraene contains a conjugated system of double bonds, and this is supported by X-ray analysis, which has also shown that the ring is non-planar.

According to Treibs (1950), cyclo-octatetraene is 'tub-shaped', there being comparatively small strain in this conformation.

If the double bonds were delocalised, the molecule would be flat and consequently would experience large strain, too large to be compensated by the gain in resonance energy.

Cyclo-octatetraene is a reactive compound and many of its reactions are accompanied by rearrangement to benzene and bicyclo [4,2,0]-ring compounds,. These are the result of transannular reactions.

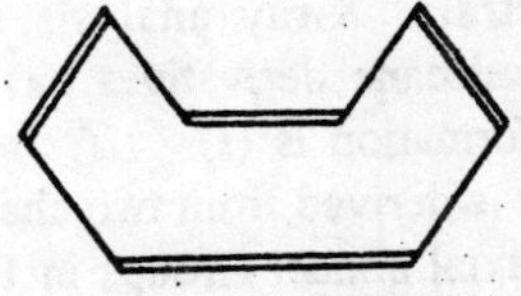

Fig. 1.24

Cyclononane, cyclodecane, and cyclo-undecane may be prepared by the methods used for large rings.

Medium-ring compounds have many properties which are quite different from those of the other classes of ring compounds.

Reference to Table 1.1 shows that medium rings have large strain relative to all other rings except the small rings.

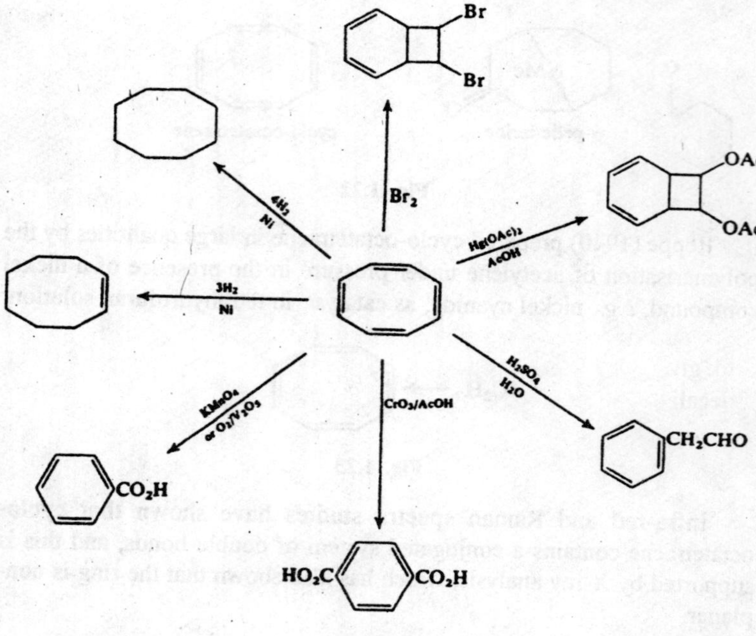

Fig. 1.25

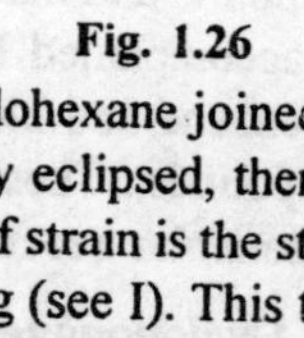

Fig. 1.26

The contributing factors to strain in cycloalkanes are bond-angle deformation, steric repulsion, and bond-opposition forces. In medium rings, however, there is *another* source of strain. X-ray analysis studies on some cyclodecane derivatives have shown that the conformation is (I).

This is derived from two chair conformations of cyclohexane joined by 1,3-axial bonds. Groups in 1,2-positions are partially eclipsed, thereby producing steric strain, but a more important source of strain is the steric repulsion between atoms on opposite sides of the ring (see I). This type of interaction has been designated **transannular interaction,** and this produces **trans-annular strain.** At the same time, it is this close proximity of *opposite* sides of the ring that gives rise to **transannular reactions,** *e.g.,* the formolysis of epoxycyclodecane (II), followed by hydrolysis gives, among other products, cyclodecane-1,6-diol (these structures are drawn in the conventional manner).

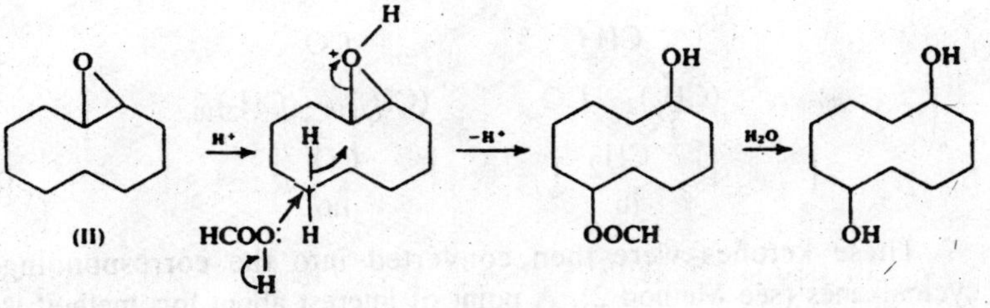

Fig. 1.27

Another example is the catalytic reduction of cyclodecane 1,6-dione to give a mixture of 1,6-dihydroxycyclodecane and 9,10-dihydroxy-decalin.

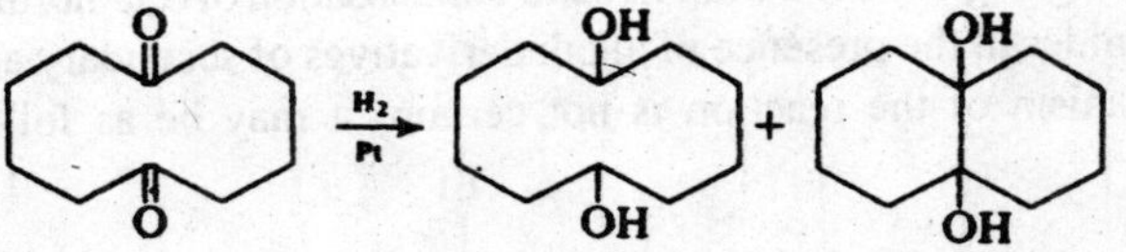

Fig. 1.28

Large-ring Compounds

Preparations of large-ring compounds. Up to 1926, the largest ring compound known contained eight carbon atoms. Ruzicka and his co-workers (1926, onwards) prepared large rings containing up to thirty-four carbon atoms. Their first method was to distil the calcium salts of dibasic acids, this method being limited to the preparation of cyclopentanone, cyclohexanone and cyclo-heptanone; the yields were poor. Ruzicka and his co-workers increased the yields, and also were able to obtain larger rings, by distilling in vacua at about 300°C the thorium, cerium or yttrium salt mixed with copper powder (which aids heat conduction). Usually a number of products was obtained by the distillation of a particular acid, *viz.,* a cyclic hydrocarbon, a cyclic monoketone and a cyclic diketone; *e.g.,* the yttrium salt of the dibasic acid $(CH_2)_{10}(CO_2H)_2$ gave a cyclic hydrocarbon, the cyclic monoketone (I), and the cyclic diketone (II).

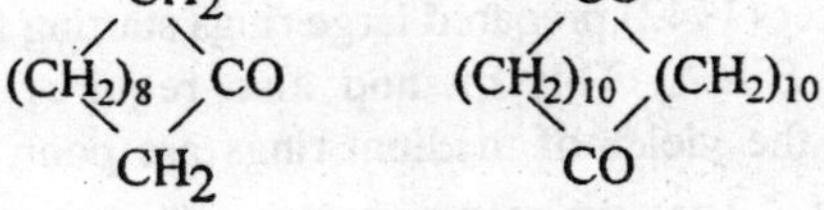

(I) (II)

$$\underset{\text{(I)}}{\text{cyclo-}[CH_2-(CH_2)_8-CO-CH_2]} \qquad \underset{\text{(II)}}{\text{cyclo-}[CO-(CH_2)_{10}-CO-(CH_2)_{10}]}$$

These ketones were then converted into the corresponding cycloalkanes (see Method 2). A point of interest about this method is that 9- and 10- rings cannot be prepared this way. The structure of the ring compound (the cyclic ketone) was establised by oxidation to the α, ω-dicarboxylic acid.

Ziegler *et al.* (1933) made use of the *high-dilution principle,* obtaining large by the intramolecular condensation of α,ω-normal aliphatic dicyanides in the presence of alkali derivatives of secondary amines. The mechanism of the reaction is not certain, it may be as follows.

$$(CH_2)_n\begin{cases}CH_2CN\\CH_2CN\end{cases} \xrightarrow{LiN(C_6H_5)(C_2H_5)} (CH_2)_n\begin{cases}CH(Li)CN\\CH_2CN\end{cases} + C_6H_5NHC_2H_5 \longrightarrow$$

$$\text{cyclo-}[CHCN-(CH_2)_n-CH_2-C{=}NLi] \xrightarrow{\text{acid}} \text{cyclo-}[CHCN-(CH_2)_n-CH_2-C{=}NH] \longrightarrow$$

$$\text{cyclo-}[CHCO_2N-(CH_2)_n-CH_2-CO] \xrightarrow{\text{heat}} \text{cyclo-}[CH_2-(CH_2)_n-CH_2-CO]$$

The method may be regarded as in extension of the Thorpe nitrlie condensation.

Hunsdiecker (1942) prepared large rings starting from ω-bromoacyl chlorides and E.A.A. This method also requires the high-dilution technique, but the yields of medium rings are poor.

$$Br(CH_2)_nCOCl + [CH_3COCHCO_2C_2H_5]^-Na^+ \longrightarrow$$

$$\mathrm{Br(CH_2)_nCO\underset{}{\overset{COCH_3}{\overset{|}{C}}}HCO_2C_2H_5 + NaCl}$$

$$\xrightarrow[CH_3OH]{CH_3ONa} \mathrm{Br(CH_2)_nCOCH_2CO_2CH_3} \xrightarrow[acetone]{NaI}$$

$$\mathrm{I(CH_2)_nCOCH_2CO_2CH_3} \xrightarrow[butan\,one]{K_2CO_3}$$

$$\mathrm{(CH_2)_n}\begin{cases}\mathrm{CO}\\ \mathrm{CHCO_2CH_3}\end{cases} \xrightarrow{H_2SO_4} \mathrm{(CH_2)_n}\begin{cases}\mathrm{CO}\\ \mathrm{CH_3}\end{cases}$$

Leonard *et al.* (1958), using the high-dilution technique, extended the Dieckmann method to the preparation of large rings.

$$\mathrm{(CH_2)_n}\begin{cases}\mathrm{CO_2Et}\\ \mathrm{CO_2Et}\end{cases} \xrightarrow{t\text{-}BuOK} \mathrm{(CH_2)_n}\begin{cases}\mathrm{CO}\\ \mathrm{CHCO_2Et}\end{cases} \xrightarrow{H_2SO_4} \mathrm{(CH_2)_n}\begin{cases}\mathrm{CO}\\ \mathrm{CH_2}\end{cases}$$

The most useful method of preparing large rings is the intramolecular acyloin condensation of α,ω-dicarboxylic esters (Prelog, 1947, cf. This method does not require the high-dilution technique.

$$\mathrm{(CH_2)_n}\begin{cases}\mathrm{CO_2R}\\ \mathrm{CO_2R}\end{cases} \xrightarrow[Et_2O:N_2]{Na} \mathrm{(CH_2)_n}\begin{cases}\mathrm{CO}\\ \mathrm{CHOH}\end{cases}$$

The corresponding cycloalkane is obtained by reduction of the acyloin by the Clemmensen method.

Larger rings have also been prepared by the reduction of cyclopolynes.

Larger ring compounds are very stable and those containing more than 20 carbon atoms consist of two approximately parallel chains joined at both ends by methylene bridges.

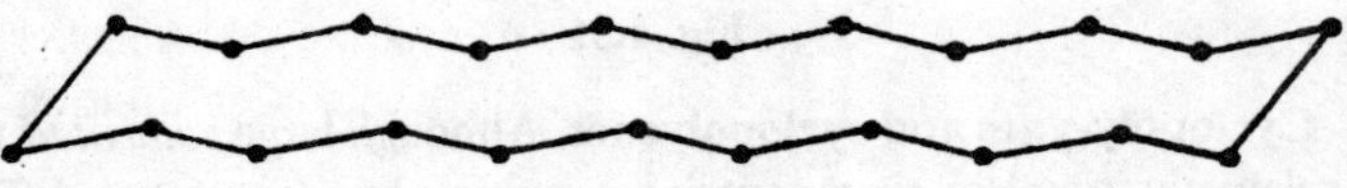

Fig. 1.29

Cyclophanes. These are ring systems which contain benzene rings, *e.g.*,

Fig. 1.30

Benzocyclenones, which contain one benzene ring, have also been prepared and have been converted into the corresponding cyclophanes.

Catenanes. These are large-ring compounds consisting of interlocked ring. Wasserman (1960) prepared the ketone (I) which, in the form of its hydrocarbon, may be represented as (II), (III), which is a *trefoil,* is an example of a single chain that is knotted. No trefoil has yet been synthesised.

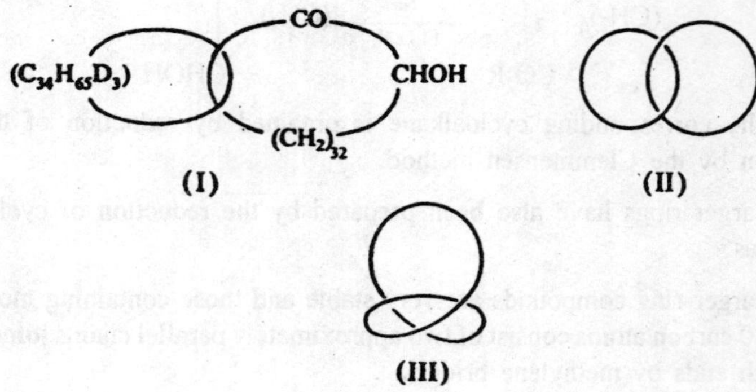

Fig. 1.31

Cyclopoly-ynes and cyclopolyenes. Although large *saturated* rings are relatively strainless, rings containing multiple bonds present a different picture. So far, the smallest ring prepared that contains, *cis-* double bond is (I), a *trans-* double bond, (II), and a triple bond, (III).

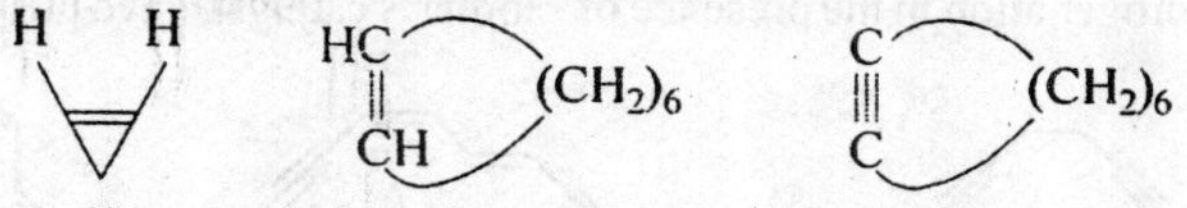

H H

HC=CH (CH₂)₆

C≡C (CH₂)₆

(I) (II) (III)

Fig. 1.32

All are highly strained and, in consequences, show abnormal reactivity. Strain in (II) is due to the fact that the molecule is twisted about the double bond.

Large-ring cyclopoly-ynes and cyclopolyenes. Many of these have been prepared and they are relatively stable. The following example illustrates the method used. Tetradeca-l,13-di-yne, on treatment with excess of cupric acetate in methanolic pyridine, in high dilution, forms cyclotetradeca-1,3-di-yne and also some of the cyclic dimer cyclo-octacosa-l,3,15,17-tetfa-yne.

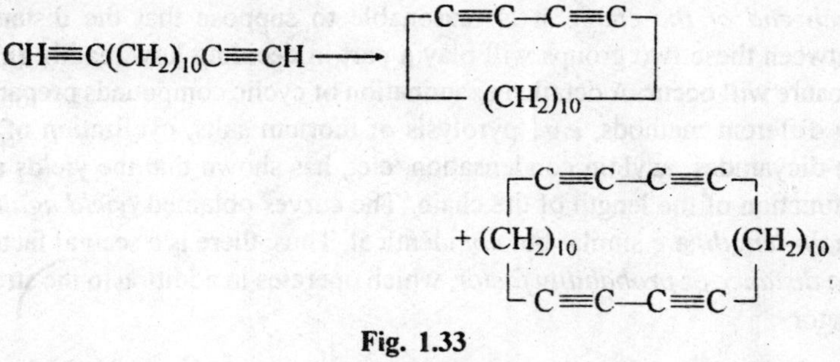

$$CH{\equiv}C(CH_2)_{10}C{\equiv}CH \longrightarrow \text{C}{\equiv}\text{C—C}{\equiv}\text{C} \;/\; (CH_2)_{10}$$

$$+ \; (CH_2)_{10} \;\; \text{C}{\equiv}\text{C—C}{\equiv}\text{C} \;\; (CH_2)_{10} \;\; \text{C}{\equiv}\text{C—C}{\equiv}\text{C}$$

Fig. 1.33

These, on catalytic hydrogenation (platinum), gave respectively cyclotetradecane ($C_{14}H_{28}$) and cyclo-octacosane ($C_{28}H_{56}$). Sondheimer *et al* (1959) have carried out oxidative coupling with octa-l,7-di-yne and obtained 16-, 24-, 32- and 40-membered polyalkynes which, on hydrogenation, give cycloalkanes. These authors have now prepared up to a 54-membered ring system, the largest ring hitherto contained 34 carbon atoms.

The above method was used by Sondheimer to prepare **annulenes** (cyclopolyenes, *e.g.*, [18] annulene was prepared by the oxidative coupling of hexa-l,5-di-yne to give the cyclic trimer (plus the tetramer and pentamer), and this, on heating with potassium t-butoxide in t-butanol, isomerises (by prototropic rearrangement) to a conjugated enyne which, on hydrogenation in the presence of Lindlar's catalyst, gave [18] annulene.

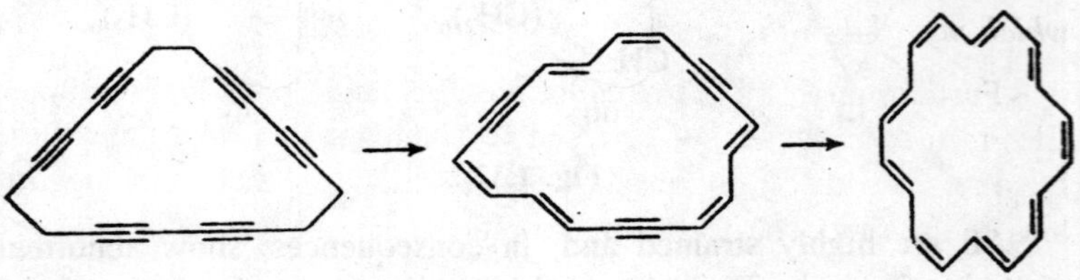

Fig. 1.34

Factors influencing the ease of ring formation. Inspection of Table 19.1 shows that if strain were the only factor, ease of ring-closure would be greatest for the six-membered ring and large rings. In practice, the ease of formation is greatest for common rings (5–7), and it therefore follows that factors other than strain must be operating. In view of the fact that rings are formed by reaction between the functional groups *at each end of the chain,* it is reasonable to suppose that the distance between these two groups will play a part in deciding how readily ring-closure will occur. A detailed examination of cyclic compounds prepared by different methods, *e.g.*, pyrolysis of thorium salts, cyclisation of α, ω)-dicyanides, acyloin condensation, etc., has shown that the yields are a function of the length of the chain. The curves obtained *(yield against chain-length)* are similar but *not* identical. Thus, there is a second factor, the *distance or probability factor,* which operates in addition to the strain factor.

Some large-ring compounds occur in nature, two of which are *civetone* and *muscone.* The following account of the elucidation of their structures is typical of the 'classical' methods used in the examination of natural products.

Civetone, m.p. 31°C, occurs in the civet cat. Ruzicka (1926, 1927) showed that its molecular formula is $C_{17}H_{30}O$, and that the oxygen is in an oxo group (formation of an oxime), and that an olefinic bond is present (addition of bromine). When catalytically reduced, civetone absorbs one molecule of hydrogen to form **dihydrocivetone.** Thus, civetone contains one double bond and therefore its parent (saturated) hydrocarbon is $C_{17}H_{34}$ (add two hydrogens for one double bond, and replace the O of the C = O by two hydrogens). This formula corresponds to a *monocyclic* hydrocarbon.

Dihydrocivetone condenses with benzaldehyde; this shows the presence of a — CH_2 — CO — group. Also, since the oxidation of dihydrocivetone with chromic acid gives a dicarboxylic acid $C_{17}H_{32}O_4$, *without* loss of carbon, the oxo group is confirmed to be a keto group *which must be in the ring.*

Furthermore, the same dicarboxylic acid is obtained by the Clemmensen reduction of civetone to **civetene** (in which the double bond remains intact), followed by ozonolysis. Thus, the double bond is also in the ring. The facts obtained so far may be formulated.

$$C_{15}H_{28}\left\}\begin{matrix}-CO\\ |\\ -C{=}CHPh\end{matrix}\right. \xleftarrow{PhCHO} C_{15}H_{28}\left\}\begin{matrix}-CO\\ |\\ -CH_2\end{matrix}\right. \xrightarrow{H_2} C_{15}H_{30}\left\}\begin{matrix}-CO\\ |\\ -CH_2\end{matrix}\right. \xrightarrow{CrO_3} C_{15}H_{30}\left\}\begin{matrix}-CO_2H\\ |\\ -CO_2H\end{matrix}\right.$$

$$C_{15}H_{28}\left\}\begin{matrix}-CO\\ |\\ -CH_2\end{matrix}\right. \xrightarrow{Zn/Hg,\ HCl} \begin{matrix}CH-\\ |\\ CH-\end{matrix}\Big\}C_{13}H_{26}\left\{\begin{matrix}-CH_2\\ |\\ -CH_2\end{matrix}\right. \xrightarrow{O_3} \begin{matrix}HO_2C-\\ |\\ HO_2C-\end{matrix}\Big\}C_{13}H_{26}\left\{\begin{matrix}-CH_2\\ |\\ -CH_2\end{matrix}\right. \equiv C_{15}H_{30}\left\}\begin{matrix}-CO_2H\\ |\\ -CO_2H\end{matrix}\right.$$

Ruzicka now showed that the dicarboxylic acid, $C_{17}H_{32}O_4$, is identical with pentadecane-1,15-dicarboxylic acid (by comparison with a *synthetic* specimen).

Hence, civetone is a seventeen-membered ring containing a keto group and one double bond, but there is no information which gives their relative positions in the ring. This was elucidated as follows. Civetone, on oxidation with cold permanganate, gave a ketodicarboxylic acid, $C_{17}H_{30}O_5$, and as the keto group is intact, this is further proof that the double bond is in the ring.

When this ketodicarboxylic acid was oxidised with sodium hypobromite, a mixture of succinic, pimelic, suberic, and azelaic acids was obtained. Since azelaic acid is $HO_2C(CH_2)_7CO_2H$, the double bond in civetone must be separated from the keto group by at *least* seven methylene groups, and the absence of higher acids suggests a symmetrical structure. The structure of civetone which fits these facts is (1), and is further supported by synthesis of the ketodicarboxylic acid from the methyl acid ester of azelaic acid.

$$\begin{array}{l} CH(CH_2)_7 \\ \| \quad \backslash CO \\ CH(CH_2)_7 \end{array} \xrightarrow{KMnO_4} \begin{array}{l} HO_2C(CH_2)_7 \\ \quad \backslash CO \\ HO_2C(CH_2)_7 \end{array} \xrightarrow[\text{(ii) hydrolysis}]{\text{(i) Fe; heat}} 2(CH_2)_7\begin{array}{l} CO_2Me \\ CO_2H \end{array}$$

Final confirmation of the structure of civetone comes from its synthesis.

Several syntheses have been carried out, that given here is due to Stoll (1948), and points to note are the use of the acyloin synthesis, protection of the keto group, and the formation of *cis*- and *trans*-isomers.

$$OC\begin{array}{l}(CH_2)_7CO_2Me \\ (CH_2)_7CO_2Me\end{array} \xrightarrow[H_3O^+]{HOCH_2CH_2OH} \left[\begin{array}{l} O \\ O \end{array}\right]C\begin{array}{l}(CH_2)_7CO_2Me \\ (CH_2)_7CO_2Me\end{array} \xrightarrow[\text{xylene}]{Na}$$

$$\left[\begin{array}{l} O \\ O \end{array}\right]C\begin{array}{l}(CH_2)_7-CO \\ (CH_2)_7-CHOH\end{array} \xrightarrow{H_2/Ni} \left[\begin{array}{l} O \\ O \end{array}\right]C\begin{array}{l}(CH_2)_7-CHOH \\ (CH_2)_7-CHOH\end{array} \xrightarrow[AcOH]{HBr}$$

cis and trans

$$OC\begin{array}{l}(CH_2)_7-CHOAc \\ (CH_2)_7-CHBr\end{array} \xrightarrow[EtOH]{Zn} C\begin{array}{l}(CH_2)_7-CH \\ \qquad\qquad \| \\ (CH_2)_7-CH\end{array}$$

cis and trans   cis and trans

Because of the large ring, *cis*= and *trans*-isomers are possible, and both were present in the product.

$$OC\begin{array}{l}(CH_2)_7-C-H \\ \qquad\qquad \| \\ (CH_2)_7-C-H\end{array} \qquad OC\begin{array}{l}(CH_2)_7-C-H \\ \qquad\qquad \| \\ (CH_2)_7\ HC\end{array}$$

cis   trans

The two forms, referred to as α,- and β-forms, were separated by fractional crystallisation of their dioxalans, and the b-form was then treated as shown.

(CH₂)₇ ... β-civetone → (i) $HOCH_2CH_2OH/H^+$ (ii) Br_2 → ketal dibromide → KOH/EtOH → ketal alkyne → (i) H_2/Ni (ii) H_3O^+ → cis-civetone

Since catalytic hydrogenation of a triple bond normally gives cis-addition, the configuration of the product is *cis (i.e.,* α-civetone). This was shown to be identical with natural civetone.

Muscone *(muskone)* occurs in natural musk (from the musk deer). It is a thick colourless oil and is optically active. Its molecular formula is $C_{16}H_{30}O$, and it was shown to be a ketone. Since it was also shown to be saturated, it must therefore be cyclic, since if it were an open-chain compound, its formula would be $C_{16}H_{32}O$, if it had contained more than one ring, the number of hydrogen atoms would have been less than thirty.

The investigation of the odours of cyclic compounds was then used to elucidate the structure of muscone. Ruzicka was led to adopt this procedure because he had found that civetone and dihydro-civetone had about the same odour, *i.e.,* he assumed that similar structures gave rise to similar odours (this might be termed a *physiologico-chemical* method). Ruzicka found that the odour of muscone was identical with that of synthetic cyclopentadecanone and its methyl derivatives. He believed that muscone was therefore a methyl derivative of the fifteen-membered cyclic ketone.

This was proved by preparing cyclopentadecanone (by distillation of the thorium salt of tetradecane-1,14-dicarboxylic acid), treating it with methylmagnesium iodide, and dehydrating the tertiary alcohol so produced to the alkene which was then catalytically reduced to methylcyclopentadecane.

$$\begin{array}{l} CH_2\text{—}CO \\ |\qquad\quad | \\ (CH_2)_{12}\text{—}CH_2 \end{array} \xrightarrow[\text{(ii) } H_2O]{\text{(i) } CH_3MgI} \begin{array}{l} \qquad\quad CH_3 \\ \qquad\quad | \\ CH_2\text{—}CO \\ |\qquad\quad | \\ (CH_2)_{12}\text{—}CH_2 \end{array} \xrightarrow{-H_2O}$$

$$\begin{array}{l} \qquad\quad CH_3 \\ \qquad\quad | \\ CH_2\text{—}C \\ |\qquad\quad | \\ (CH_2)_{12}\text{—}CH \end{array} \xrightarrow{H_2/Ni} \begin{array}{l} \qquad\quad CH_3 \\ \qquad\quad | \\ CH_2\text{—}CH \\ |\qquad\quad | \\ (CH_2)_{12}\text{—}CH_2 \end{array}$$

Methylcyclopentadecane is also obtained when muscone is reduced by the Clemmensen method.

The problem now was to determine the position of the methyl group with respect to the carbonyl group. When oxidised with chromic acid, muscone yields two acids with the formula $C_{14}H_{28}(CO_2H)_2$, and also a mixture of even-numbered dicarboxylic acids from succinic to decane-dicarboxylic acid. It therefore follows that there is a straight chain of at least ten methylene groups in muscone. Hence, muscone must be (I), (II), or (III).

$$\underset{\text{(I)}}{\begin{array}{l} CH_3CH(CH_2)_3\text{—} \\ \qquad\qquad\qquad CO \\ \quad (CH_2)_{10}\text{—} \end{array}} \qquad \underset{\text{(II)}}{\begin{array}{l} CH_3CH(CH_2)_2\text{—} \\ \qquad\qquad\qquad CO \\ \quad (CH_2)_{11}\text{—} \end{array}} \qquad \underset{\text{(III)}}{\begin{array}{l} CH_3CHCH_2\text{—} \\ \qquad\qquad\quad CO \\ \quad (CH_2)_{12}\text{—} \end{array}}$$

Fig. 1.35

Although, Ruzicka failed to separate the mixture of the two C14 dicarboxylic acids (see above) into two pure specimens, he obtained a partially purified specimen of one of them which he showed was identical with a *synthetic* specimen of 1-methyltridecane-1,13-dicarboxylic acid. Thus, muscone is (III), and this was confirmed by synthesis, *e.g.*, Ziegler *et al.* (1934)

$$\begin{array}{l} \quad CH(CH_3)\text{—}CH_2CN \\ (CH_2)_{10} \\ \quad CH_2CH_2CN \end{array} \xrightarrow[\text{(ii) } H_3O^+]{\text{(i) LiN(Ph)(Et)}} \underset{(\pm)\text{-muscone}}{\begin{array}{l} \quad CH(CH_3)CH_2 \\ (CH_2)_{10} \qquad\qquad CO \\ \quad CH_2CH_2 \end{array}}$$

Civetone and muscone are both used in perfumery. Investigation of the relationship between ring size and odours of cyclic ketones is as follows, C_5, bitter almonds, C_6, mint, $C_7 - C_9$, transition to camphor, $C_{10} - C_{13}$, transition to cedar, $C_{14} - C_{16}$, transition to must with maximum musk odour at C_{15}, C_{17}, civetone, C_{18}, very weak civetone, from C_{19} onwards the odour decreases very rapidly.

Cyclopentadecanone is used in perfumery under the name of *exaltone.*

EXERCISES

1. Formulate the following reaction.

$$CH_3CO_2Et + (CH_2)_5(MgBr)_2 \xrightarrow[\text{(ii) } H^+]{\text{(i) } Et_2O} \text{(A) } (C_7H_{14}O \xrightarrow{H^+}$$

$$\text{(B) } C_7H_{12} \xrightarrow[\text{reagent}]{\text{Lemieux}} \text{(C) } C_7H_{12}O_3$$

2. Complete the following equations.

(i) cyclohexanone $\xrightarrow{SeO_2}$?

(ii) cyclohexanone + $HCO_2Et \xrightarrow{EtONa}$

(iii) $ClCH_2CH_2CH_2CN \xrightarrow{NaOH}$ (cyclic compound) ?

(iv) $PhCHBrCH_2CH_2Br \xrightarrow{Zn-Cu}$?

(iv) 2-methylcyclohexanol (Me, OH) $\xrightarrow[\text{warm}]{CrO_3-H_2SO_4}$?

(vi) $C_3H_7CHClCOCH_2Me \xrightarrow{EtONa}$?

(vii) $MeCH(CH_2CO_2Et)CH_2CH_2Et \xrightarrow{EtONa}$?

(viii) $MeCO(CH_2)_3CO_2Et \xrightarrow{EtONa}$?

(ix) cyclohexanone $\xrightarrow[H^+]{H^+}$? $\xrightarrow{H^+}$? $\xrightarrow[H^+]{HN_3}$

3. Treatment of $Me_2C = CHCH_2CH_2Cl$ with $CaCO_3 - H_2O$ gives cyclopropyldimethylmethanol. Suggest a mechanism.

4. Convert cyclohexanone into cyclopentanone.

5. Would you expect the NMR spectra of *cis-* and *trans* 1,2-dibromocyclopropane to be identical ? Explain.

6. Using i.r., spectroscopy, differentiate between 2-methylcyclopentanone, cyclohexanone, cyclohex-2-enone, and cyclohex-3-enone.

7. Given the conformation shown, state (a) the nature (e or a) of each group, and (b) the geometrical relationship (*cis* or *trans*) of, (i) Cl and Br, (II) I and Et, (III) Cl and Et, (iv) ET and Br (v) Me and 1, (vi) Cl and Me, (vii) 1 and Br, (viii) Et and Me, (ix) Cl and I, (x) I and Me.

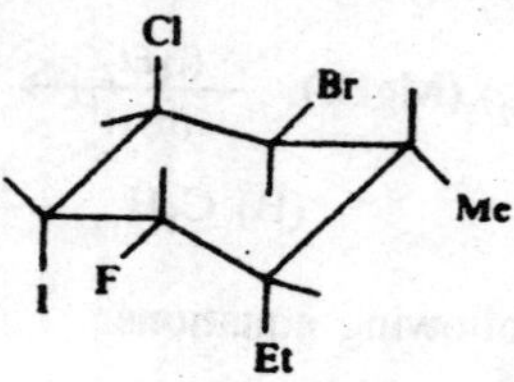

Fig. 1.36.

8. Which halogen atom in Prob. 7 would be most readily eliminated by the E2 mechanism to form the ene compound and why ?

9. Draw all possible conformations for the disubstituted cyclohexanes.

   (i) 1-Br-2-Cl, (ii) 1,3-Br-2, (iii) 1-Cl-4I, (iv) 1,3-$(OH)_2$.

   Which conformation is the most stable one in each set ? Which are optically active compounds ?

10. *cis*-and *trans*-Cyclohexane-1,2-diol are each treated with.

    (i) 1 molecule of Ac_2O, (ii) a second molecule of Ac_2O.

    Are the rates of the first and second reactions the same or different? Explain.

11. How many conformations are possible for, (i) *cis*-decal-2-ol ? Draw them.

12. Camphor contains *two* asymmetric carbon atoms, and exists as *one* pair of enantiomers. Explain.

13. Draw the equation (conformational) for the pyrolysis of cyclohexyl acetate and explain.

14. *One* isomer of hexachlorocyclohexane undergoes dehydrochlorination with great difficulty. Suggest its conformation. Explain. How many NMR signals would this isomer show ?

15. Menthyl and neomenthyl chlorides are stereoisomers of 1-methyl-3-chloro-4-isopropylcyclohexane and in both the alkyl groups are *trans*. On treatment with EtINa, menthyl chloride gives menth--2-ene (100%) and neomenthyl chloride gives menth-2-ene (25%) and menth-3-ene (75%). Deduce the conformation of the chlorine atom in each chloride, and formulate the reactions.

16. Discuss the data of the heats of formation (kJ mol^{-1}) for the following methylcyclohexanols, *cis*-2-Me-n– 396.6, *trans*–2-Me-, – 422.2 *cis*-3-Me-, – 422.6, *trans*-3-Me-, – 400.8.

17. *cis*-Cyclohexane-1,2-diol is oxidised by periodic acid more rapidly than the corresponding *trans*- isomer. Discuss.

18. Convert cyclohexanol into *cis*-and *trans*-cyclohexane-1,2-diol.

19. Discuss, *cis*-2-chlorocyclohexanol with aqueous alkali gives cyclohexanone, whereas the *trans* isomer gives cyclohexene oxide.

20. The deamination (with nitrous acid) of *trans*-2-aminocyclohexanol gives a cyclopentane derivative, $C_6H_{10}O$, as the only product and that of the *cis*- isomer gives two products, one the same as that from the *trans*-isomer and the other a cyclohexane derivative, $C_6H_{10}O$. Formulate the reactions and explain them.

21. Starting from D.E.M. and any other compounds you like, synthesis cyclohexane-1,4-dicarboxylic acid.

22. Convert 1,2,4-trihydroxycyclohexane into 3,4-dihydroxycyclohexanone.

23. Expoxidation of 4-methylcyclopentene gives two isomeric epoxides, with one predominating. Discuss.

24. $Ph\,N_2^+Cl^-$ couples with $PhNMe_2$, but not with 2,6-Me_2; C_6H_3-NMe_2. Explain

25. Convert:

(i) H_2N–C_6H_4–C_6H_4–NH_2 ⟶ F–C_6H_4–C_6H_4–F ;

(ii) m-dinitrobenzene ⟶ m-chloronitrobenzene;

(iii) benzaldehyde ⟶ m-chlorobenzaldehyde;

(iv) o-chloronitrobenzene ⟶ o-bromochlorobenzene;

(v) p-bromoaniline ⟶ p-bromoiodobenzene;

(vi) 2-Br-4-Me-aniline ⟶ m-bromotoluene;

(vii) m-aminobenzaldehyde ⟶ m-hydroxybenzaldehyde;

(viii) o-toluidine ⟶ o-methylphenyl cyanide;

(ix) p-nitroaniline ⟶ p-dinitrobenzene;

(x) o-nitrotoluene ⟶ o-cresol;

(xi) nitrobenzene ⟶ pentabromo-benzene.

26. p-Toluidine reacts with benzenediazonium chloride to form a compound which, on boiling with H_2SO_4, gives four products (excluding nitrogen). Discuss.

27. Convert:

(i) PhH ⟶ m-dibromobenzene;

(ii) PhMe ⟶ 3-Br-4-NO_2-toluene;

(iii) PhH ⟶ m-difluorobenzene;

(iv) PhMe ⟶ 3,4-dinitrotoluene;

(v) $PhNH_2$ ⟶ 1,3,5-tribromobenzene.

28. Discuss, including the use of thermochemical data, the; reaction:

$$\text{Ph}\,\text{N}_2^+\text{Cl}^- + \text{CH}_2(\text{CN})\text{CO}_2\text{Et} \xrightarrow{\text{AcONa}} \text{PhNHN}{=}\text{C(CN)CO}_2\text{Et}$$

29. Complete the following equations:

(i) $\text{PhN} = \text{NNHPh} + \text{PhOH} \xrightarrow{\text{HCl}}$

(ii) $\text{PhN}_2^+\text{BF}_4^- + \text{PhNO}_2 \xrightarrow{\text{heat}} ?$

(iii) $\text{p-ClC}_6\text{H}_4\text{H}_2^+\text{Cl}^- + \text{CH}_2{=}\text{CHCHO} \xrightarrow{\text{Cu}^{2+}} ?$

2

PHENOLS AND QUINONES

INTRODUCTION

Phenols are aromatic compounds containing hydroxyl groups directly attached to the nucleus, and they are classified as monohydric, dihydric, trihydric phenols, etc., according as they contain one, two, three, etc., hydroxyl groups.

MONOHYDRIC PHENOLS

General Methods of Preparation

1. A number of monohydric phenols occur in coal-tar and their extraction from this source is very important commercially.

   (i) *Middle oil* is cooled, whereupon naphthalene crystallises out (43 per cent). The oil is pressed free from the naphthalene and then treated with aqueous sodium hydroxide, which dissolves the phenols. The alkaline liquor is drawn off, boiled, and air is blown through; this removes naphthalene (that remained after cooling the oil), pyridine, etc. The liquid is allowed to cool and then carbon dioxide is blown through, thereby decomposing the sodium phenoxides, into the free phenols and sodium carbonate, the latter dissolving in the aqueous layer. This aqueous layer is drawn off, and the crude phenols (the yield of which is about 12 per cent of the middle oil) are fractionated. Three fractions are collected: *phenol*, b.p. 182°C (20%), *cresols*, b.p. 190-203°C (43%) and *xylenols*, b.p. 211-225°C (26%); the residue is *pitch*.

   (ii) *Heavy oil* is treated in the same manner as above. After being pressed to remove naphthalene, the residual oil contains cresols, higher phenols, naphthol, etc. (heavy oil contains about 7 per cent of the phenols).

2. Phenols may be prepared by fusion of sodium arylsulphonates with sodium hydroxide:

$$ArSO_3^- Na^+ + 2NaOH \longrightarrow AtO^-Na^+ + Na_2SO_3 + H_2O$$

3. When a diazonium sulphate solution is steam distilled, a phenol is produced

$$ArN_2^+ HSO_4^- + H_2O \longrightarrow ArOH + N_2 + H_2SO_4$$

4. Phenols are formed when compounds containing an activated halogen atom are heated with aqueous sodium hydroxide; *e.g.*, p-nitrophenol from p-chloronitrobenzene:

$$p\text{-}ClC_6H_4NO_2 + 2NaOH \longrightarrow p\text{-}O^-Na^+C_6H_4NO_2 + NaCl + H_2O$$

5. Distillation of phenolic acids with soda-lime produces phenols; *e.g.*, sodium salicylate gives phenol:

$$o\text{-}HOC_6H_4CO_2Na + NaOH(CaO) \longrightarrow C_6H_5O^-Na^+ + Na_2CO_3$$

6. Phenols may be prepared by means of a Grignard reagent:

$$ArMgBr \xrightarrow{O_2} ArO_2MgBr$$

$$\xrightarrow{ArMgBr} 2ArOMgBr \xrightarrow{H^+} 2ArOH$$

Alternatively, the Grignard reagent may be converted into a phenol via a boronic or bromic acid (these need not be isolated; Hawthorn, 1957):

$$ArMgX \xrightarrow[(ii)H_2O]{(i)B(OMe)_3} ArB(OH)_2 \text{ or}$$

$$Ar_2BOH \xrightarrow{H_2O_2} ArOH\ (60\text{–}70\%)$$

Phenol (*carbolic acid, hydroxybenzene*) may be prepared by any of the general methods; commercially, it is prepared from coal-tar. The supply from this source is now insufficient to give the amount of phenol required for industry, and so it is also prepared synthetically. Various methods are used.

The oldest synthetic method is the alkaline fusion of sodium benzenesulphonate. Two other methods start from chlorobenzene, one conversion being effected by heating with aqueous sodium carbonate or sodium hydroxide under pressure at about 300°C, and the other conversion being effected by heating with steam at 425°C in the presence of a catalyst (in the latter, the chlorobenzene is prepared 'by the Raschig process;). Another method is the oxidation of toluene by air in the presence of manganous and cupric salts as catalyst:

$$C_6H_5CH_3 \xrightarrow[\text{cat.}]{O_2} C_6H_5OH + CO_2$$

The *cumene-phenol* process appears to be the most important commercial method and is carried out by the oxidation of cumene to its hydroperoxide, which is then decomposed by acid into phenol and acetone (which is sold as a by-product). The mechanism of the conversion of cumene hydroperoxide into phenol is not certain, but it most probably involves a 1,2-shift (cf. the Baeyer-Villager oxidation).

$$C_6H_6 + CH_3CH{=}CH_2 \xrightarrow[250\,^\circ C\,;\ \text{press}]{H_3PO_4} \text{C6H5CH(CH3)2} \xrightarrow[OH^-\,;130\,^\circ C]{O_2}$$

$$(CH_3)_2\underset{\displaystyle|}{\overset{C_6H_5}{C}}\!-\!O\!-\!OH \xrightarrow[100\,^\circ C]{H_2SO_4} (CH_3)_2\overset{C_6H_5}{C}\!-\!O\!-\!\overset{}{O}H_2 \xrightarrow{-H_2O}$$

$$(CH_3)_2\overset{C_6H_5}{C}\!-\!O^+ \longrightarrow (CH_3)_2\overset{+}{C}\!-\!O\!-\!C_6H_5 \xrightarrow[(ii)-H^+]{(i)H_2O}$$

$$(CH_3)_2\overset{:OH}{C}\!-\!OC_6H_5 \longrightarrow (CH_3)C{=}OH^+ + C_6H_5O^-$$

$$\longrightarrow (CH_3)_2C{=}O + C_6H_5OH$$

Properties of phenol. These are characteristic of monohydric phenols. Phenol is a colourless crystalline solid, m.p. 43°, b.p. 182°C, which turns pink on exposure to air and light. It is moderately soluble in cold water, but is readily soluble in ethanol and ether. Phenol undergoes the Liebermann reaction, when phenol is dissolved in concentrated sulphuric acid and a few drops of aqueous sodium nitrite added, a red colour is obtained on dilution, and turns blue when made alkaline with aqueous sodium hydroxide.

Phenol is used as an antiseptic and disinfectant, and in the preparation of dyes, drugs, bakelite, etc.

The infrared absorption region of the phenolic hydroxyl group (O—H stretch) is the same as that for alcohols, but the nucleus may be identified by its own absorption spectrum. At the same time, there will also be present the pattern due to a substituted benzene. Furthermore, there may be complications due to intermolecular hydrogen bonding, or intramolecular hydrogen bonding in various o-derivatives.

Mass Spectra

The molecular ion of phenols is usually very intense and fragments in various ways according to whether other substituent groups are present, *e.g.*, phenol itself:

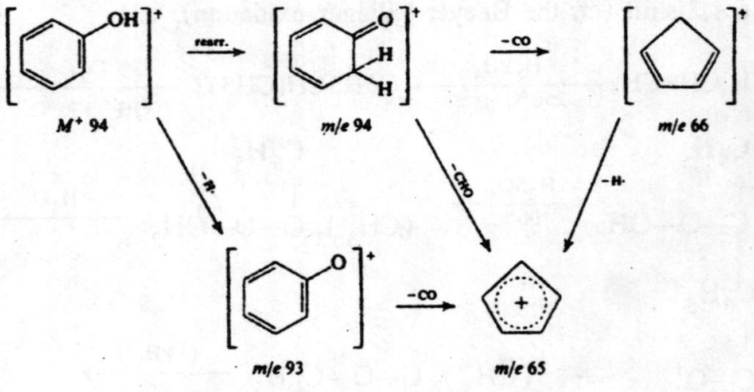

OH; rearr.; O; H; H; -CO; M^+ 94; m/e 94; m/e 66; -H·; -CHO; -H·; O; -CO; +; m/e 93; m/e 65

Cresols form the hydroxytropylium ion *e.g.*,

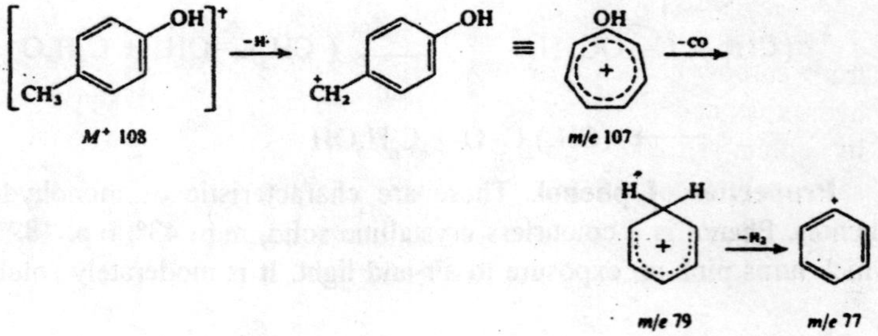

OH; CH_3; -H·; OH; $\overset{+}{C}H_2$; ≡; OH; +; -CO; M^+ 108; m/e 107

H; H; +; $-H_2$; m/e 79; m/e 77

Phenolic ethers give a strong molecular ion peak and undergo several fragmentation patterns. One follows that of phenol, *e.g.*.

$[C_6H_5OCH_3]^{+\cdot}$ $\xrightarrow{-CH_3\cdot}$ $[C_6H_5O]^+$ $\xrightarrow{-CO}$ $C_5H_5^+$

M^+ 108 — *m/e* 93 — *m/e* 65

An alternative path involves a rearrangement

$[C_6H_5OCH_3]^{+\cdot}$ $\xrightarrow{-CH_2O}$ $[C_6H_6]^{+\cdot}$ $\xrightarrow{-H\cdot}$ $C_6H_5^+$ $\xrightarrow{etc.}$

m/e 78 — *m/e* 77

If the alkyl group contains two or more carbon atoms, then rearrangement can occur with elimination of an alkene (cf. alkyl ethers):

$[C_6H_5O\text{—}CH_2\text{—}CHR\text{—}H]^{+\cdot}$ $\longrightarrow$ $[C_6H_6O]^{+\cdot}$ + RCH=CH$_2$

m/e 94

Reactions of phenols. The hydroxyl group in phenol behaves in some respects like that in alcohols, but the majority of the reactions of phenol are those connected with the benzene ring.

1. Phenol gives a violet colour with ferric chloride; this reaction is characteristic of all compounds containing the grouping—C(OH)=C (cf. enols).

2. Phenol behaves as a weak acid, forming *phenoxides* with strong alkalis:

$$C_6H_5OH + NaOH \longrightarrow C_6H_5O^- Na^+ + H_2O$$

Since phenol is a weaker acid than carbonic acid, it may be separated from carboxylic acids by making the solution alkaline with sodium hydroxide, and then passing in carbon dioxide. Phenol is liberated from its sodium salt and so may be extracted with ether; the carboxylic acid salts are not decomposed by carbon dioxide.

Phenols are stronger acids than alcohols, and the explanations offered are similar to those for the carboxylic acids.

(i) Because of resonance, the oxygen atom of the OH group acquires a positive charge and so proton release is facilitated:

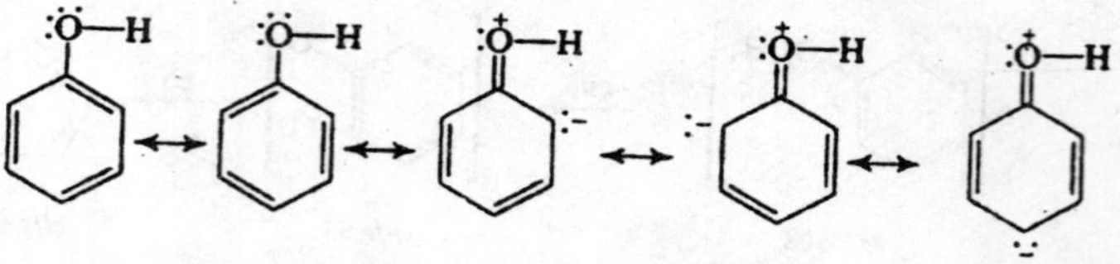

It might be noted that since the carbon atom of the C—OH group is sp^2 hybridised, it is more electron-attracting than an sp^2 carbon atom (as in alcohols), and consequently there is an increased –I effect which facilitates proton release. This inductive effect alone, however, is not large enough to explain the increased acid strength of phenols as compared with alcohols.

(ii) When phenol ionises, the phenoxide ion is also a resonance hybrid, but it is more stabilised by resonance than is the un-ionised phenol molecule because of spreading of a negative charge only. In the un-ionised molecule, unlike charges are spread and so is less stable.

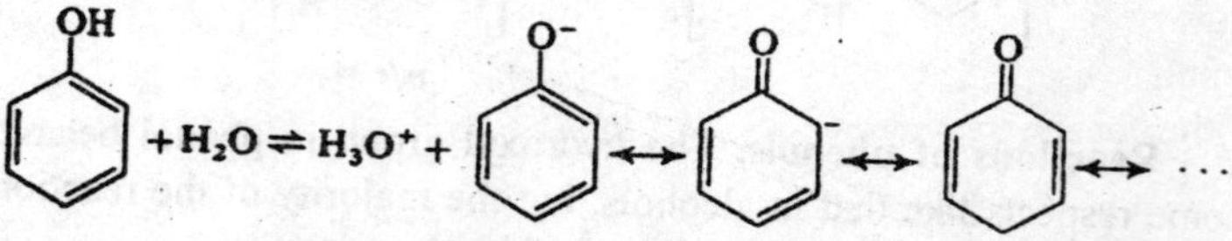

$$+ H_2O \rightleftharpoons H_3O^+ +$$

The effect of a ring substituent on the acid strength of phenols depends on whether the group is electron attracting or releasing, its ability to enter into resonance with the hydroxyl group and its position (cf. the anilines). Effects in the o-position are similar to those in the p-position but there are added complications due to the stearic effect and to hydrogen bonding (when this is possible).

The methyl group decreases the acid strength of phenol from all ring positions, Table 2.1.

The methyl group is electron releasing and so release of a lone pair from oxygen (of the OH group in the un-ionised phenol or from the O^- in the phenoxide ion) into the ring is opposed (cf. the anilines). This results in diminished resonance in the contributing structures, and consequently the phenoxide ion is more resonance-stabilised with respect to phenol than is the methylphenoxide ion with respect to methylphenol. Hence phenol is a stronger acid than methylphenol.

Table 26.1

Substituent	*pKa*		
	o	*m*	*p*
None: 9.98	10.28	10.08	10.14
NO_2	7.23	8.40	7.15
F	8.81	9.28	9.95
Cl	8.48	9.02	9.38
MeO	9.98	9.65	10.21
OH	9.48	9.44	9.96
NH_2	9.71	9.87	10.30
2,4,6,-tri-NO_2:			
0.71			

Since the nitro-group can enter into resonance (— R) with the hydroxyl group from both the o- and p-positions, there is increased resonance in the contributing structures, and so o- and p-nitrophenol are stronger acids than phenol (see also the Hammett equation). The cumulative effect of three nitro-groups in the 2,4,6-positions is shown by the fact that picric acid

is a very strong acid , Table 2.1. A m-nitro-group cannot enter into resonance with the hydroxyl group, but it can exert a –I effect from this position (cf. the anilines). Hence m-nitrophenol is a stronger acid than phenol, but is not so strong as the o- and p-isomers.

Groups with a +R effect, *e.g.*, NH_2, MeO, weaken the phenol when in the p-position, but strengthen it when in the m-position. In the latter case, only the –I effect can operate. All the halogenophenols are stronger acids than phenol, and so it follows that the –I effect of a halogen atom is greater than its + R effect.

A study of Table 2.1 shows that the effect of a given group in the o-position can be opposite to that when in the p-position. Intramolecular hydrogen bonding, when this is possible, may play a part; this would stabilise the un-ionised form. The stearic effect, because of the small size of a hydrogen atom, would be expected to be very small or completely absent; if it is the phenoxide ion that is being considered, then the stearic effect is absent.

This line of argument is supported by the following observations:

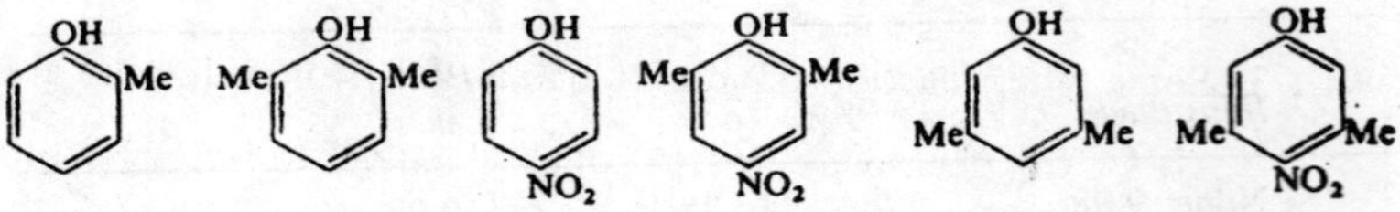

The decrease in acid strength from (I) to (II) can be explained by the additional +I effect in (II). (III) and (IV) have close PK, values, and so the stearic effect is absent, (IV) being slightly weaker due to the +1 effect of the two o-methyl groups. In (V), the +1 effect of the methyl group is operating, but in (VI) there is now stearic inhibition of resonance. Thus, (VI) is much weaker than (III), (cf. the anilines).

A point of interest in connection with the acid strength of phenols is that ring deuteration of phenol lowers the acidity by about 12 percent (Kresge *et al.*, 1961). This is an example of the secondary isotope effect, and it therefore appears that deuterium is more electron-releasing than protium.

The thermodynamic study of the ionisation of phenols may be carried out by considering the following equations (cf. carboxylic acids):

$$\text{ArOH (aq.)} \rightleftharpoons \text{ArO}^- \text{ (aq.)} + \text{H}^+ \text{ (aq.)}$$

and $-RT\ln K_a = \Delta G \text{ (ionisation)} = \Delta H \text{ (ionisation)} - T\Delta S\text{(ionsiation)}$

These thermodynamic functions have been evaluated for a number of phenols (Biggs, 1956; Hepler *et al.*, 1959). Table 2.2 shows values for Me and NO_2 substituents.

Table 2.2

Substituent	*ΔG (kJ)*	*ΔH (kJ)*	*ΔS (J/K⁻¹)*
none	56.1	23.6	−111.7
o-Me	58.6	29.8	− 97.06
m-Me	57.7	20.5	−128.5
p-Me	58.6	17.7	−137.2
o-NO_2	41.0	19.5	−72.8
w-NO_2	47.7	19.7	− 91.14
p-NO_2	40.6	19.7	−70.7

3. Some other reactions involving mainly the hydroxyl group are:

   (i) Alkali phenoxides react with alkyl halides to form phenolic ethers.

   (ii) Phenol reacts with phosphorus pentachloride to give only a small amount of chlorobenzene; the main product is triphenyl phosphate. On the other hand, the presence of nitro- groups in the o- and or p-positions leads to good yields of chloronitrobenzenes, *e.g.*, picric acid (2,4,6-trinitrophenol) gives picryl chloride.

   In this case, activated nucleophilic substitution occurs at one stage in the multistage reaction.

   (iii) Phenol forms phenyl esters when warmed with acid chlorides or acid anhydrides, *e.g.*,

$$C_6H_5OH + (CH_3CO)_2O \xrightarrow{CH_3CO_2Na} CH_3CO_2C_6H_5 + CH_3CO_2H$$

These esters, however, because of the presence of the benzene ring, can undergo rearrangement under suitable conditions (cf. the allylic rearrangement). This is known as the **Fries rearrangement** (1908), and is the conversion of a phenyl ester into an o- or -hydroxyketone, or a mixture of both, by treatment with anhydrous aluminium chloride:

$$C_6H_5OCOR \xrightarrow{AlCl_3} o\text{-}HOC_6H_4COR + p\text{-}HOC_6H_4COR$$

Generally low temperatures (60°C or less) favour the formation of the p-isomer, whereas high temperatures (above 160°C), favour the o-isomer. This rearrangement is a useful synthetic route to phenolic ketones.

Many theories have been proposed for the Fries rearrangement, but none is certain. There is evidence for an intermolecular mechanism, (*e.g.*, Baltzly *et al.*, 1948), but more recently evidence has been obtained to show that the rearrangement is at least partly intramolecular (Crawford, 1959).

It would appear that the rearrangement is best regarded as a combination of the two mechanisms occurring simultaneously:

(i) Intermolecular (cf. the Friedel-Crafts acylation):

$$\text{PhOCOR} + \text{AlCl}_3 \rightleftharpoons \text{PhO}^+\overset{\bar{\text{A}}\text{lC}_3}{—}\text{CORR}$$

$$\rightleftharpoons \text{PHO}\bar{\text{A}}\text{lCl}_3 + \text{RCO}^+ \rightleftharpoons$$

$\overset{+}{O}$—$\bar{A}lCl_3$, H, COR $\xrightarrow{-HCl}$ $OAlCl_2$, COR $\xrightarrow{H_2O}$ OH, COR

(ii) Intramolecular:

RCO, $\bar{A}lCl_3$, $:\overset{+}{O}$ ⟶ [RCO, H, $\overset{+}{O}$—$\bar{A}lCl_3$] $\xrightarrow{-HCl}$ RCO, $OAlCl_2$ $\xrightarrow{H_2O}$ RCO, OH

Just as alcohols undergo ammonolysis, so can phenol be converted into aniline by ammonia either when heated under pressure or by heating in the presence of zinc chloride.

$$C_6H_5OH + NH_3 \xrightarrow{ZnCl_2} C_6H_5NH_2 + H_2O$$

4. Many of the reactions of phenol are those which are examples of electrophilic aromatic substitution, and since the hydroxyl group is o-lp-orienting, these are always the products formed (see below).

5. Phenol undergoes alkylation to form mainly the p-derivative and a small amount of the o- :

OH + CH_3Cl $\xrightarrow{AlCl_3}$ OH, CH_3 (para) + OH, CH_3 (ortho)

6. Phenol is chloromethylated so readily that usually polymers are obtained.

The presence of –I group, however, decreases the activating effect of the hydroxyl group *e.g.*, p-nitrophenol may be successfully chloromethylated to give 2-hydroxy-5-nitrobenzyl chloride.

On the other hand, phenol may be chloromethylated successfully by first converting it into an ester (usually the ethyl phenyl carbonate by means of chloroformic ester), and chloromethylating this.

7. Phenol can be readily mercurated, *e.g.*, when refluxed with aqueous mercuric acetate, o-acetoxymercuriphenol is formed (together with some dimercurated compound):

$$C_6H_5OH + (CH_3CO_2)_2Hg \longrightarrow o\text{-}HOC_6H_4HgOCOCH_3 + CH_3CO_2H \quad (44\%)$$

8. Phenol couples in the p-position with diazonium salts in alkaline solution to form hydroxy as compounds.
9. Phenol condenses with aliphatic and aromatic aldehydes in the o- and p-positions, the most important example being the condensation with formaldehyde. At low temperature, in the presence of dilute acid or alkali, and using formalin (40 per cent aqueous formaldehyde), the main product is o-hvdroxvbenzvl alcohol, together with a small amount of the o-isomer.

$$C_6H_5OH + HCHO \xrightarrow[\text{(6 days)}]{NaOH} p\text{-}HOC_6H_4CH_2OH + o\text{-}HOC_6H_4CH_2OH \quad (80\%\ o+p)$$

This is known as the **Lederer-Manasse reaction** (1894); its mechanism may be:

Acid catalysis

$$CH_2{=}O \ H^+ \rightleftharpoons {}^+CH_2{-}OH$$

$$C_6H_5OH + {}^+CH_2{-}OH \longrightarrow \left[\text{(protonated cyclohexadienone, } \overset{+}{O}H, \text{ H, } CH_2OH) \right] \longrightarrow o\text{-}HOC_6H_4CH_2OH + H^+$$

Base catalysis

$$C_6H_5O^- + H_2C{=}O \longrightarrow \left[\text{(cyclohexadienone, O, H, } CH_2{-}O^-) \right] \longrightarrow o\text{-}{}^-OC_6H_4CH_2OH$$

When larger amounts of formaldehyde are used, bishydroxymethylphenol and p,p′-dihydroxy-diphenylmethane are obtained:

$$C_6H_5OH + 2HCHO \rightarrow C_6H_3(OH)(CH_2OH)_2 \text{ (o,p)} + C_6H_3(OH)(CH_2OH)_2 \text{ (o,o')}$$

$$2C_6H_5OH + HCHO \rightarrow HO{-}C_6H_4{-}CH_2{-}C_6H_4{-}OH$$

These condensations are the basis of the preparation of phenol-formaldehyde resins (**Bakelites**). Phenol and excess of formaldehyde, in the presence of dilute sodium hydroxide, slowly form a three-dimensional polymer, the structure of which is uncertain. One possibility is:

$$-CH_2-C_6H_2(OH)(CH_2-)-CH_2-C_6H_2(OH)(CH_2-)-CH_2-C_6H_2(OH)(CH_2-)_2$$

10. Phenol undergoes the *Reimer-Tiemann reaction* and the *Gattermann reaction*.
11. When sodium or potassium phenoxide is heated with carbon dioxide, a phenolic acid is formed, *e.g.*, salicylic acid:

$$C_6H_5ONa + CO_2 \rightarrow o\text{-}HOC_6H_4CO_2Na$$

12. Phenol can be hydrogenated in the presence of a nickel catalyst at 160°C to cyclohexanol:

$$C_6H_5OH + 3H_2 \xrightarrow{Ni} C_6H_{11}OH(ex.)$$

When distilled with zinc dust, phenol is converted into benzene. This is of very little preparative value, but is useful for the identification of various nuclei in natural products, *e.g.*, alkaloids.

When phenol is oxidised with potassium permanganate, the ring is broken down (most phenols behave in a similar manner).

Homologues of phenol can, however, be oxidised to the corresponding phenolic acid *provided the hydroxyl group is protected by alkylation or acylation.*

The best means of protection is the formation of the benzene-sulphonate, *e.g.*:

$$p\text{-}CH_3C_6H_4OH + C_6H_5SO_2Cl \xrightarrow{NaOH} p\text{-}CH_3C_6H_4OSO_2C_6H_5 \xrightarrow{KMnO_4} p\text{-}CO_2HC_6H_4OSO_2C_6H_5 \xrightarrow{H^+} p\text{-}CO_2HC_6H_4OH$$

Hydrogen of the hydroxyl group of phenol is readily removed by '*one-electron transfer*' oxidising agents, *e.g.*, ferric ions ($FeCl_3$), alkaline potassium ferricyanide, etc.

$$C_6H_5OH + Fe^{3+} \longrightarrow Fe^{2+} + H^+ + C_6H_5O^{\bullet} \longleftrightarrow \ldots \longleftrightarrow \ldots \longleftrightarrow \ldots$$

This phenoxyl radical, because of its high reactivity, quickly disappears, mainly by coupling) (dimerisation), which is predominantly of the C—C type (o-o, p-p, o-p).

The other possible modes of coupling, C—O and O—O, occur less frequently, particularly the latter.

These combination reaction have useful synthetic applications and are also important in biosynthetic studies.

Coupling between aroxyl radicals is sterically hindered by the presence of large substitutents, *e.g.* 2,4,6-tri-t-butylphenol forms a stable free radical when oxidised in benzene solution (the solution is dark blue).

This radical has been isolated as a dark blue solid, m.p. 96°C. In the presence of oxygen, the colour of the solution becomes light yellow due to the formation of the peroxide.

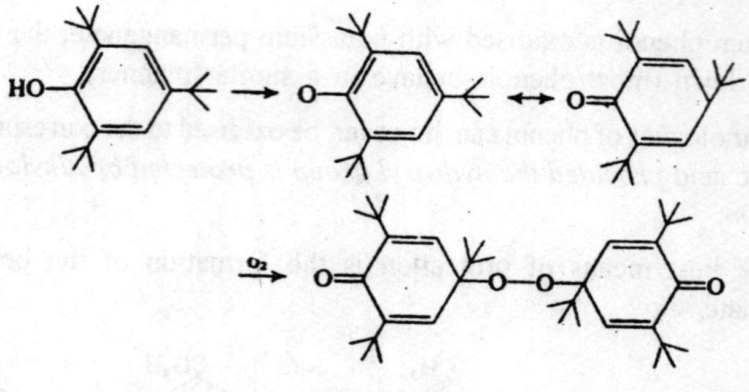

Fig. 2.1

On the other hand if the p-t-butyl group is absent, p-coupling occurs, and the dimmer is then oxidised to a red compound.

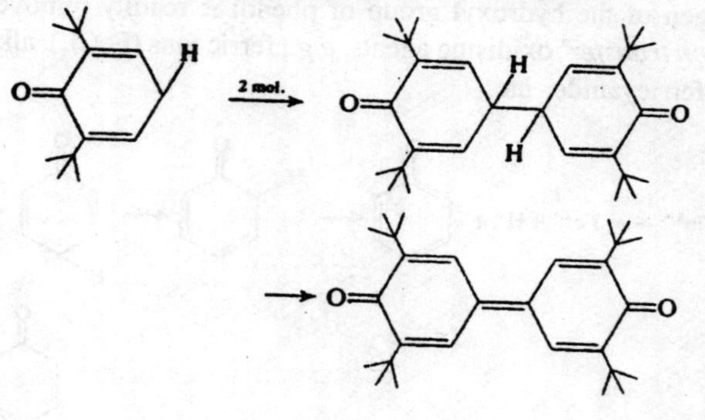

Fig. 2.2

Phenols also undergo the **Elbs persulphate oxidation** (1890). In this reaction, monohydric phenols are oxidised to dihydric phenols by alkaline potassium persulphate. Hydroxylation normally occurs in the p-position but, if this is blocked, then in the o-position. According to Behrman *et al.* (1963), the mechanism is:

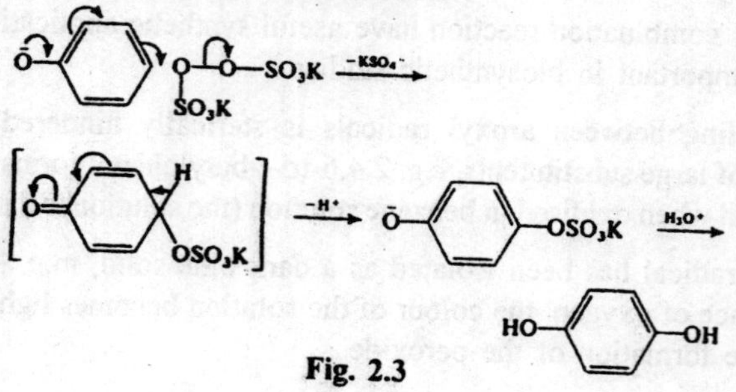

Fig. 2.3

Substituted Phenols

Halogenated phenols. Phenol, on treatment with chlorine or bromine water, gives an immediate precipitate of the 2,4,6-trihalogen derivative:

$$C_6H_5OH + 3Br_2 \xrightarrow{H_2O} 2,4,6\text{-}Br_3C_6H_2OH + 3HBr$$

This reaction may be used to estimate phenol quantitatively.

Monohalogenated phenols may be prepared from the corresponding halogenoanilines (p-iodophenol below) or, *e.g.*, p-bromophenol, as follows:

$$C_6H_5OH + Br_2 \xrightarrow[0°C]{CS_2} p\text{-}BrC_6H_4OH + HBr \quad (80\text{–}84\%)$$

The reason for the formation of different products when water and carbon disulphide are used as solvents is believed to be due to the fact that in the former the substrata is the phenoxide ion.

This is produced by ionisation in water, but is not formed in carbon disulphide (in the phenoxide ion, the benzene ring is more activated than in phenol itself).

o-Bromophenol may be prepared by protecting one o-position and the p-position by sulphonation:

$$C_6H_5OH \xrightarrow{H_2SO_4} 2,4\text{-}(SO_3H)_2C_6H_3OH \xrightarrow{Br_2/NaOH} 6\text{-}Br\text{-}2,4\text{-}(SO_3Na)_2C_6H_2OH \xrightarrow[boil]{acid} o\text{-}BrC_6H_4OH \quad (40\text{–}43\%)$$

o-*Iodophenol* is best prepared by heating phenol with mercuric acetate, converting the o-acetoxymercuriphenol into the corresponding chlormerodrin-derivative by heating with aqueous sodium chloride, and replacing the chloromercuri-group by treatment with iodine in chloroform solution:

$$C_6H_5OH \xrightarrow{(CH_3CO_2)_2Hg} o\text{-}HOC_6H_4HgOCOCH_3 \xrightarrow[aq.]{NaCl} o\text{-}HOC_6H_4HgCl\ (44\%) \xrightarrow{I_2/CHCl_3} o\text{-}HOC_6H_4I\ (63\%)$$

p-*Iodophenol* may be prepared as follows:

$$p\text{-}HOC_6H_4NH_2 \xrightarrow[H_2SO_4]{NaNO_2} p\text{-}HOC_6H_4N_2^+HSO_4^- \xrightarrow{KI} p\text{-}HOC_6H_4I \quad (69\text{–}72\%)$$

Dihalogenophenols are prepared by methods depending on the positions of the halogen atoms, *e.g.*, 2,6-dichlorophenol may be prepared from ethyl p-hydroxybenzoate:

$$p\text{-}HOC_6H_4CO_2Et \xrightarrow{SO_2Cl_2} 3,5\text{-}Cl_2\text{-}4\text{-}HOC_6H_2CO_2Et\ (83\text{–}85\%) \xrightarrow[(ii)\ HCl]{(i)\ KOH} 3,5\text{-}Cl_2\text{-}4\text{-}HOC_6H_2CO_2H\ (93\text{–}97\%) \xrightarrow[PhNMe_2]{heat\ in} 2,6\text{-}Cl_2C_6H_3OH\ (80\text{–}91\%)$$

o-, p-, 2,4-Dichloro-, and 2,4,6-trichlorophenols are manufactured by the action of the requisite amount of chlorine on molten phenol. When monochlorination in the p-position is required, sulphur-chloride is preferred as the chlorinating agent.

Nitrophenols. Treatment of phenol with cold dilute nitric acid gives a mixture of o- and p-nitrophenols, the latter predominating; oxidation products are also obtained. These isomers may be separated by steam distillation. As we have seen, solubility in hydroxylic solvents depends on, among other things, the power to form hydrogen bonds with the solvent. Phenol can form these bonds and hence a certain solubility in water can be expected. This argument also applies to substituted phenols since the hydroxyl group is still present, but in the o-compounds, however, because chelation is possible, hydrogen bonding with the solvent water molecules is hindered and hence the solubility is lowered. Furthermore since chelation causes the o-compound to behave as a 'monomer', this isomer will be more volatile than the corresponding m- and p-isomers.

Thus, the effects of chelation are lower solubility and greater volatility in the o-compounds, thereby enabling these to be separated from their m- and p-isomers by steam distillation. The o-isomer may also be separated from the p- by crystallisation or by chromatography.

[Structure: o-nitrophenol showing intramolecular hydrogen bond O–H···O⁻ to $\overset{+}{N}$=O]

It has been found that the nitration of phenol (and of aniline) is accelerated by the presence of nitrous acid, and it appears that the mechanism of the reaction is different from that of ordinary nitration. Evidence for this is based on the observation that when phenol is nitrated in the presence of very little nitrous acid, o- and p-nitrophenols are formed in the ratio of 7 : 3. When nitrated in the presence of a large amount of nitrous acid, the ratio becomes 1 : 9, which is the ratio in which o- and p-nitrosophenols are formed if the nitric acid is omitted. It is therefore believed that the nitroso-compound is formed first, and this is then oxidised to the nitro-compound (Hughes *et al.*, 1946). This scheme may be used to prepare pure p-nitrophenol in very good yield, phenol is treated with nitrous acid and the product, p-nitrosophenol, is oxidised to p-nitrophenol.

$$HNO_2 + 2H_2SO_4 \rightleftharpoons NO^+ + H_3O^+ + 2HSO_4^-$$

$$HOC_6H_5 + NO^+ \longrightarrow HO\text{-}C_6H_4\text{-}NO \xrightarrow{HNO_3} HO\text{-}C_6H_4\text{-}NO_2$$

Since nitrosation occurs only with activated compounds such as phenols, the nitrosonium ion is not so reactive an electrophilic reagent as the nitronium ion.

o- and p-Nitrophenols are prepared commercially by heating o- and p-chloronitrobenzenes with aqueous sodium hydroxide under pressure.

o- and p-Nitrophenols are also formed when nitrobenzene is heated with solid potassium hydroxide.

m-Nitrophenol may be prepared from m-dinitrobenzene:

$$C_6H_4(NO_2)_2 \xrightarrow{NH_4HS} NH_2\text{-}C_6H_4\text{-}NO_2 \xrightarrow[H_2SO_4]{NaNO_2} N_2^+HSO_4^-\text{-}C_6H_4\text{-}NO_2 \xrightarrow{H_2O} OH\text{-}C_6H_4\text{-}NO_2$$

(81–86% on the nitroaniline)

o-Nitrophenol is a yellow solid, m.p. 45°C; the m- and p-isomers are colourless solids, m.p. 97° and 114°C, respectively. All three are readily reduced to the corresponding amino-phenols, and the nitro-group in the o- and p-compounds is displaced on treatment with bromine water, 2,4,6-tribromophenol being formed.

o- and p-Nitrophenols give rise to two series ethers, one coloured and the other colourless. The colourless series contains the benzenoid structure, whereas the coloured series contains the chromophore quinone, *e.g.*, colourless coloured

The colourless ethers are stable but the coloured ethers are unstable and easily hydrolysed.

When o- and p-nitrophenols are nitrated, 2,4-dinitrophenol is formed and this, in turn, can be further nitrated to 2,4,6-trinitrophenol (*picric acid*). The yield of picric acid is poor due to large losses by oxidation. Picric acid is prepared commercially by first sulphonating phenol and then nitrating the product. Another commercial method is as follows:

$$C_6H_5Cl \xrightarrow[H_2SO_4]{HNO_3} C_6H_3Cl(NO_2)_2 \xrightarrow{aq.\ Na_2CO_3} C_6H_3(OH)(NO_2)_2 \xrightarrow[H_2SO_4]{HNO_3} C_6H_2(OH)(NO_2)_3$$

The interesting point to note about this method is that the presence of the nitro-group protects, to a large extent, the hydroxyl group from oxidation. It is in the nitration of phenol to the o- and p-nitrophenols that the loss by oxidation is greatest.

Picric acid may be obtained .in the laboratory by oxidising 2,4,6-trinitrobenzene with potassium ferricyanide.

Picric acid is a yellow crystalline solid, m.p. 122°C, with a bitter taste (Greek: *pikros*, bitter), and is a strong acid, decomposing carbonates. It forms charge-transfer complexes known as *pirates*, with aromatic hydrocarbons, amines and phenols; their picrates are frequently used to identify these classes of compounds. When treated with bleaching powder, picric acid forms chloropicrin as one of the products and forms *picryl chloride* when treated with phosphorus pentachloride. The chlorine atom in picryl chloride is very reactive (owing to the presence of the three nitro-groups in the p- and two o-positions): *e.g.*, when boiled with water, picryl chloride forms picric acid, and when shaken with concentrated ammonia, *picramide*:

$$C_6H_2(OH)(NO_2)_3 \xleftarrow{H_2O} C_6H_2Cl(NO_2)_3 \xrightarrow{NH_3} C_6H_2(NH_2)(NO_2)_3$$

Picramic acid (*2-amino-4,6-dintrophenol*) is formed when picric acid is reduced with sodium sulphide:

$$\text{2,4,6-}(NO_2)_3C_6H_2OH \xrightarrow{Na_2S} \text{2-}NH_2\text{-4,6-}(NO_2)_2C_6H_2OH \quad (90\%)$$

Aminophenols. These may be prepared by reducing the corresponding nitrophenols with metal and acid, or catalytically.

m-Amino-phenol (used in the manufacture of dyes) is prepared commercially by fusion of metanilic acid (m-aminobenzenesulphonic acid) with sodium hydroxide or by the reduction of m-nitrophenol with Fe/HCl.

o- and p-Aminophenols are more weakly acidic than phenol; thus they do not form phenoxides with alkalis. On the other hand, they form salts with strong inorganic acids.

The o- and p-derivatives are readily oxidised to the corresponding quinones; the m-compound is not easily oxidised (and does not give a quinone). o-Aminophenol has a marked tendency to form cyclic compounds (cf. o-phenylenediamines); *e.g.*, with acetic acid it forms 2-methyl-benzoxazole:

$$o\text{-}HOC_6H_4NH_2 + HO{-}C(=O){-}CH_3 \longrightarrow \text{2-methylbenzoxazole}$$

The amino-group in aminophenols is more readily acetylated than the hydroxyl group; *e.g.*, when p-aminophenol hydrochloride is acetylated with one equivalent of acetic anhydride in the presence of aqueous sodium acetate, p-acetamidophenol is formed (it is the *free* amino-group and not the ammonium ion that is attacked):

$$p\text{-}HOC_6H_4NH_2{\cdot}HCl + (CH_3CO)_2O \xrightarrow{CH_3CO_2Na} p\text{-}HOC_6H_4NHCOCH_3$$

p-*Aminophenol*, m.p. 186°C, is very important as a photographic developer. It may be prepared by boiling phenylhydroxylamine with sulphuric acid, or by reducing hydroxyazobenzene with sodium dithionite:

$$C_6H_5N_2^+Cl^- + C_6H_5OH \xrightarrow{NaOH} C_6H_5\text{-N=N-}C_6H_4OH \xrightarrow{Na_2S_2O_4} C_6H_5NH_2 + H_2NC_6H_4OH$$

It is prepared industrially by the electrolytic reduction of nitrobenzene in sulphuric acid

Two other important photographic developers are amidol, m.p. 78°C, and metol, m.p. 87°C:

OH, NH_2, NH_2 — amidol; OH, $NHCH_3$ — metol

Phenolsulphonic acids. When phenol is treated with concentrated sulphuric acid, o- and p-phenolsulphonic acids are formed, the former being the main product at ordinary temperatures, and the latter at higher temperatures (110°C); the o-compound rearranges to the p- on heating: thus, the o-isomer is the kinetically and the -isomer is the thermodynamically controlled product. m-Phenolsulphonic may be obtained by the controlled potassium hydroxide fusion of benzene-m-disulphonic acid at about 180°C:

$$C_6H_4(SO_3K)_2 \xrightarrow{KOH} C_6H_4(OK)SO_3K \xrightarrow{H^+} C_6H_4(OH)SO_3H$$

When o- or p- to a hydroxyl group (or an amino-group), a sulphonic acid group is often displaced by halogen when the sulphonic acid is halogenated in aqueous solution (cf. nitrophenols):

$$HOC_6H_4SO_3H \xrightarrow{Br_2/H_2O} C_6H_2Br_3OH + H_2SO_4$$

Nitrosophenols. When treated with nitrous acid, phenol forms mainly p-nitrosophenol, and a small amount of the o-compound:

OH → $\xrightarrow[(NO^+)]{HNO_2}$ p-nitrosophenol (90%)

p-Nitrosophenol is also formed when p-nitrosodimethylaniline is boiled with alkali.

p-Nitrosophenol crystallises from hot water in pale yellow needles which readily turn brown. On the other hand, it crystallises from ether in brownish-green flakes. This colour suggests that p-nitrosophenol may have a quinonoid structure, *i.e.*, the following tautomeric system is present:

Phenol $\xrightarrow{HNO_2}$ p-nitrosophenol ⇌ p-benzoquinone monoxime (NOH) $\xleftarrow{NH_2OH}$ p-benzoquinone

This is supported by the fact that p-nitrosophenol has been shown to be identical with the monoxime of p-benzoquinone, which, may be prepared by the action of hydroxylamine on p-benzoquinone.

HOMOLOGUES OF PHENOL

Cresols (*hydroxytoluenes*). These occur in the middle and heavy oil fractions of coal-tar. The mixture of the three cresols (together with a little phenol) is known as *cresylic acid* or *creosote*, and is used for preserving purposes, *e.g.*, timber, railway sleepers, etc.

A solution of cresols in soapy water is known as *lysol*, which is used as a disinfectant.

The boiling points of the cresols are: o-, 191° ,m-, 201°-, p-, 202-5°C. By means of a very good fractionating column it is possible to separate the o-isomer from the other two. Each isomer can be obtained pure from the corresponding toluidine.

When either o- or m-chlorotoluene is heated with aqueous sodium hydroxide at about 300-320°C under pressure, a certain amount of m-cresol is obtained.

m- and p-Cresols are used in the manufacture of resins, plasticisers, etc.

Two higher homologues of phenol are the isomers *thymol* (3-hydroxy-4-isopropyltoluene), and *carvacrol* (2-hydroxy-4-isopropyltoluene):

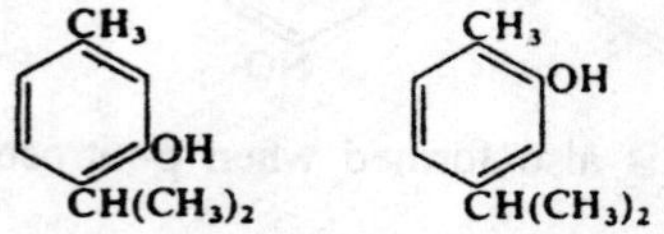

thymol, m.p. 51°C carvacrol, b.p. 237°C

Thymol occurs in the essential oil, oil of thyme, but is prepared commercially by heating m-cresol and isopropanol with sulphuric acid; it is used in perfumery and as an antiseptic. Carvacrol also occurs in some essential oils, but is prepared by heating camphor with iodine:

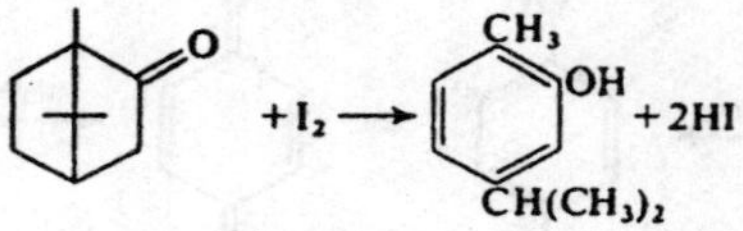

$$+ I_2 \longrightarrow + 2HI$$

It is used in perfumery and as an antiseptic.

A number of monochloro-derivatives of monohydric phenols are also very good disinfectants, *e.g.*, 2-chloro-5-hydroxytoluene.

DIHYDRIC PHENOLS

Catechol (o-*dihydroxybenzene*) occurs in certain plants. It may be prepared by the alkaline fusion of o-phenolsulphonic acid; commercially.

It is prepared by heating o-chlorophenol or o- dichlorobenzene with 20 per cent aqueous sodium hydroxide and a trace of copper sulphate at 190°C under pressure.

It may be conveniently prepared in the laboratory by the action of alkaline hydrogen peroxide on salicylaldehyde:

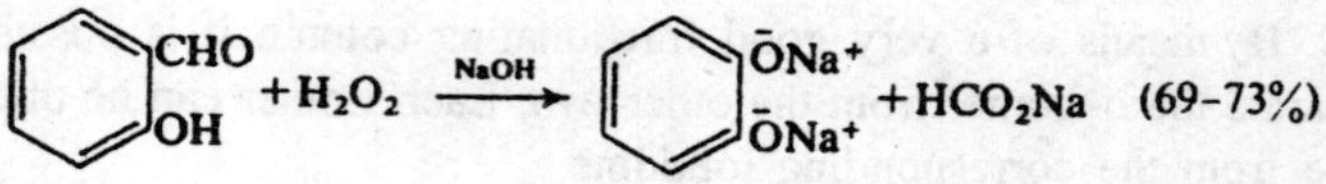

$$C_6H_4(CHO)(OH) + H_2O_2 \xrightarrow{NaOH} C_6H_4(\bar{O}Na^+)_2 + HCO_2Na \quad (69\text{–}73\%)$$

This reaction is characteristic of o- and p-hydroxyaldehydes; it is known as the **Dakin reaction** (1909).

The mechanism of the Dakin reaction is uncertain; a strong possibility is one that involves a 1,2-shift (cf. the cumene-phenol process).

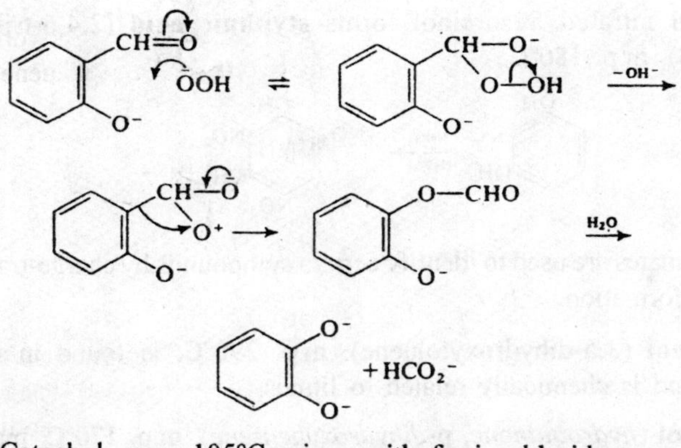

$$\text{o-}(CH{=}O)C_6H_4O^- + \ddot{O}OH \rightleftharpoons \text{o-}(CH(O^-)-O-OH)C_6H_4O^- \xrightarrow{-OH^-}$$

$$\text{o-}(CH{-}O^-)C_6H_4-O^+ \longrightarrow \text{o-}(O{-}CHO)C_6H_4O^- \xrightarrow{H_2O}$$

$$C_6H_4(O^-)_2 + HCO_2^-$$

Catechol, m.p. 105°C, gives with ferric chloride a green colour which turns red on the addition of sodium carbonate. Catechol is a powerful reducing agent; its aqueous solution darkens on exposure to air due to oxidation; it reduces cold silver nitrate and warm Fehling's solution. It is used as a photographic developer. Catechol is oxidised by silver oxide in ether solution to o-benzoquinone, and it condenses with many compounds, *e.g.*, with phthalic anhydride in the presence of sulphonic acid to form alizarin .

Important derivatives of catechol are guaiacol and adrenaline (the hormone secreted by the adrenal glands):

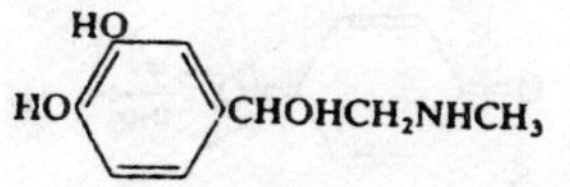

$$(HO)_2C_6H_3CHOHCH_2NHCH_3$$

adrenaline

Resorcinol (m-*dihydroxybenzene*), m.p. 110°C, is prepared industrially by the alkaline fusion of benzene-m-disulphonic acid.

Resorcinol is also formed when benzene-p-disulphonic acid or all three bromobenzenesulphonic acids are fused with alkali. Its aqueous solution gives a violet colour with ferric chloride. It is not so powerful a reducing agent as the o- and -isomers, but it will reduce silver nitrate and Fehling's solution on waning. With nitrous acid it forms dinitrosoresorcinol:

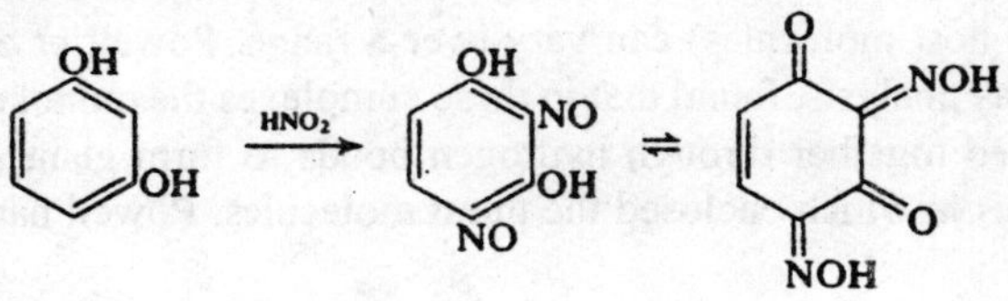

$$C_6H_4(OH)_2 \xrightarrow{HNO_2} C_6H_2(OH)_2(NO)_2 \rightleftharpoons \text{(quinone oxime form: }=O,\ =NOH,\ =O,\ =NOH)$$

When nitrated, resorcinol forms **styphnic acid** (2.4,6-trinitroresorcinol), m.p. 180°C:

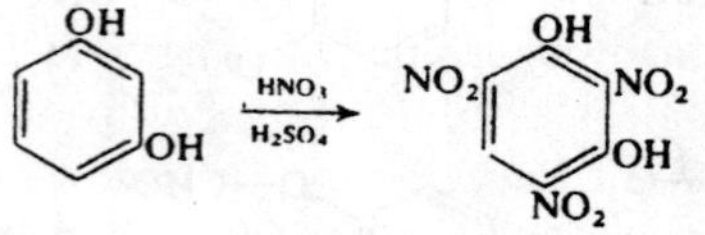

OH, OH $\xrightarrow[H_2SO_4]{HNO_3}$ OH, NO_2, NO_2, OH, NO_2

Styphnates are used to identify certain compounds by charge-transfer complex formation.

Orcinol (3,5-dihydroxytoluene), m.p. 290°C. is found in many lichens, and is chemically related to litmus.

Quinol (*hydroquinone*, p-*dihydroxybenzene*), m.p. 170°C, may be prepared by diazotising p-aminophenol (yield: 30%), or by reducing p-benzoquinone with sulphurous acid:

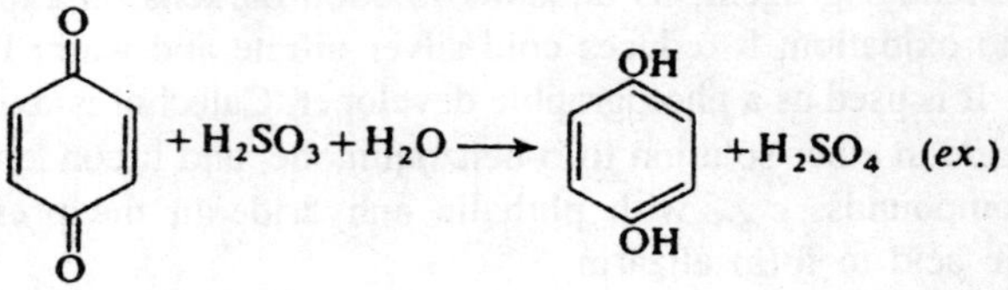

$$\text{O=C}_6\text{H}_4\text{=O} + H_2SO_3 + H_2O \longrightarrow \text{HO-C}_6\text{H}_4\text{-OH} + H_2SO_4 \quad (ex.)$$

It is made commercially as follows:

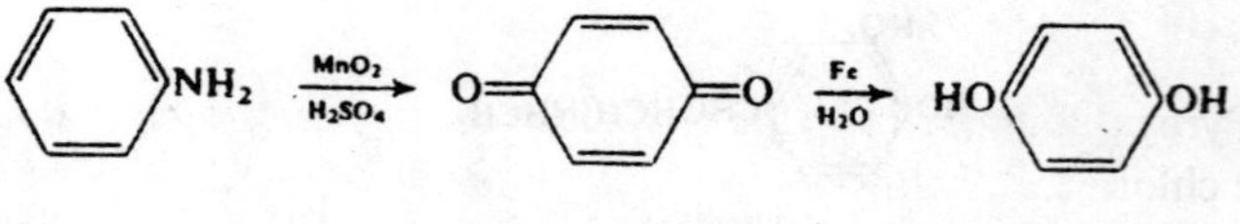

$$C_6H_5NH_2 \xrightarrow[H_2SO_4]{MnO_2} O{=}C_6H_4{=}O \xrightarrow[H_2O]{Fe} HO{-}C_6H_4{-}OH$$

Quinol is a powerful reducing agent and hence is used as a photographic developer. It is oxidised by ferric chloride to -benzoquinone; it is also oxidised by diazonium salts, no coupling taking place at all.

Clathrates. When quinol is crystallised from a solution in water saturated with sulphur dioxide, a quinol-sulphur dioxide complex is obtained. These inclusion complexes are also formed with hydrogen sulphide, methanol, etc., and have a molecular formula of the type

$nC_6H_5(OH)_2.Z$ (where Z is one guest molecule), and the value of n (for the host molecules) can vary over a range. Powell *et al.* (1948), using X-ray analysis, found that in these complexes the quinol molecules were linked together through hydrogen bonds to form giant molecules the cavities in which enclosed the guest molecules. Powell named these complexes clathrates, and showed that the 'cages' must be large enough to contain the guest molecule and that they must be so arranged that the guest molecules do not escape. These clathrates are stable, but the imprisoned molecules may be released either by melting or by means of an organic solvent which dissolves quinol.

Clathrates differ from channel complexes in that each guest molecule is enclosed in separate *molecular* cage of limited size, and, thus the possible variations of the guest molecule depend on molecular side. Hence clathrates may be used to separate certain homologues. *e.g.*, quinol forms a clathrate with methanol but not with any other homologue of this alcohol series

TRIHYDRIC PHENOLS

Pyrogallol (1,2,1-*trihydroxybenzene*) may be prepared by heating solid gallic acid in a stream of carbon dioxide, or by heating an aqueous solution of gallic acid at 210°C under pressure:

$$HO_2C{-}C_6H_2(OH)_3 \xrightarrow[210°C]{(solid)} C_6H_3(OH)_3 + CO_2 \quad (40\%)$$

Pyrogallol, m.p. 133°C, in aqueous solution gives a red colour with ferric chloride. Alkaline solutions of pyrogallol oxidise very rapidly on exposure to air, and hence are used in gas analysis for the absorption of oxygen (and carbon dioxide); the pyrogallol is oxidised to a complex mixture containing, among other things, carbon monoxide, carbon dioxide, acetic and oxalic acids. Pyrogallol also reduces the salts of silver, gold, platinum and mercury to their metals. It is used as a photographic developer.

Hydroxyquinol (1,2,4-*trihydroxybenzene*), m.p. 140°C, may be prepared by the alkaline fusion of quinol in air. It is best prepared by hydrolysing its triacetate, which is obtained by heating p-benzoquinone with acetic anhydride and concentrated sulphuric acid.

Phloroglucinol (1,3,5-trihydroxybenzene), m.p. 218°C, is obtained when many plant resins are fused with alkalis. It may be prepared by fusing resorcinol with sodium hydroxide in the air, but a convenient laboratory method is to reduce 2,4,6-trinitrobenzoic acid, and heat the resulting amino-derivative with hydrochloric acid (yield: 46-53% as dihydrate):

$$\text{2,4,6-}(NO_2)_3C_6H_2CO_2H \xrightarrow{Sn/HCl} \text{2,4,6-}(NH_2)_3C_6H_2CO_2H \xrightarrow{HCl} \text{1,3,5-}C_6H_3(OH)_3 + CO_2 + 3NH_4Cl$$

Alkaline solutions of phloroglucinol rapidly darken on exposure to air due to oxidation.

Since it reacts with certain reagents as if it were triketocyclohexane, it has been suggested that phloroglucinol is a tautomeric compound, *e.g.*,

$$\text{1,3,5-}C_6H_3(OH)_3 \rightleftharpoons \text{cyclohexane-1,3,5-trione} \xrightarrow{NH_2OH} \text{trioxime (HON=, =NOH, NOH)}$$

$$\text{cyclohexane-1,3,5-trione} \xrightarrow[KOH]{MeI} \text{2,2,4,4,6,6-hexamethyl (}Me_2, Me_2, Me_2\text{) cyclohexane-1,3,5-trione}$$

Infra-red studies, however, have shown that all phenols are entirely enolic in the solid state and also in solution (Thomson, 1956). This author has suggested that C-alkylation may be explained by attack of alkyl carbonium ions at negative sites in the system. Thus, since alkylations are carried out in alkaline solution, the phenoxide ion may be regarded as a resonance hybrid:

$$\text{1,3,5-}C_6H_3(OH)_3 \xrightarrow{OH^-} \text{1,3,5-}C_6H_3(O^-)_3 \longleftrightarrow \text{trione with negative charges on ring carbons}$$

In protic solvents, *e.g.*, water, which are strongly hydrogen-bonding, the more electronegative oxygen atoms are solvated and so C-attack is favoured. For highly reactive halides in non-protic solvents (or weakly protic solvents, *e.g.*, ethanol), it appears that C-alkylation is favoured when the solvent is non-polar, *e.g.*.

$$o\text{-}HOC_6H_4CH_2CH{=}CH_2 \xleftarrow[C_6H_6]{CH_2{=}CHCH_2Br} C_6H_5^-.N^+$$

$$\xrightarrow[Me_2CO]{CH_2{=}CHCH_2Br} C_6H_5OCH_2CH{=}CH_2$$

AROMATIC ETHERS

The aromatic ethers may be divided into two groups, the *phenolic ethers* (alkaryl ethers), which are of the type ArOR, and the ethers of the type Ar_2O (the diary l ethers).

The infra-red absorption region of diaryl and alkaryl ethers (=C—O— stretch) is 1270-1230 cm^{-1} (s). Thus, these types of ethers may readily be distinguished from dialkyl ethers, but not always from epoxides (the molecular formula will help here; for mass spectra).

Phenolic ethers. These may be prepared by heating sodium phenoxide with alkyl halide in ethanol solution, or by treating an alkaline solution of a phenol with alkyl sulphate:

$$ArO^-Na^+ + RX \longrightarrow ArOR + NaX \ (e.g.)$$

$$ArO^-Na^+ + R_2SO_4 \longrightarrow ArOR + RNaSO_4 \ \text{(v.g. –ex)}$$

Phenolic methyl ethers may be obtained in excellent yield by the action of diazomethane on a phenol:

$$ArOH + CH_2N_2 \longrightarrow ArOCH_3 + N_2 \ \text{(ex)}$$

Two important phenolic ethers are **anisole** (*methyl phenyl ether, methoxybenzene*), b.p. 155°C and **phenetole** (*ethyl phenyl ether, ethoxybenzene*), b.p. 172°C.

These are prepared industrially by the alkylation of phenol with methyl or ethyl sulphate, or with the methyl and ethyl esters of p-toluenesulphonic acid.

Anisole and phenetole are unaffected by most acids and alkalis, but are decomposed by concentrated hidrotic acid (or hydrobromic acid) into phenol and alkyl iodide (or bromide):

$$R—O—C_6H_5 + HI \rightleftharpoons I^- \ R—\overset{+}{\underset{H}{O}}—C_6H_5 \longrightarrow I—R + C_6H_5OH$$

The alkyl iodide can be absorbed by an ethanolic solution of silver nitrate, and the silver iodide so formed, weighed. This is the basis of the **Zeisel method** for the estimation of methoxyl an ethoxyl groups.

Phenolic ethers may also be decomposed by heating with pyridine hydrochloride at 170-240°C. This method is particularly useful for ethers that are sensitive to other demethylating reagents.

$$PhOR + C_5H_5NH^{+}Cl^{-} \longrightarrow PhOH + C_5H_5\overset{+}{N}MeCl^{-} + H_2O$$

Formation of an ether provides a means of protecting the hydroxyl group in many reactions; *e.g.*, in the Friedel-Crafts reaction, due to the acidic character of the hydroxyl-group, aluminium chloride is attacked with the liberation of hydrogen chloride:

$$C_6H_5OH + AlCl_3 \longrightarrow C_6H_5OAlCl_2 + HCl$$

Thus, to carry out a successful Friedel-Crafts reaction with phenol, it is necessary to use a large amount of aluminium chloride. This, however, may be avoided by carrying out the reaction with the methyl ether, but the temperature must be kept as low as possible to prevent the catalyst decomposing the ether.

Alternatively, alkylation of phenol may be carried out with an alkene or alcohol, and hydrogen fluoride.

In certain cases, phenolic ethers rearrange to the o- or p-alkyl compound:

$$C6H5OR \xrightarrow{AlCl_3} \text{o-or p-}RC_6H_4OH$$

Catechol ethers. Guaiacol (o-*hydroxyanisole*), m.p. 32°, b.p. 205°C, occurs in beech-wood tar from which it may be obtained by fractional distillation. Guaiacol is prepared synthetically from o-anisidine:

$$o\text{-}C_6H_4(OCH_3)NH_2 \xrightarrow[\text{boil}]{NaNO_2/H_2SO_4} o\text{-}C_6H_4(OCH_3)OH$$

When refluxed with constant boiling hydrobromic acid, or heated with hydriodic acid at 130°C, guaiacol is converted into catechol.

Veratrole (1,2-*dimethoxybenzene*), b.p. 207°C, may be prepared by methylating catechol with methyl sulphate in alkaline solution:

$$o\text{-}C_6H_4(OH)_2 \xrightarrow[NaOH]{(CH_3)_2SO_4} o\text{-}C_6H_4(OCH_3)_2 \quad (95\%)$$

Eugenol (4-*allylguaiacol*), b.p. 254°C, occurs in oil of cloves and in many other essential oils. When heated with ethanolic potassium hydroxide, or better, with potassium hydroxide in diethylene glycol at 180°C, eugenol isomerises to **isoeugenol** (4-*propenylguaiacol*):

OH, OCH_3, $CH_2CH{=}CH_2$ → OH, OCH_3, $CH{=}CHCH_3$

eugenol isoeugenol

This migration of the double bond in the ally) side-chain, under the influence of alkali, is general. The rearranged product contains phenyl-vinyl conjugation, and this produces a more stable system (for mechanism).

Isoeugenol, b.p. 267-5°C, also occurs naturally, and gives vanillin on gentle oxidation.

Anethole (p-*methoxypropenylbenzene*), m.p. 22-23°, b.p. 235°C, is one of the chief constituents of aniseed oil, from which it is obtained. It is also prepared synthetically by the interaction of p-methoxy-phenylmagnesium bromide and allyl bromide, and isomerising the product, **estragole** (*methylchavicol*), with alkali:

OCH_3, MgBr $+ CH_2{=}CHCH_2Br$ → OCH_3, $CH_2CH{=}CH_2$ $\xrightarrow{KOH}$ OCH_3, $CH{=}CHCH_3$

estragole, b.p. 215°C anethole

Safrole *(4-allyl-l,2-methylenedioxybenzene)*, b.p. 232°C, occurs in camphor oil and sassafras oil. When heated in ethanolic potassium hydroxide solution, or better, in a concentrated solution of potassium hydroxide in cellosolve, safrole isomerises to **isosafrole**:

O—CH_2—O, $CH_2CH{=}CH_2$ $\xrightarrow{KOH}$ O—CH_2—O, $CH{=}CHCH_3$

safrole isosafrole

Isosafrole, b.p. 252°C, also occurs naturally, and gives piperonal on gentle oxidation.

Diaryl ethers. These are conveniently prepared by means of the *Ullmann reaction*; *e.g.*, *diphenyl ether*:

$$C_6H_5Br + C_6H_5OH + KOH \xrightarrow{Cu} (C_6H_5)_2O + KBr + H_2O$$

(90%)

Diphenyl ether, m.p. 28°C, is not decomposed by hydriodic acid, and is valuable as a high temperature heat transfer medium.

Diaryl ethers may be prepared by the action of diaryliodonium salts on phenoxides.

Diaryl ethers are extremely weak bases, and this may be explained on the basis that they are resonance hybrids:

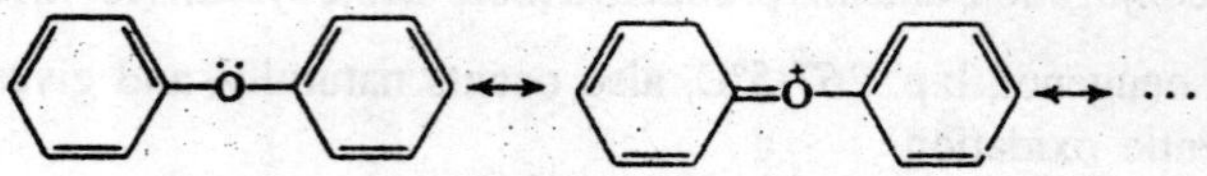

The availability of the lone pair for protonation is decreased and the extended resonance stabilisation is lost in the protonated molecule.

Ethers of aminophenols. The methyl ethers of o- and p-aminophenol are known as o-anisidine and p-anisidine, respectively, and each may be prepared by reducing its corresponding nitroanisole, *e.g.* o-anisidine from o-nitroanisole:

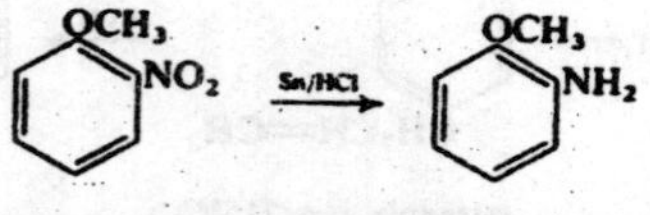

$$\text{o-}OCH_3C_6H_4NO_2 \xrightarrow{Sn/HCl} \text{o-}OCH_3C_6H_4NH_2$$

The anisidines are used in the preparation of azo-dyes.

The ethyl ethers of the aminophenols are known as phenetidines; these are also used in the preparation of azo-dyes. The acetyl derivative of p-phenetidine is known as *phenacetin*, and is prepared commercially as follows:

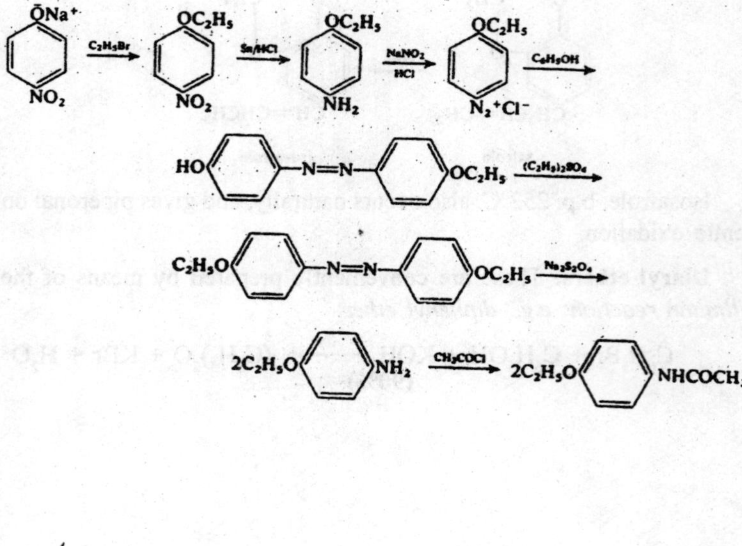

$$p\text{-}NO_2C_6H_4O^-Na^+ \xrightarrow{C_2H_5Br} p\text{-}NO_2C_6H_4OC_2H_5 \xrightarrow{Sn/HCl} p\text{-}NH_2C_6H_4OC_2H_5 \xrightarrow[HCl]{NaNO_2} p\text{-}N_2^+Cl^-C_6H_4OC_2H_5 \xrightarrow{C_6H_5OH}$$

$$HO\text{-}C_6H_4\text{-}N{=}N\text{-}C_6H_4\text{-}OC_2H_5 \xrightarrow{(C_2H_5)_2SO_4}$$

$$C_2H_5O\text{-}C_6H_4\text{-}N{=}N\text{-}C_6H_4\text{-}OC_2H_5 \xrightarrow{Na_2S_2O_4}$$

$$2C_2H_5O\text{-}C_6H_4\text{-}NH_2 \xrightarrow{CH_3COCl} 2C_2H_5O\text{-}C_6H_4\text{-}NHCOCH_3$$

This preparation is a very good example of the *recycling procedure*, diazotisation of part of the p-phenetidine *produced* then permits *phenol* to be used as the starting material instead of p-nitrophenol (which is more expensive than phenol).

Phenacetin is replacing acetanilide in medicine; it is a very good analgesic (*i.e.*, promotes relief from pain) and antipyretic (*i.e.*, fever-reducing).

Dulcin is the carbamyl derivative of p-phenetidine, and is prepared by heating p-phenetidine with urea:

$$p\text{-}H_2NC_6H_4OC_2H_5 + CO(NH_2)_2 \longrightarrow p\text{-}C_2H_5OC_6H_4NHCONH_2 + NH_3$$

Dulcin, m.p. 171°C, is about 200 times as sweet as sugar: it is used commercially as a sweetening agent.

Claisen rearrangement (1912). This is the rearrangement of allyl aryl ethers to allylphenols. No catalyst is required and it occurs when the substrate is heated alone to about 200°C or in some inert solvent, *e.g.*, diphenyl ether. The ally I group migrates to the o-position if one is free, but if *both* o-positions are occupied, then migration occurs to the p-position, *e.g.*,

$$C_6H_5OCH_2CH{=}CH_2 \xrightarrow{200^\circ C} o\text{-}HOC_6H_4CH_2CH{=}CH_2 \quad \textbf{(ex.)}$$

$$2,6\text{-}(CH_3)_2C_6H_3OCH_2CH{=}CH_2 \xrightarrow{200^\circ C} 2,6\text{-}(CH_3)_2\text{-}4\text{-}(CH_2CH{=}CH_2)C_6H_2OH \quad \textbf{(85\%)}$$

There are, however, some exceptions to these rules, *e.g.*, Schmid *et al.* (1963) found that crotyl 3,5-dimethylphenyl ether, $MeCH{=}CHCH_2OC_6H_3Me_2$, gave both the o- and p-rearrangement, with the former predominating.

The mechanism of the Claisen rearrangement has been the subject of a great deal of work. The intramolecular nature of the rearrangement has been demonstrated by the fact that if two different ethers are heated together, they rearrange independently, *e.g.*, Hurd *et al.* (1937) heated a mixture of allyl 2-naphthyl ether and cinnamyl phenyl ether and did not obtain cross products:

$$\text{C}_{10}\text{H}_7\text{OCH}_2\text{CH}{=}\text{CH}_2 + \text{C}_6\text{H}_5\text{OCH}_2\text{CH}{=}\text{CHPh} \longrightarrow$$

$$\text{(1-allyl-2-naphthol) } \text{CH}_2\text{CH}{=}\text{CH}_2,\ \text{OH} + \text{(o-substituted phenol) OH},\ \text{CHPhCH}{=}\text{CH}_2$$

Kincaid el al. (1939) showed the rearrangement was of the first order, and other workers have showing that when the allyl group migrates to the ring in the o-rearrangement, the α, γ-bonding is reversed (above example). Schmid *et al.* (1953), using the allyl group labelled with ^{14}C at the γ-carbon atom, also showed that the rearranged product contained the tracer atom attached to the nucleus:

$$\text{C}_6\text{H}_5\text{OCH}_2\text{CH}{=}\overset{*}{\text{C}}\text{H}_2 \longrightarrow o\text{-HOC}_6\text{H}_4\overset{*}{\text{C}}\text{H}_2\text{CH}{=}\text{CH}_2$$

These examples suggest that the reversal of the bonding in the allyl group is a necessary requirement in the rearrangement. The following intramolecular cyclic mechanism is in agreement with the foregoing facts:

$$\text{C}_6\text{H}_5\text{O–CH}_2\text{–CH}{=}\text{CH}_2 \underset{\text{slow}}{\rightleftharpoons} [\text{T.S.}] \rightleftharpoons \left[\text{2-(CH}_2\text{CH}{=}\text{CH}_2\text{)-2-H-cyclohexa-3,5-dienone}\right] \xrightarrow{\text{fast}} o\text{-HOC}_6\text{H}_4\text{CH}_2\text{CH}{=}\text{CH}_2$$

The formation of a cyclic transition state is supported by the fact that the entropies of activation are negative; thus, the transition state is more 'ordered' than the substrate. It will also be noted that the formation of a dienone is postulated in this mechanism, evidence for its existence has been obtained in the *para* rearrangement (see below). Support for the migration of the o-hydrogen in the dienone given by the following experiment of Kistiakowsky *et al.* (1942):

$OCH_2CH{=}CH_2$ (2,4,6-trideuterio) → OD, $CH_2CH{=}CH_2$ (4,6-dideuterio)

The mechanism of the p-rearrangement is believed to occur in two stages, each one with reversal, and thus the final product will *not* be reversed. Schmid et at (1953) showed the following rearrangement occurred with labelled allyl 2,6-dimethylphenyl ether:

$OCH_2CH{=}\overset{*}{C}H_2$, CH_3, CH_3 ⟶ OH, CH_3, CH_3, $\overset{*}{C}H_2CH{=}CH_2$

Conroy *et al.* (1954, 1956) have demonstrated the existence of the dienone by rearranging allyl 2,6-dimethylphenyl ether in the presence of maleic anhydride; a small amount of adduct was isolated. Thus a Diels-Alder reaction has occurred, and this is strong evidence for the formation of the dienone (the dienophile in this system). The reversibility of both stages in the p-arrangement has also been demonstrated be Schmid *et al.* (1956) using allyl groups labelled with ^{14}C:

$\overset{\circledast}{C}H_2{=}CHCH_2O$, Me, Me, $CH_2CH{=}\overset{*}{C}H_2$ (I) ⇌ O=, Me, $CH_2CH{=}\overset{\circledast}{C}H_2$, Me, $CH_2CH{=}\overset{*}{C}H_2$ ⇌

O=, Me, Me, $CH_2CH{=}\overset{\circledast}{C}H_2$, $CH_2CH{=}\overset{*}{C}H_2$

Compound (I), in which labelling was only in the C-allyl group (C), was recovered *chemically* unchanged after being heated to 160°C, but was now also labelled in the O-allyl group (C), the isotope being equally distributed between the terminal methylene groups.

A two-stage mechanism for this p-rearrangement, each stage involving a cyclic transition state, accounts for all of the facts. In the second stage, o- to p-, the cyclic transition state will be highly deformed; this is probably the reason why the o-rearrangement is preferred when it is possible.

The Claisen rearrangement is the best known example of the class of reactions called **sigmatropic rearrangements**. These are uncatalysed intramolecular (unimolecular one-step) processes whereby a σ-bond, flanked by one or two π-electron systems, migrates to a new position. In the Claisen rearrangement, the migrating α-bond is the O—C_α bond (in the ether) which forms the new α-bond, o-C—Cγ, in the dienone (o-C is the *ortho*-carbon atom in the dienone)

QUINONES

Quinones are cyclic unsaturated diketones in which the double bonds and the keto groups are conjugated. Thus they are not aromatic compounds, but they are readily prepared from, and readily converted into, aromatic compounds. Two quinones derived from benzene are possible, o- and p-benzoquinone:

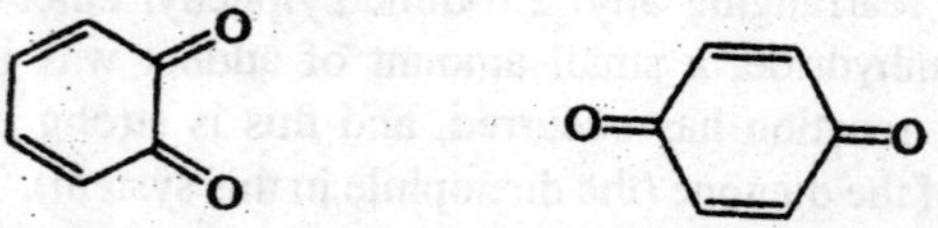

A characteristic property of quinones is their colour, the p-compounds being yellow and the o-compounds being (usually) red. These quinonoid structures are powerful chromospheres.

The infra-red absorption region of the carbonyl group (C=O str.) in quinones (with both carbonyl groups in the same ring) is 1690-1660 cm^{-1} (s). This is similar to that in α, β-unsaturated ketones. The molecular formula, however, will help here.

p-Benzoquinone (p-*quinone*) may be prepared by the oxidation of quinol with ferric chloride, manganese dioxide and sulphuric acid, or acid dichromate: the best oxidising agent is sodium chlorate in dilute sulphuric acid in the presence of vanadium pentoxide as catalyst:

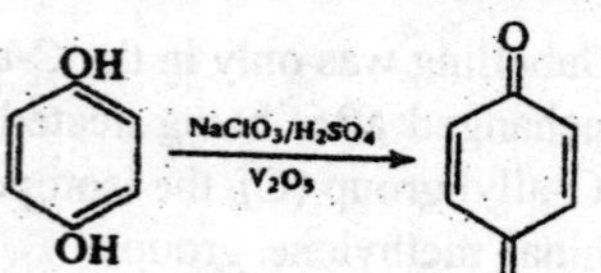

$NaClO_3/H_2SO_4$, V_2O_5 (92–96%)

p-Benzoquinone is usually prepared in the laboratory by the oxidation of aniline with potassium dichromate and sulphuric acid.

In general, p-quinones may be prepared by the oxidation (using dichromate-sulphuric acid) of p-dihydroxy-, p-diamino- or -amino-hydroxy-compounds, *e.g.*,

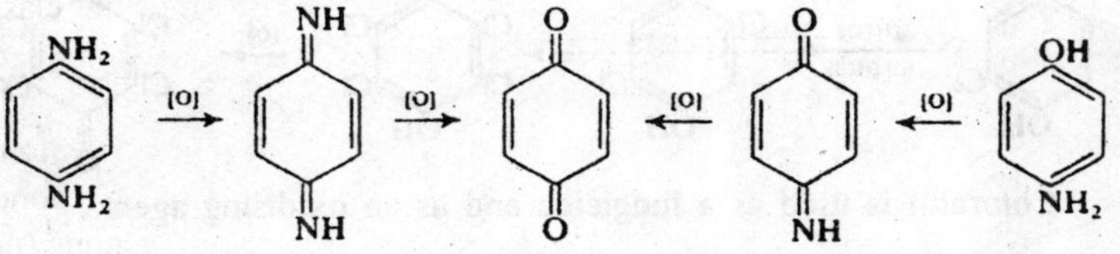

The intermediate products (1) (p-*benzoquinonedi-imine*) and (Il) (p-*benzoquinoneimine*) can be isolated under certain conditions.

p-Benzoquinone crystallises in yellow prisms, m.p. 116°C, it has a sharp smell, sublimes when heated, is slightly soluble in water, and is steam-volatile. On exposure to light, p-benzoquinone turns brown. The structure of p-benzoquinone is known as a *crossed conjugated system*, it contains three (or more) conjugated double bonds which are not arranged in a continuous chain.

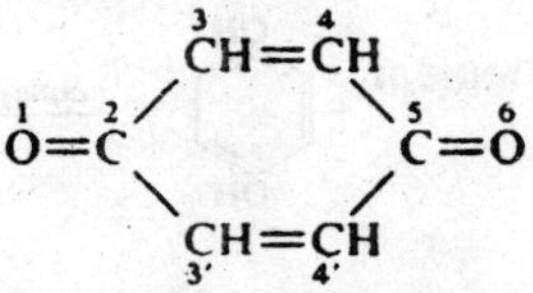

As might be expected from what has been said above, p-benzoquinone behaves as an α, β-unsaturated ketone, most of its addition reactions being typical 1,4-additions, followed by aromatisation. The driving force of these additions is largely due to the fact that the resonance energy of the quinone is 20-9 kJ mol^{-1}, and when it is converted into the aromatic compound, the product has gained an increased stabilisation energy of at least 125-5 kJ mol. p-Benzoquisnone adds on hydrogen chloride to form mainly *chloroquinol* (*chlorohydroquinone*) m.p. 106°C:

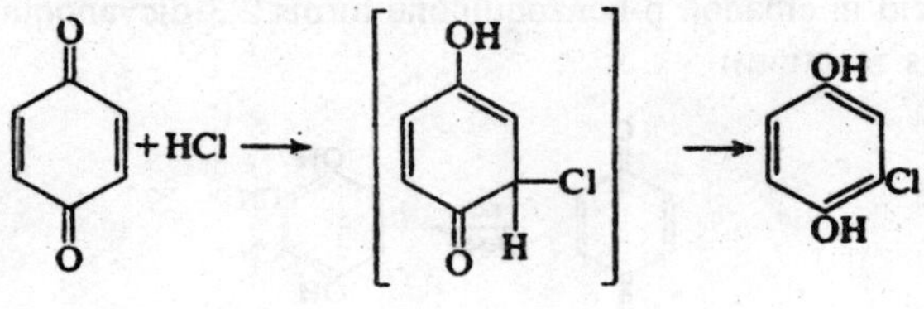

Chloroquinol, on oxidation, gives chloro-p-benzoquinone. By repeating the process of adding hydrogen chloride and then oxidising, the final product obtained is *tetrachloro*-p-*benzoquinone* (*chloranil*):

OH, Cl, OH —(i) [O], (ii) [HCl]→ OH, Cl, Cl, OH ····→ OH, Cl, Cl, Cl, Cl, OH —[O]→ O, Cl, Cl, Cl, Cl, O

Chloranil is used as a fungicide and as an oxidising agent.

In the same way, p-benzoquinone forms addition compounds with primary or secondary amines in ethanolic solution.

In this case the quinone and not the quinol derivative, is obtained, *e.g.*, with aniline, the final product is 2,5-dianilino-p-benzoquinone :

O, O $+ C_6H_5NH_2$ ⟶ OH, NHC_6H_5, OH —p-O=C_6H_4=O→

O, NHC_6H_5, O + OH, OH —$C_6H_5NH_2$→

OH, NHC_6H_5, C_6H_5NH, OH —p-O=C_6H_4=O→ O, NHC_6H_5, C_6H_5NH, O

In a similar way, primary alcohols, in the presence of zinc chloride, form 2,5-dialkoxy-p-benzoquinones with p-benzoquinone.

When treated with aqueous potassium cyanide in the presence of sulphuric acid in ethanol. p-benzoquinone forms 2,3-dicyanoquinol (cf. the previous reaction):

O, O —KCN, H_2SO_4→ OH, CN, CN, OH

With acetic anhydride in the presence of sulphuric acid, p-benzoquinone forms hydroxyquinol triacetate; this reaction is known as the **Thiele acetylation** (1898).

$$+ (MeCO)_2O \xrightarrow[1,4\text{-addn.}]{H^+} \left[\text{...H, OCOMe} \longrightarrow \text{...OCOMe} \right] \xrightarrow{(MeCO)_2O} \text{(OCOMe)}_3$$

p-Benzoquinone is easily reduced to quinol (1,6-addition) by sulphurous acid, hydrogen sulphide, or sodium hyposulphite (yield: 80%).

An intermediate in this reduction is quinhydrone (green prisms, m.p. 171°C). **Quinhydrones** are a group of coloured substances formed from quinones and another aromatic compound. They are believed to be charge-transfer complexes, *e.g.*, phenoquinone (I); red); phenol acts as the electron donor and the quinone as the electron acceptor. (II) is a 2:1 molar charge transfer complex, but 1:1 molar adducts are also known, *e.g.*, (II). In either case, the quinhydrone is stabilised by hydrogen bonding. Quinhydrones dissociate into their components in solution.

$C_6H_5OH \cdots O{=}\langle\ \rangle{=}O \cdots HOC_6H_5$

(I)

O·····H—O

O·····H—O

(II)

p-Benzoquinone is so easily reduced that it is used as an oxidising agent generally chloranil is more satisfactory–in reactions where inorganic oxidising agents must be avoided. It liberates iodine from acidified potassium iodide solution and is oxidised by silver oxide to maleic acid (and other products). It forms a monoxime and a dioxime and can act as; dienophile in the Diels-Alder reaction, *e.g.*, with butadiene it forms a hydrogenated anthraquinone derivative.

The oxidising property of benzoquinone is due to the fact that when the quinone adds on two electrons, the benzenoid structure (with a large resonance energy) is obtained:

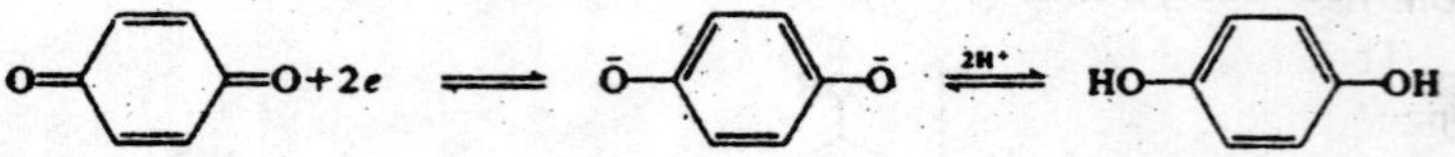

$$O{=}C_6H_4{=}O + 2e \rightleftharpoons {}^{-}O{-}C_6H_4{-}O^{-} \overset{2H^+}{\rightleftharpoons} HO{-}C_6H_4{-}OH$$

This *reversible* oxidation-reduction is characteristic of all quinones and a measure of the oxidising power is given by the *oxidation-reduction* or *redox potential* of that system. The system with the greater positive redox potential oxidises a system with a less positive redox potential, and the higher the redox potential the higher is the energy content of the quinone. Thus, the value of a redox potential depends on the number of nuclei in the quinone and on the relative positions of the two carbonyl groups. The potential of the o-quinone is higher than that of the corresponding p-isomer, and the larger the number of rings fused in a linear fashion, the lower is the redox potential, *e.g.*,

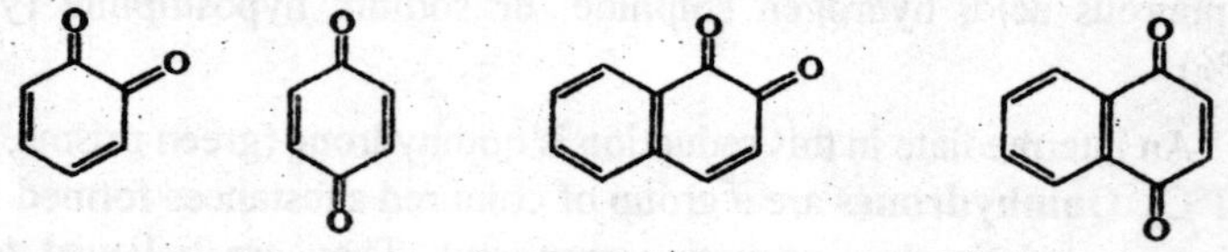

volts 0·79 0·70 0·56 0·47

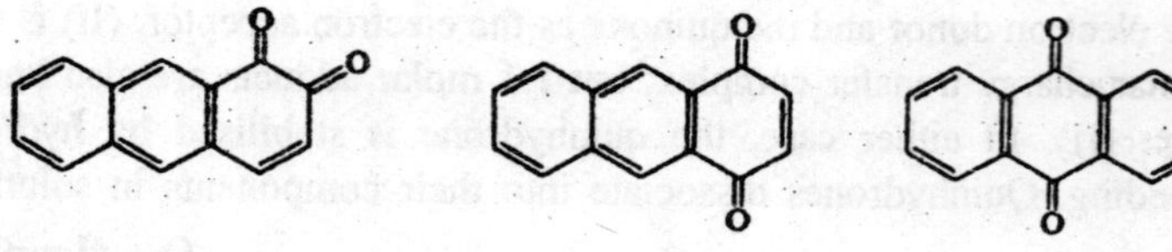

volts 0·49 0·40 0·15

The property of being good electron-acceptors is generally due to the low energy level of the lowest empty M.O. in quinones. Apart from the factors mentioned above that control the oxidation-reduction potentials of quinones, it has also been found that substituent groups often have a marked effect. If the nucleus contains a strongly electron-attracting group, the quinone has a higher oxidation-reduction potential than that of the parent quinone. When the substituent is electron-releasing, the potential is lower than that of the parent quinone, *e.g.*, Cl (which has a –I effect) raises the potential of chloro-p-benzoquinone above that of p-benzoquinone. Thus, the former quinone-quinol system will oxidise the latter system, *i.e.*, the latter will have its equilibrium position mainly on the quinone side. Hence in the case, when HCl adds to p-benzoquinone, the chloroquinol produced is *not* oxidised by unchanged quinone. On the other hand, the anilino and ethoxyl groups are electron-releasing (both have a +I effect).

Therefore the oxidation-reduction potentials of these substituted p-benzoquinones are lower than that of p-benzoquinone. Hence, when aniline or ethanol adds to p-benzoquinone to form the corresponding quinol, this is then oxidised by unchanged quinone, and the reaction can proceed again (see above for the equations).

In the electrolytic reduction of benzoquinone, one electron is transferred to produce an anionic free radical, known as semiquinone, this is stabilised by resonance.

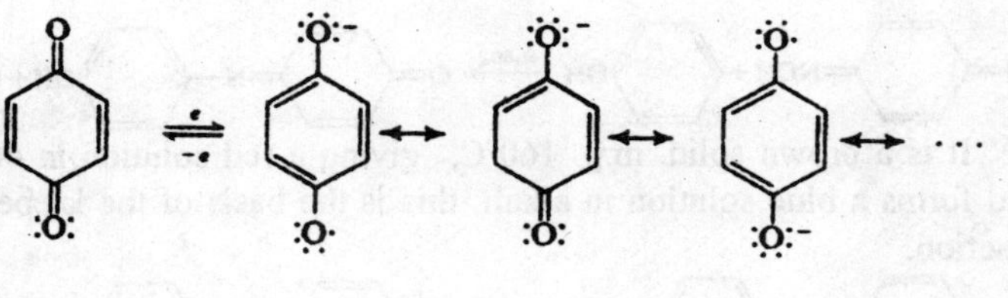

Fig. 2.4

Corresponding anionic free radicals derived from o-benzoauinone are also known.

When derivatives of p-phenylenediamine are oxidised with bromine in acetic acid, coloured solutions containing **Wurster salts** are produced. These salts are cationic free radicals and, like semi-quinone, are stabilised by resonance, *e.g.*, the tetramethyl bromide (blue):

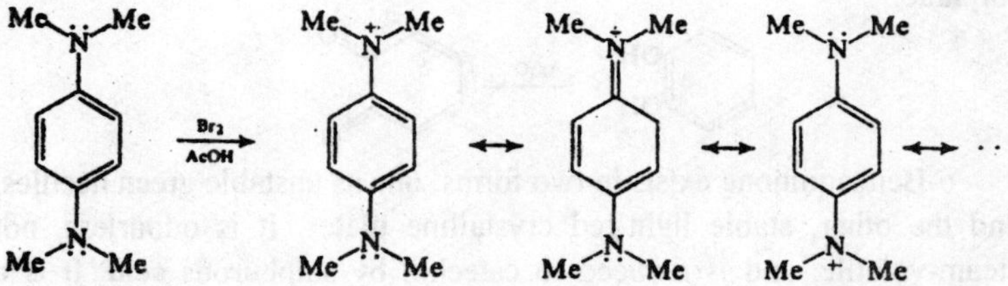

Fig. 2.5

Most Wurster salts are stable in the pH range 3-5-6.

p-Benzoquinone oximes. p-Benzoquinone forms two oximes with hydroxylamine. With one equivalent of hydroxylamine the monoxime is formed, and this is tautomeric with p-nitrosophenol. The dioxime is a yellow crystalline solid.

p-Benzoquinoneimine, $NH{=}C_6H_4{=}O$, and **p-benzoquinone-imine**, $NH{=}C_6H_4{=}NH$, may be prepared by the careful oxidation of p-aminophenol and p-phenylenediamine, respectively, with silver oxide in ethereal solution. The mono-imine is a bright yellow crystalline solid, the di-imine is colourless and unstable. When warmed with inorganic acids, both compounds form p-benzoquinone and ammonia.

Indophenol may be prepared by condensing p-nitrosophenol with phenol in the presence of 70 percent sulphuric acid or concentrated hydrochloric acid:

O=C₆H₄=NOH + C₆H₅OH $\xrightarrow{H_2SO_4}$ O=C₆H₄=N—C₆H₄—OH + H_2O

It is a brown solid, m.p. 160°C, giving a red solution in ethanol, and forms a blue solution in alkali; this is the basis of the Liebermann reaction.

O=C₆H₄=N—C₆H₄—OH $\xrightarrow{NaOH}$ O=C₆H₄=N—C₆H₄—O⁻ ⟷ ⁻O—C₆H₄—N=C₆H₄=O

o-Benzoquinone may be prepared by the oxidation of catechol with silver oxide in dry ethereal solution in the presence of anhydrous sodium sulphate.

$C_6H_4(OH)_2 \xrightarrow{Ag_2O}$ o-$C_6H_4O_2$

o-Benzoquinone exists in two forms, one as unstable green needles, and the other, stable light-red crystalline plates. It is odourless, not steam-volatile, and is reduced to catechol by sulphurous acid. It is a strong oxidising agent, *e.g.*, it liberates iodine from acidified potassium iodide.

Many benzoquinone compounds occur naturally, and are the cause of the colour in the pigments in which they are found.

Quinone methods. These occur naturally, and are formally derived from quinones by replacement of one of the carbonyl oxygen atoms by a methylene, or substituted methylene, group. Only p-quinone methods have been obtained in the pure state, *e.g.*. (I) is obtained by the oxidation of 2,6-di-t-butyl-4-isopropylphenol. The most characteristic property of quinone methods is their ready reaction with nucleophiles at the methylene group, *e.g.*

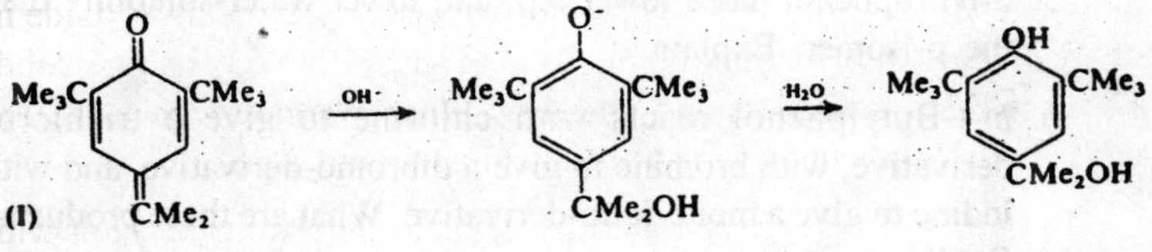

(I) $\xrightarrow{OH^-}$ $\xrightarrow{H_2O}$

Quinols (quinoles) may be regarded as derivatives of p-benzoquinone in which one carbonyl group has been converted into a tertiary alcohol. Quinols may be prepared by oxidising p-alkylphenols with Caro's acid; *e.g.*, p-cresol gives p-*toluqinol* (4-*methylquinol*):

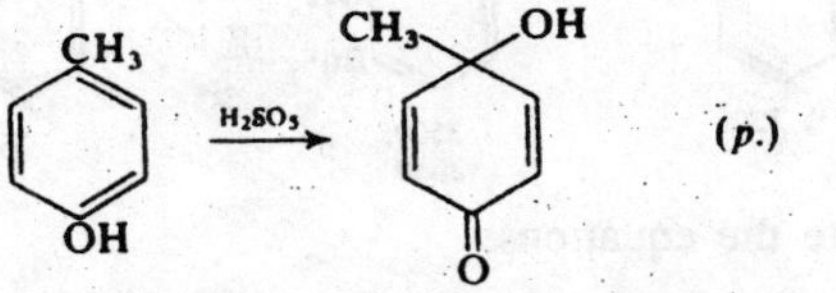

$\xrightarrow{H_2SO_5}$ (*p.*)

The most characteristic reaction of quinols is their ready rearrangement to p-dihydric phenols (the **dienone-phenol rearrangement**), *e.g.*, p-toluquinol, when heated in aqueous methanol containing sulphuric acid, rearranges to 2-methylquinol.

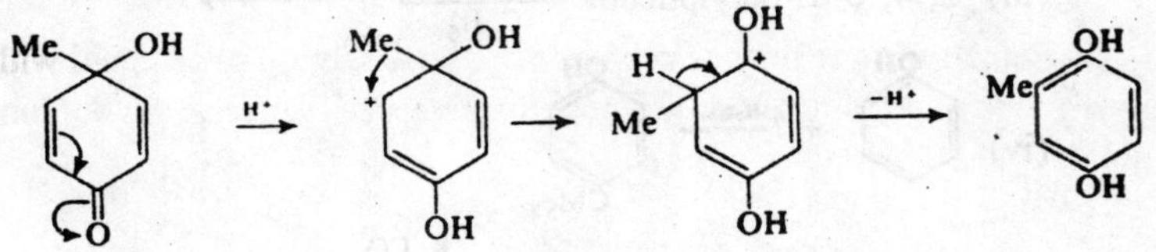

$\xrightarrow{H^+}$ $\longrightarrow$ $\xrightarrow{-H^+}$

Fig. 2.6

EXERCISES

1. How would the i.r. spectra of o- and p-hydroxybenzaldehyde differ?
2. How could you separate chemically a mixture of benzene, aniline, phenol, and benzoic acid?
3. Alcohols are weaker acids than phenols but are stronger nucleophiles. Explain.

4. Which in each pair is the more stable and why?

   (i) $Ph\overset{+}{C}HMe$ and $PhCH_2CH_2C_2^+$

   (ii) PhO^- and p-$NO_2C_6H_4O$.

5. o-Nitrophenol has a lower b.p. and lower water-solubility than the p-isomer. Explain.

6. m-t-Butylphenol reacts with chlorine to give a trichloro-derivative, with bromine to give a dibromo-derivative, and with iodine to give a mono-iodo-derivative. What are these products? Explain.

7. 2,6-Di-t-butylphenol is a much weaker acid than phenol. Explain.

8. Discuss the following data:

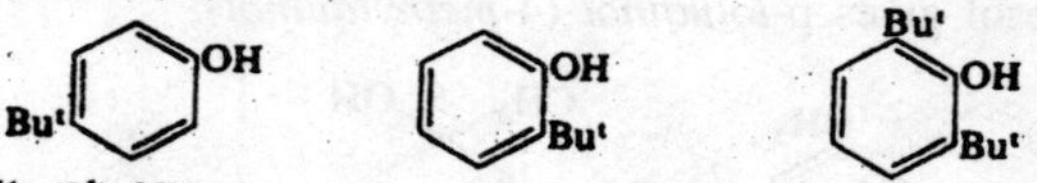

   OH, Bu^t — OH(cm^{-1}) 3500

   OH, Bu^t — 3500, 3640

   Bu^t, OH, Bu^t — 3640

9. Complete the equations:

   (i) $PhOH + (AcO)_2Hg \xrightarrow{heat} ? \xrightarrow[H_2O]{NaCl}$

   (ii) $PhOH + ClCH_2CO_2H \xrightarrow[heat]{NaOH}$

   (iii) 2, 4, 6-tri-butylphenol $\xrightarrow[C_6H_6]{PbO_2} ? \xrightarrow{Br_2}$

   (iv) 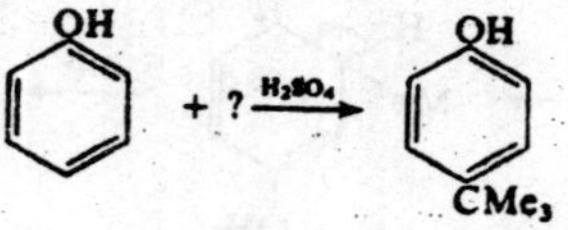 OH (phenol) $+ ? \xrightarrow{H_2SO_4}$ OH, CMe_3

   (v) $PhOH + CH_2 = CHCH_2Br \xrightarrow{K_2CO_3} ?$

10. Phenyl triphenylmethyl ether, on treatment with acid, gives p-triphenylmethylphenol. Discuss.

11. How many N.M.R. signals would you expect from Ph—O—CH_2CH_2—O—Ph?

12. Convert:

    (i) phenol ⟶ p-fluorophenol;

    (ii) phH ⟶ m-$HOC_6H_4NO_2$;

(iii) PhCl ⟶ 2, 4-$(NO_2)_2$-phenol;

(iv) phOH ⟶ 2,6-Br_2-4-NO_2-phenol;

(v) o-hydroxyacetophenone ⟶ 2, 5-dihydroxyacetophenone;

(vi) o-MeC_6H_4OH ⟶ o-$HOC_6H_4CO_2H$.

13. The Wurster salt derived from p-$Me_2NC_6H_4NMe_2$ is very stable, whereas that derived from the corresponding 2, 3, 5, 6-tetramethyl compound (durene derivative) cannot be detected. Explain.

14. Account for the following ions in the mass spectra of:

(i) o-ethylphenol; 122 (100%), 79 (9%), 77 (18%), 51 (5%);

(ii) n-butyl phenyl ether: 150 (19%), 107 (2%), 94 (100%), 93 (1%), 66 (4%), 65(4%).

3

HETEROCYCLIC COMPOUNDS

INTRODUCTION

Heterocyclic compounds are cyclic compounds with the ring containing carbon and other elements. The commonest being oxygen, nitrogen and sulphur. There are a number of heterocyclic rings which are easily opened and do not possess any aromatic properties, *e.g.*, ethylene oxide, γ- and δ-lactones etc. These are not considered to be heterocyclic compounds. Heterocycles are those compounds with five or six-membered heterocyclic rings which are stable, contain conjugated double bonds, and exhibit aromatic character.

Nomenclature

Many heterocyclic systems have trivial names. The following is the systematic method of nomenclature.

(i) The names of monocyclic compounds are derived by a prefix (or prefixes) indicating the nature of the hetero-atoms present, and eliding the 'a' where necessary, *e.g.*, oxygen, **oxa**; sulphur, **thia**, nitrogen, **aza**: silicon, sila, phosphorus, **phospha**. When two or more of the same hetero-atoms are present, the prefixes di, tri, etc. are used, *e.g.*, dioxa, triaza. If the hetero-atoms are *different*, their order of citation starts with the hetero-atom of as high a group in the periodic table and as low an atomic number in that group. Thus, the order of naming will be O, S, N, P, Si, *e.g.*, thiaza (S then N).

(ii) The size of a monocyclic ring from 3 to 10 is indicated by a stem: 3, ir (tri), 4. et (*tetra*); 5, ol; 6, in; 7, ep (*hepta*)', 8, oc (*octa*); 9, on (*nona*); 10, ec (*dcca*) [see Table 3.1].

(iii) The state of hydrogenation is indicated in the suffix (see Table 3.1) or by the prefixes dihydro, tetrahydro, etc., or by

prefixing the name of the parent unsaturated compound with the symbol H preceded by a number indicating the position of saturation.

(iv) (a) In a monocyclic compound containing only one hetero-atom numbering starts at this atom.

(b) The ring is numbered to give substituents or other hetero-atoms the lowest numbers possible. If the hetero-atoms are different, then numbering starts at the atom cited first according to the rule in (i) and proceeds round the ring in order of precedence.

Table 3.1

No. of members in the ring	*Rings containing nitrogen*		*Rings containing no nitrogen*	
	Unsaturation	*Saturation*	*Unsaturation*	*Saturation*
	(a)		(a)	
3	-irine	-iridine	-iren	-iran
4	-ete	-etidine	-et	-etan
5	-ole	-olidine	-ole	-olan
6	-ine	(b)	-in	-ane
7	-epine	(b)	-epin	-epan
8	-ocine	(b)	-ocin	-ocan
9	-onine	(b)	-onin	-onan
10	-ecine	(b)	-ecin	-ecan

(a) Corresponding to the maximum number of non-cumulative double bonds.

(b) Expressed by prefixing 'perhydro' to the name of the corresponding unsaturated compound.

Examples (see text for the various trivial names).

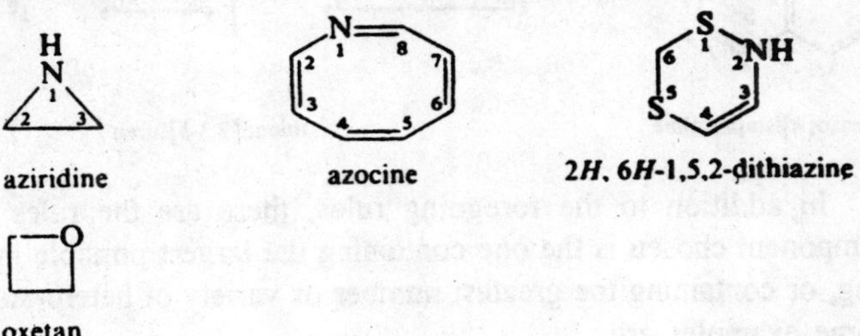

Fig. 3.1

Fused heterocyclic systems. Only a very elementary account is given here. When one heterocyclic ring is present, this is chosen as the parent compound. If more than one heterocyclic ring is present, the order of preference is given to the nitrogen-containing component (nitrogen rings are the most common).

For a component containing a hetero-atom other than nitrogen, the order of preference is that in (i) above (O before S, etc.). When the parent compound has been chosen, its name is prefixed by the name of the fused ring attached, *e.g.*, benz(o), naphth(o). Also, the parent compound chosen is the component containing the largest number of rings and has a simple name.

For the purpose of numbering, the structure is written with the greatest number of rings in a horizontal position and a maximum number of rings above and to the right of the horizontal row. Numbering is then carried out (usually) in a clockwise direction starting with the uppermost ring farthest to right and omitting atoms at ring junctions. To distinguish isomers, *the peripheral sides of the parent compound* are lettered a, b, c, etc., beginning with a for the side 1,2, *b* for 2,3, etc.

To the letter as early in the alphabet as possible, denoting the side where fusion occurs, are prefixed, if necessary, the numbers indicating the positions of fusion of the other component, their order conforms to the *direction of lettering* of the base component.

It should be noted that these numbers apply to the *prefixed component* (as a separate entity) and *not* to the combined system (which is numbered according to the usual rules). Two examples are:

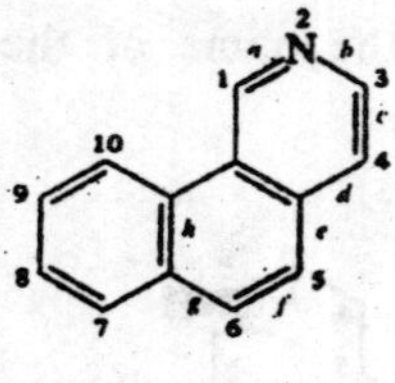

benzo[*h*]isoquinoline

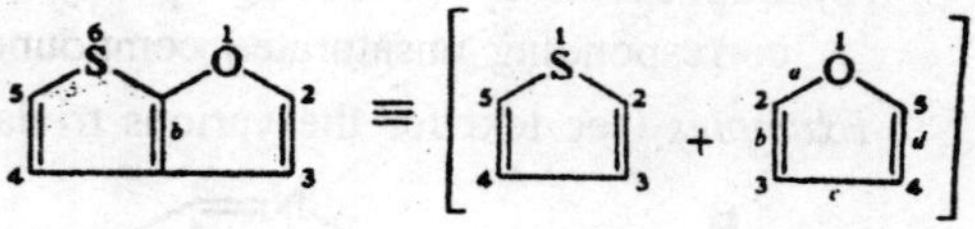

thieno[2,3-*b*]furan

In addition to the foregoing rules, there are the rules that the component chosen is the one containing the largest possible individual ring, or containing the greatest number or variety of hetero-atoms, etc. Some examples are:

2*H*-furo[3,2-*b*]pyran

1*H*-pyrazolo[4,3-*d*]oxazole

5*H*-pyrido[2,3-*d*]-*o*-oxazine

Five-membered rings

FURAN AND ITS DERIVATIVES

Furan (*furfuran*) contains one oxygen atom in its ring, and its structure is as shown. There are two monosubstituted derivatives of furan, 2 (or α) and 3 (or β): there are four disubstitution products: 2,3 (α, β), 2,4 (α, β'), 2.5 (α, α') and 3,4 (β, β').

Furan is obtained when wood, especially pine-wood, is distilled. It may be prepared by the dry distillation of mucic acid, and heating the product, *furoic acid*, at its b.p.:

$$HO_2C(CHOH)_4CO_2H \longrightarrow \text{(furoic acid)} + CO_2 + 3H_2O \longrightarrow \text{(furan)} + CO_2$$

Furan is most conveniently prepared by decarboxylating furoic acid in quinoline in the presence of copper powder.

It is manufactured by the catalytic decomposition of furfural in steam in the presence of an oxide catalyst:

$$\text{(furfural, CHO)} \longrightarrow \text{(furan)} + CO$$

A general method of preparing furan derivatives is to dehydrate 1,4-diketones or dialdehydes with, *e.g.*, phosphorus pentoxide, sulphuric acid, etc.:

$$RCO\,CH_2-CH_2\,COR \xrightarrow[-H_2O]{P_2O_5} \text{(2,5-dialkylfuran, R, R)}$$

Alternatively, furan derivatives may be prepared from ethyl acetoacetate as follows:

$$2CH_3CO\overset{-}{C}H\overset{+}{N}aCO_2C_2H_5 + I_2 \longrightarrow 2NaI + \begin{array}{c} CH_3COCHCO_2C_2H_5 \\ | \\ CH_3COCHCO_2C_2H_5 \end{array}$$

When diacetosuccinic ester is heated with dilute sulphuric acid, 2,5-dimethylfuran-3,4-dicarboxylic acid is formed:

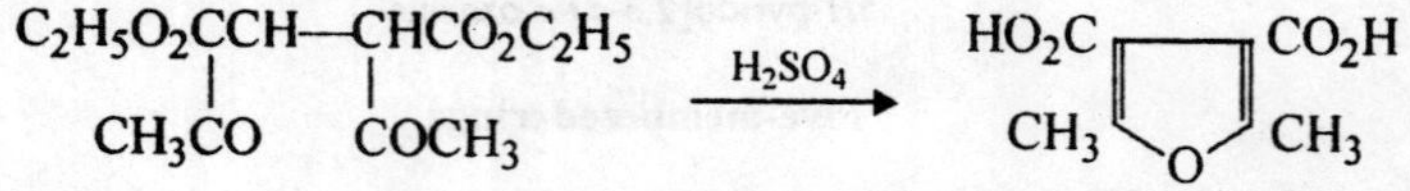

$$\begin{array}{cc} C_2H_5O_2CCH & —CHCO_2C_2H_5 \\ | & | \\ CH_3CO & COCH_3 \end{array} \xrightarrow{H_2SO_4}$$ 2,5-dimethylfuran ring bearing HO_2C and CO_2H at positions 3,4 and CH_3 groups at positions 2,5 (O in ring)

If ammonia is used instead of sulphuric acid, 2,5-dimethylpyrrole-3,4-dicarboxylic ester is obtained.

Furan derivatives may also be prepared by the **Feist-Benary synthesis** (1902, 1911); an α-chloro-ketone is condensed with a β-keto-ester in the presence of pyridine (cf. the Hantzsch synthesis, for a possible mechanism):

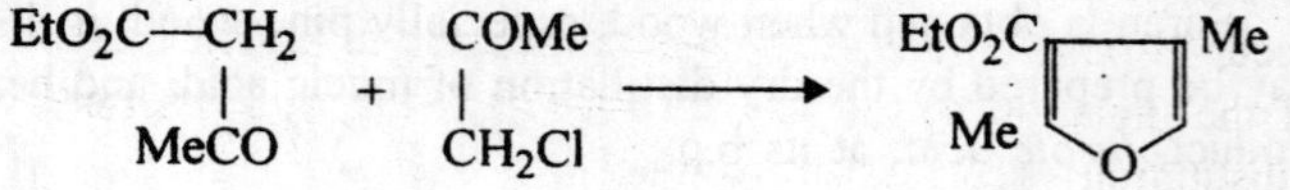

$$\begin{array}{c} EtO_2C—CH_2 \\ | \\ MeCO \end{array} + \begin{array}{c} COMe \\ | \\ CH_2Cl \end{array} \longrightarrow$$ furan ring with EtO_2C, Me (top) and Me (at C-2, next to O)

STRUCTURE OF FURAN

Furan has a resonance energy of 71.1-96.2 kJ mol^{-1}, and is best represented as a resonance hybrid of contributing structures (I)-(V). As we have seen, to be aromatic, a monocyclic molecule must be a planar $(4n + 2)$ π-electron molecule. Hence, to provide six π-electrons for the ring system, one of the lone pairs of electrons of the oxygen atom is involved.

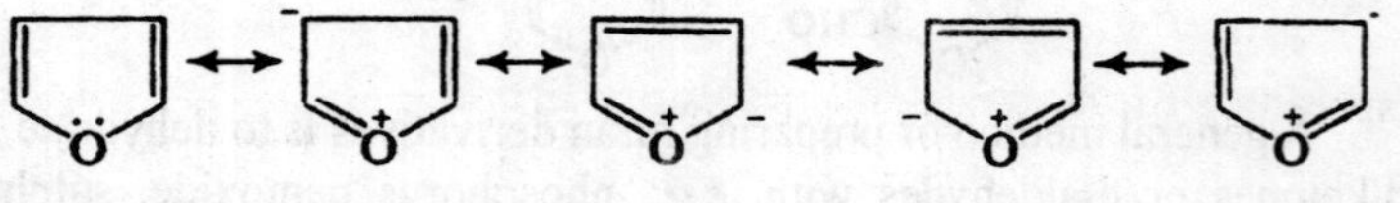

(I) ⟷ (II) ⟷ (III) ⟷ (IV) ⟷ (V)

The ring current in furan, evaluated from N.M.R. data (Elvidge, 1965: de Jongh et at., 1965), is less than that of benzene and so furan is less aromatic than benzene.

N.M.R. spectra of heterocyclic compounds. Because of the inductive effect of the hetero-atom (O, S, and N) ring protons have τ-values very much shifted downfield compared with the corresponding aromatic compound, and this downfield shift is more pronounced for hydrogen atoms in the α-positions, *e.g.*, (τ-values):

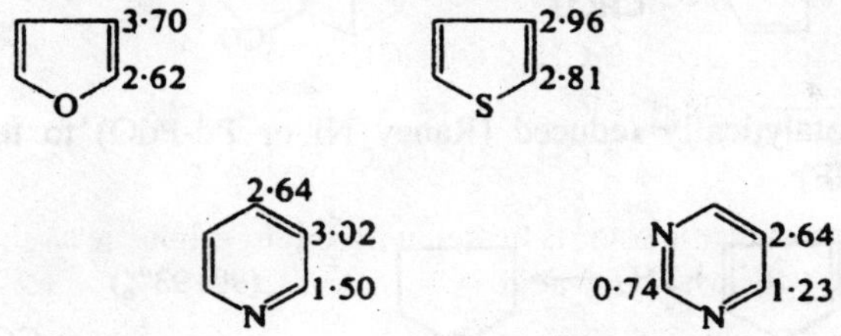

3·70 2·62 O   2·96 2·81 S   3·91 3·40 N H

2·64 3·02 1·50 N   N 2·64 0·74 1·23 N

Fig. 3.2

SUBSTITUTION IN FURAN

As we have seen, canonical structures carrying unlike charges are not so stable as un-charged structures, and the larger the charge separation, the less stable is that structure. Hence, (II) and (V) contribute less than (III) and (IV), and consequently the resonance hybrid has a larger electron density at position 2 (or 5) than at 3 (or 4). Thus, on the basis of charge distribution, electrophilic substitution would be expected to occur at position 2 (or 5). This is the case in practice. Also, because, of the donation of the oxygen lone-pair, the ring becomes activated, substitution occurring faster than in benzene.

Alternatively, if we consider the contributing resonating structures of the intermediate carbonium ion (σ-complex), then there are three for 2- and two for 3-substitution.

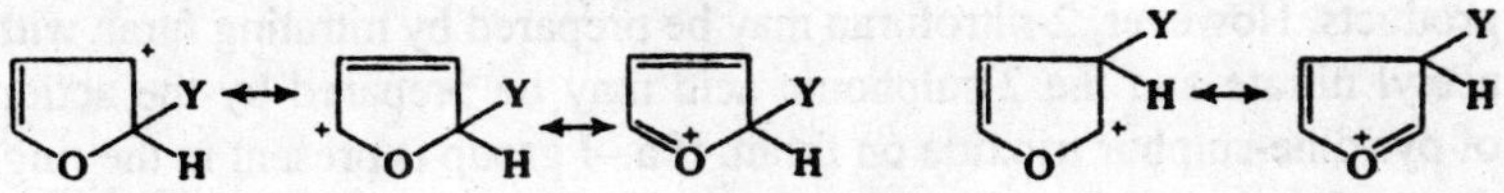

Fig. 3.3

Hence 2-substitution will be favoured, since there is also greater spreading of charge in this resonance hybrid.

Reactions of Furan

Furan, b.p. 32°C, turns green a pine splint moistened with hydrochloric acid. It behaves chemically as an aromatic compound, but in some ways it behaves as a 1,3-diene. Thus, it is less aromatic than thiophen and pyrrole, and is the only one of these to undergo the Diels-Alder reaction.

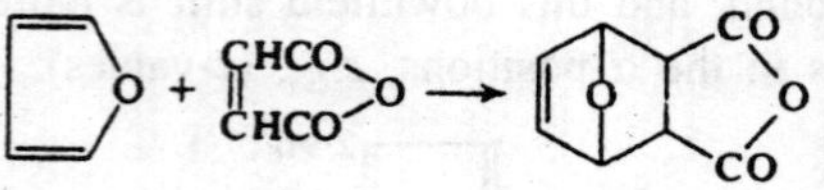

$$\text{Furan} + \begin{matrix}CHCO \\ \| \\ CHCO\end{matrix}\!\!>O \longrightarrow \text{adduct (with } \begin{matrix}CO \\ CO\end{matrix}\!\!>O\text{)}$$

Furan is catalytically reduced (Raney Ni or Pd-PdO) to tetrahydrofuran (THF):

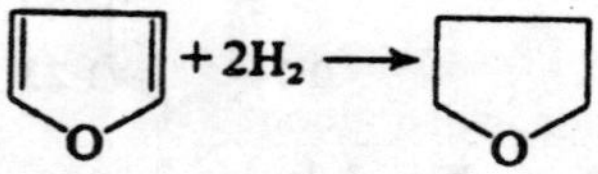

$$\text{Furan} + 2H_2 \longrightarrow \text{THF} \quad (90–93\%)$$

Tetrabydrofuran is manufactured synthetically from butyne-1,4-diol:

$$HOCH_2C{\equiv}CCH_2OH \xrightarrow[Ni]{H_2} HOCH_2CH_2CH_2CH_2OH \xrightarrow[(heat)]{H_2O} \text{THF}$$

Tetrahydrofuran is a valuable solvent (*e.g.*, for the preparation of Grignard reagents). With ammonia it forms pyrrolidine, and with hydrogen chloride, tetramethylene chlorohydrin:

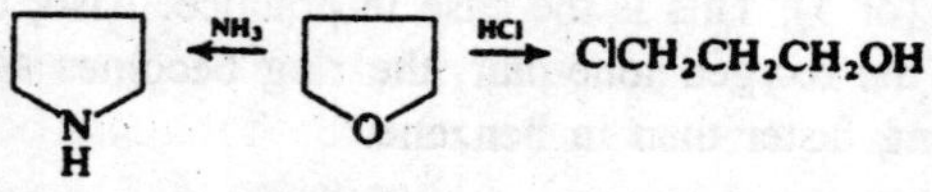

$$\text{Pyrrolidine (NH)} \xleftarrow{NH_3} \text{THF} \xrightarrow{HCl} ClCH_2CH_2CH_2OH \quad (54–57\%)$$

Furan is readily polymerised by concentrated acids, and probably the oxonium salt is involved (furan is very weakly basic; cf. pyrrole). Thus, direct nitration (mixed acid) and sulphonation result in resinified products. However, 2-nitrofuran may be prepared by nitrating furan with acetyl nitrate and the 2-sulphonic acid may be prepared by the action of pyridine-sulphur trioxide on furan. If a –I group is present in the ring, then sulphonation can be carried out directly, *e.g.*, furoic acid gives furoic-5-sulphonic acid. Furan reacts readily with halogens, but the liberated halogen acid causes polymerisation. Because of this, halogenated furans are usually prepared by indirect methods, *e.g.*, by brominating furoic acid and decarboxylating the product, 5-bromofuroic acid, to 2-bromofuran by heating in quinoline in the presence of copper powder.

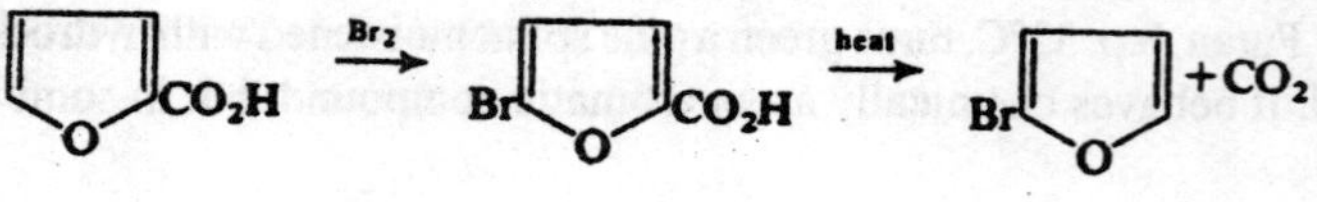

$$\text{Furan-}CO_2H \xrightarrow{Br_2} Br\text{-Furan-}CO_2H \xrightarrow{heat} Br\text{-Furan} + CO_2$$

However, chlorination of furan at about –40°C gives 2-chloro- and 2,5-dichlorofuran as the main products.

Furan undergoes the Gattermann reaction to form furfural. Since aluminium chloride attacks the ring, Friedel-Crafts reactions are best carried out with stannic chloride as catalyst. In any case, alkylation reactions with furan lead to polymers, but acylation with acid chlorides or anhydrides with stannic chloride as catalyst gives 2-acyl derivatives. A better catalyst for the reaction with anhydrides is boron trifluoride in ether (the reagent is the complex Et_2O^+ —BF_3^-).

Furan is very readily mercurated in the 2-position, *e.g.*, when heated with mercuric chloride in aqueous sodium acetate, 2-chloromercurifuran is obtained. Since the mercuri-group is readily replaced by bromine or iodine, and by an acyl group, these mercuri-compounds are useful intermediates:

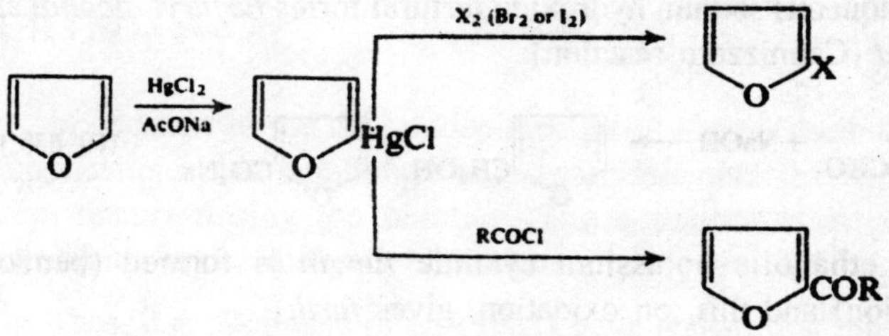

Fig. 3.4

Furan reacts with n-butyl-lithium to form the 2-lithium derivative. This undergoes many of the usual reactions of organolithium compounds, *e.g.*. with carbon dioxide, furoic acid is obtained:

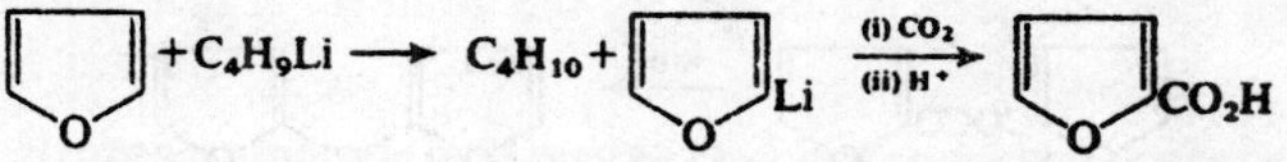

Furan undergoes the Gomberg reaction when treated in alkaline solution with diazonium salts:

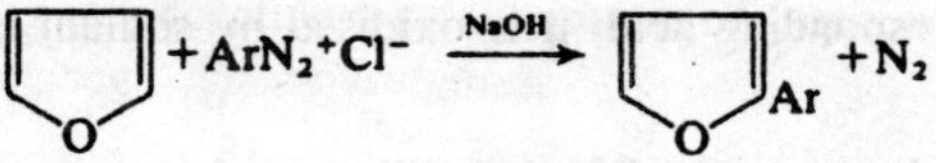

The furan ring is readily opened under suitable conditions to give 1,4-dioxo-compounds, *e.g.*,

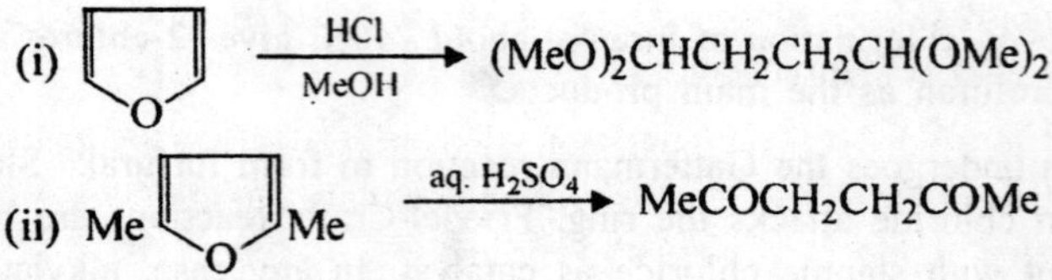

(i) [furan] $\xrightarrow[\text{MeOH}]{\text{HCl}}$ $(MeO)_2CHCH_2CH_2CH(OMe)_2$

(ii) Me[furan]Me $\xrightarrow{\text{aq. } H_2SO_4}$ $MeCOCH_2CH_2COMe$

Furfural (*furfuraldehyde*) may be prepared by distilling pentoses with dilute sulphuric acid:

$$HOCH_2(CHOH)_3CHO \longrightarrow \text{[furan-2-CHO]} + 3H_2 \quad (100\%)$$

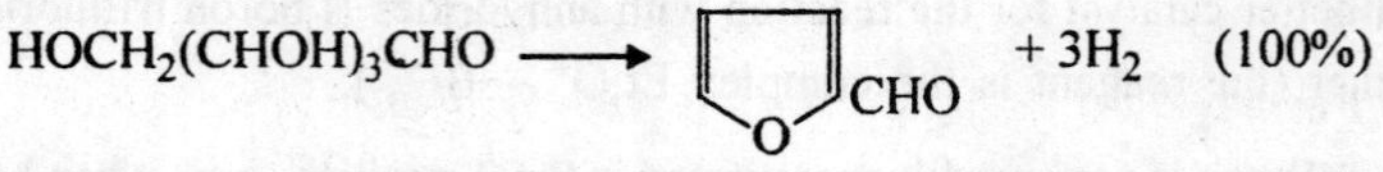

It is manufactured by treating oat husks, cotton-seed hulls or corn cobs with dilute sulphuric acid followed by steam distillation (the starting materials are rich in pentoses).

Furfural, b.p. 162°C, is chemically very similar to benzaldehyde, *e.g.*, with aqueous sodium hydroxide furfural forms *furfuryl alcohol* and *furoic acid* (Cannizzaro reaction):

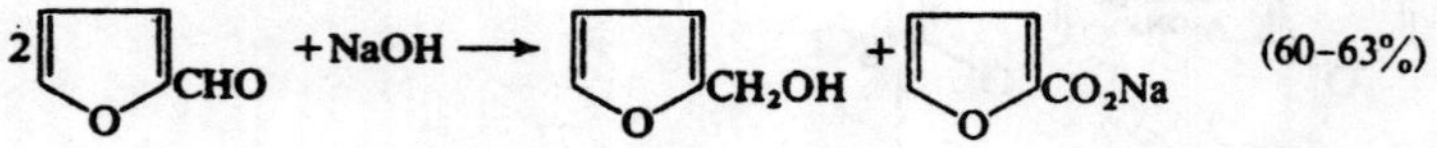

$$2\,\text{[furan-2-CHO]} + NaOH \longrightarrow \text{[furan-2-}CH_2OH] + \text{[furan-2-}CO_2Na] \quad (60\text{–}63\%)$$

With ethanolic potassium cyanide *furoin* is formed (benzoin condensation) and this, on oxidation, gives *furil:*

$$2\,\text{[furan-2-CHO]} \xrightarrow{\text{KCN}} \text{[furyl]}CH(OH)CO\text{[furyl]} \xrightarrow{[O]} \text{[furyl]}COCO\text{[furyl]}$$

When heated with aqueous potassium hydroxide, furil gives *furilic acid* (benzilic acid rearrangment):

$$\text{[furyl]}COCO\text{[furyl]} \xrightarrow{\text{KOH}} \text{[furyl]}_2C(OH)CO_2K$$

Furfural reacts with aniline to form an anal, and can undergo the Perkin reaction and the Claisen reaction. It is easily oxidised by silver oxide to the corresponding acid; it is oxidised by sodium chlorate to fumaric acid.

Furfural condenses with dimethylaniline in the presence of zinc chloride to form *furfuraldehyde green* (analogous to malachite green). A characteristic reaction of furfural is the red colour it gives with aniline and hydrochloric acid; it also turns green a pine splint moistened with hydrochloric acid (cf. furan, above).

Furfural is used for the preparation of dyes, plastics and fumaric acid. It is also used as a solvent in synthetic rubber manufacture, and as an extraction liquid in petroleum refining.

Furfuryl alcohol, b.p. 170°C, may be prepared by reducing furfural, or by means of the Cannizzaro reaction on furfural. It is manufactured by the catalytic reduction (copper chromite) of furfural.

Furoic acid (furan-2-carboxylic acid, ***pyromucic acid*****),** m.p. 133°C, may be prepared by the dry distillation of mucic acid, or by the oxidation of furfural with cuprous oxide-silver oxide in aqueous sodium hydroxide (yield: 86-90%). It behaves more like an unsaturated aliphatic acid rather than benzoic acid (cf. furfural); *e.g.*, furoic acid is readily oxidised by alkaline permanganate, brominated by bromine vapour (it adds on four bromine atoms), and oxidised by bromine-water to fumaric acid. The sodium salt of furoic acid is decarboxylated by mercuric chloride to give the mercuri-chloride (this does not occur with furan-3-carboxylic acid):

$$\text{C}_4\text{H}_3\text{O}\text{-}CO_2Na + HgCl_2 \longrightarrow \text{C}_4\text{H}_3\text{O}\text{-}HgCl + CO_2 + NaCl$$

On the other hand, when heated with mercuric acetate, furoic acid forms 3-acetoxymercurifuran.

$$\text{C}_4\text{H}_3\text{O}\text{-}CO_2H \xrightarrow{Hg(OAc)_2} \text{C}_4\text{H}_3\text{O}\text{-}CO_2HgOAc \xrightarrow{-CO_2} \text{C}_4\text{H}_3\text{O}\text{-}HgOAc$$

It has already been pointed out that the presence of a –I group stabilises the furan ring. Another example of this is that on nitration with mixed acids, furoic acid gives first the 2-nitro derivative, and then 2,5-dinitrofuran, the latter being produced by displacement of the carboxyl group.

$$\text{C}_4\text{H}_3\text{O}\text{-}CO_2H \xrightarrow[H_2SO_4]{HNO_3} O_2N\text{-C}_4\text{H}_2\text{O-}CO_2H \longrightarrow O_2N\text{-C}_4\text{H}_2\text{O-}NO_2$$

Furan-3-carboxylic acid occurs naturally.

Benzofuran (benzfuran, ***coumarone*****),** b.p. 174°C, occurs in coal-tar. It is used in the manufacture of plastics.

Dibenzofuran (dibenzfuran, *diphenylene oxide*), m.p. 87°C, may be prepared by heating phenol with lead oxide at 150°C:

$$2\,C_6H_5OH \xrightarrow{PbO} \text{dibenzofuran} \quad (28\%)$$

It may also be prepared by passing diphenyl ether through a red-hot tube.

THIOPHEN AND ITS DERIVATIVES

Thiophen occurs in coal-tar and shale oils. Its b.p. (84° C) is close to that of benzene and hence it is difficult to separate from the benzene fraction obtained from coal-tar. Thiophen can be sulphonated more readily than benzene, and this property is used to separate the two compounds by repeatedly shaking benzene (from coal-tar) with cold concentrated sulphuric acid, whereby the water-soluble thiophensulphonic acid is formed.

A better means of separation is to reflux the benzene with aqueous mercuric acetate whereupon thiophen is mercurated and benzene is not. Thiophen may be recovered from its mercurated derivative by distilling the latter with hydrochloric acid. The presence of thiophen in benzene may be detected by the *indophenin reaction*. This is the development of a blue colour when benzene is treated with isatin and sulphuric acid.

Thiophen may be prepared by passing a mixture of acetylene and hydrogen sulphide through a tube containing alumina at 400°C.

$$2C_2H_2 + H_2S \longrightarrow C_4H_4S + H_2$$

This method is used commercially. It is also manufactured by reaction between n-butane and sulphur in the vapour phase.

$$CH_3CH_2CH_2CH_3 + 4S \xrightarrow{600^\circ C} C_4H_4S + 3H_2S$$

Thiophen may also be prepared by heating sodium succinate with phosphorus trisulphide:

$$\begin{array}{l} CH_2CO_2Na \\ | \\ CH_2CO_2Na \end{array} \xrightarrow{P_2S_3} C_4H_4S \quad (30\%)$$

Derivatives of thiophen may be prepared by heating 1,4-diketones with phosphorus trisulphide (cf. furan derivatives).

Thiophen is easily sulphonated, nitrated or chlorinated. It was this close chemical similarity to benzene that masked the presence of thiophen in benzene from coal-tar. V. Meyer (1882) found that a sample of benzene prepared by heating sodium benzoate with soda-lime did not give the indophenin test.

Subsequently he showed that it was a sulphur-containing compound, which he called *thiophene* that was responsible for the indophenin reaction.

This similarity to benzene has resulted in a similar nomenclature:

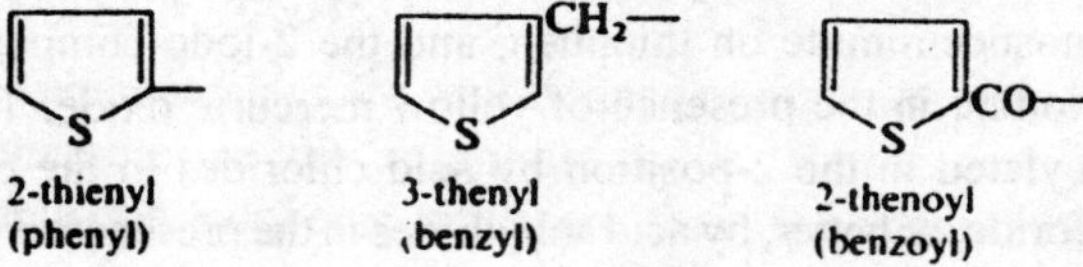

STRUCTURE OF THIOPHEN

Thiophen is a resonance hybrid (R.E. 117-130 kJ mol^{-1}), the sulphur atom contributing two electrons to form a (4n + 2) π-electron molecule (cf. furan). However, there are some complications in elucidating the electronic structure of thiophen.

Sulphur is less electronegative than oxygen (or nitrogen) and can also use 3d orbitals (oxygen and nitrogen cannot). Hence, more canonical forms are possible for thiophen than for furan and pyrrole. In group (a) the sulphur atom uses p-orbitals; in group (b) it also uses d-orbitals.

(*a*)

(*b*)

Fig. 3.5

The estimation of the ring current in thiophen has shown it to be more aromatic than furan.

Substitution in Thiophen

As with furan, electrophilic substitution would be expected, on the basis of charge distribution and stabilities of the carbonium ions, to take place at position 2 (or 5). This occurs in practice.

Reactions of Thiophen

Thiophen shows no basic properties and is stable to aqueous acids. It can be nitrated by fuming nitric acid in acetic anhydride to give mainly 2-nitrothiophen, and is readily sulphonated in the 2-position with *cold* concentrated sulphuric acid. Chlorination of thiophen results in the formation of both substitution and addition products, but 2-chloro- and 2,5-dichlorothiophen are the main products when thiophen ischlorinated at – 30°. On the other hand, 2-bromothiophen is prepared by the action of N-bromosuccinimide on thiophen, and the 2-iodo-compound by the action of iodine in the presence of yellow mercuric oxide. Thiophen is readily, acylated in the 2-position by acid chlorides in the presence of stannic chloride, or better, by acid anhydrides in the presence of phosphoric acid. *e.g.*, methyl 2-thienyl ketone:

$$\text{Thiophen} + (MeCO)_2O \xrightarrow{H_3PO_4} \text{2-COMe-thiophen}$$

Thiophen may be chloromethylated (formaldehyde and HCl) and formylated (dimethylformamide and $POCl_3$) in the 2-position. Thiophen-2-aldehyde (b.p. 198°C) undergoes the Cannizzaro reaction and the benzoin condensation (cf. furfural). Mercuration with mercuric chloride in the presence of a small amount of sodium acetate produces the 2-mercuri-chloride as the main product, and when treated with n-butyl-lithium in ether, thiophen forms the 2-lithium derivative. These compounds may be used to prepare various 2-substituted thiophens.

Various derivatives of thiophen may also be prepared from the monobromo-derivative, *e.g.*,

$$\text{2-Br-thiophen} \xrightarrow{Mg} \text{2-MgBr-thiophen} \xrightarrow[\text{(ii) HCl}]{\text{(i) }CO_2} \text{2-}CO_2H\text{-thiophen}$$

Thiophen does not undergo the Diels-Alder reaction. Reduction of thiophen with sodium in liquid ammonia produces 2,3-dihydro-(2-thiolen) and 2,5-dihydrothiophen (3-thiolen) (Birch *et al.*, 1951). Catalytic reduction to tetrahydrothiophen (*thiophan, thiolan*) may be carried out provided a large amount of the catalyst, palladium, is used to overcome the poisoning effect of the sulphur. On the other hand catalytic reduction of thiophen with Raney nickel as catalyst results in the removal of sulphur, the main product being n-butane:

$$\text{thiophen} \xrightarrow[H_2]{\text{Raney Ni}} CH_3CH_2CH_2CH_3 + NiS$$

This desulphurisation has been used as a means of preparing aliphatic mono- and dicarboxylic acids. The method is to treat a thiophen mono- or dicarboxylic acid in aqueous alkali with Ni-Al alloy *e.g.*, n-heptanoic acid from 5-ethylthiophen-2-carboxylic acid:

$$C_2H_5\text{–(thiophen)–}CO_2H \xrightarrow[\text{aq. NaOH}]{\text{Ni–Al}} CH_3(CH_2)_5CO_2H$$

Thiophen does not form sulphonium salts and cannot be oxidised to a sulphoxide or sulphone; hydrogen peroxide *opens* the thiophen ring, the sulphur being oxidised to sulphuric acid.

A number of condensed thiophen systems are known, *e.g.*, *benzothiophen* (I) (also known as *thionaphthen* because it closely resembles naphthalene), *dibenzothiophthen* (II), and *thiophthen* (III).

(I)

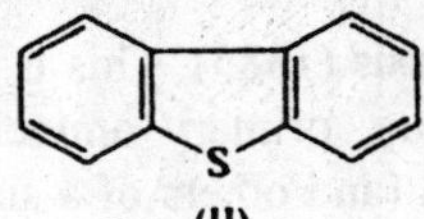

(II)

(III)

PYRROLE AND ITS DERIVATIVES

Pyrrole is a very important five-membered heterocyclic ring because its nucleus occurs in many natural compounds, *e.g.*, alkaloids, chlorophyll, haematin, etc.

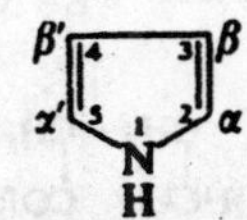

Pyrrole occurs in coal-tar and bone oil. It may be isolated from bone oil by washing the latter with dilute alkali to remove acidic substances, then with acid to remove strongly basic substances, and finally fractionating. Pyrrole distils over in the fraction boiling between 100° and 150°C, and may be purified by fusing with potassium hydroxide. Solid potassiopyrrole is formed and this, on steam distillation, gives pure pyrrole.

Pyrrole may be synthesised by passing a mixture of acetylene and ammonia through a red-hot tube:

$$2C_2H_2 + NH_3 \longrightarrow C_5H_4N + H_2$$

It is conveniently prepared by distilling a mixture of ammonium mucate and glycerol at 200°C:

$$H_4NO_2C(CHOH)_4CO_2NH_4 \longrightarrow C_4H_5N + 2CO_2 + NH_3 + 4H_2O \ (37\text{–}40\%)$$

If salts of mucic acid with primary amines are decomposed as above, then 1-substituted pyrroles are obtained, *e.g.*, aniline mucate gives 1-phenylpyrrole.

Pyrrole is also formed when succinimide is distilled with zinc dust:

$$\begin{matrix} CH_2-CH_2 \\ CO \quad CO \\ NH \end{matrix} \xrightarrow{Zn} \text{pyrrole (NH)}$$

Pyrrole is manufactured by passing a mixture of furan, ammonia and steam over heated alumina as catalyst:

$$\text{furan} + NH_3 \xrightarrow[H_2O]{Al_2O_3} \text{pyrrole (NH)}$$

Many methods are available for synthesising pyrrole derivatives, *e.g.*,

(i) **Paal-Knorr synthesis** (1885). This is carried out by treating a 1,4-diketone with ammonia, primary amines, hydrazines, etc. The mechanism of this reaction (and others of a similar nature) is not fully understood, but there is evidence that it proceeds in a similar fashion to the reaction between oxo compounds and compounds of the type Z—NH_2, *e.g.*, (proton transfers have been combined in one step).

$$\begin{matrix} CH_2-CH_2 \\ R^1CO \quad COR^1 \end{matrix} \xrightarrow{R^2NH_2} \begin{matrix} CH_2-CH_2 \\ HO\diagdown C \quad COR^1 \\ R^1 \quad NHR^2 \end{matrix} \xrightarrow{-H_2O} \begin{matrix} CH-CH_2 \\ R^1C \quad COR^1 \\ NHR^2 \end{matrix} \longrightarrow$$

$$\begin{matrix} CH-CH_2 \\ R^1C \quad C\diagup OH \\ N \quad R^1 \\ R^2 \end{matrix} \xrightarrow{-H_2O} \text{1-}R^2\text{-2,5-di}R^1\text{-pyrrole} \qquad (g.)$$

Possibly the two molecules of water are eliminated from the ring-closed intermediate and not stepwise as shown.

If succinaldehyde is used as the 1,4-dicarbonyl compound, pyrrole itself is obtained (yield is poor).

(ii) **Knorr pyrrole synthesis** (1884; 1886). This is the most general method, and involves the condensation between an α-aminoketone and a β-diketone or β-keto-ester (cf. with (iii) below).

(i) $\underset{EtO_2CCH_2}{\overset{MeCO}{|}} \xrightarrow{HNO_2} \underset{EtO_2CC=NOH}{\overset{MeCO}{|}} \xrightarrow[AcOH]{Zn} \underset{EtO_2CCHNH}{\overset{MeCO}{|}}$

(ii) $\underset{EtO_2CCHNH_2}{\overset{MeCO}{|}} + \underset{COMe}{\overset{CH_2Y}{|}} \longrightarrow$ (pyrrole ring: Me, Y; EtO_2C, Me; N–H)

Thus, if Y is CO_2Et, the product is 3, 5-dimethylpyrrole-2,4-dicarboxylic ester.

(iii) **Hantzsch synthesis** (1890). This is the condensation between chloroacetone, a β-ketoesteran and a primary amine, *e.g.*:

(i) $MeCOCH_2CO_2Et + NH_3 \longrightarrow \underset{NH_2}{\overset{MeC=CHCO_2Et}{|}}$

(ii) EtO_2C–C(H)=C(Me)–NH–H + CH_2–Cl / C=O / Me $\xrightarrow{-HCl}$ EtO_2CC——CH_2 / H / MeC–N(H)–C(=O)Me $\longrightarrow$ EtO_2CC——CH_2 / MeC–N(H)–C(OH)Me $\xrightarrow{-H_2O}$ (pyrrole: EtO_2C; Me, Me; N–H)

Fig. 3.6

Some furan derivative is also formed.

Structure of Pyrrole

Pyrrole is a resonance hybrid (R.E. 87.8–130 kJ mol^{-1})

Fig. 3.7

The estimation of ring current in pyrrole has shown that it is less aromatic than thiophen and more aromatic than furan, *i.e.*, the order of aromaticity is: thiophen > pyrrole > furan.

Substitution in Pyrrole

Electrophilic substitution in pyrrole occurs predominantly in the 2- (or 5-) position, and arguments to explain this (charge distribution and carbonium ion stabilities) are similar to those used for furan.

Basicity of Pyrrole

Because of the contribution of the lone pair of electrons of the nitrogen atom to the formation of a (4n + 2) π-electron molecule, the availability of this lone pair for protonation is very much decreased, and consequently pyrrole is a very weak base (see Table 3.2). However, in acid solution, protonation also occurs at the 2- and 3-positions, and in concentrated solution, pyrrole polymerises (to form *pyrrole-red*). On the other hand, when in contact with dilute acid for a brief period, pyrrole forms the trimer (I).

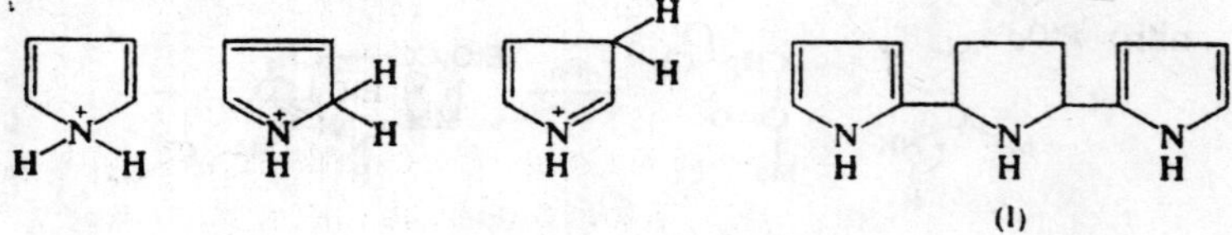

Fig. 3.8

Properties of Pyrrole

Pyrrole, b.p. 131°C, rapidly darkens on exposure to air. A characteristic reaction is the turning red of a pine splint moistened with hydrochloric acid when exposed to the vapour of pyrrole (and many of its derivatives). Pyrrole gives 2-nitropyrrole when nitrated with nitric acid in acetic anhydride at low temperatures, and sulphonation with pyridine-sulphur trioxide in ethylene dichloride gives the 2-sulphonic acid. Pyrrole reacts rapidly with halogens to form 2,3,4,5- tetrahalogenopyrroles, but with sulphuryl chloride, 2.5-dichloropyrrole is the main product.

Pyrrole shows a number of resemblances to phenols and aromatic amines. The imino-hydrogen of pyrrole is replaceable by potassium, alkyl or acyl groups. When pyrrole is heated with solid potassium hydroxide, potassiopyrrole is formed (thus pyrrole is acidic):

$$C_4H_4NH + KOH \longrightarrow C_4H_4N^-H^+ + H_2O$$

Potassiopyrrole reacts with carbon dioxide and with chloroform as do phenols in the Kolbe Schmitt and Reimer-Tiemann reactions, to form 2- and 3-pyrrolecarboxylic acids and pyrrole-2-aldehyde, respectively. In the latter case, however, some 3-chloropyridine is also formed (see also below). Potassiopyrrole also reacts with acyl chlorides, acid anhydrides, and alkyl halides to form the 1- (N-) derivatives.

At higher temperature, the 2-derivative is the main product, presumably by rearrangement of the 1-derivative formed first (cf. Hofmann-Martius rearrangement). 2-Ketones may also be prepared by the Houben-Hoesh synthesis. Pyrrole undergoes the Gattermann reaction to form the 2-aldehyde. but this compound is most conveniently prepared by formylation with dimethylformamide and phosphoryl chloride.

Pyrrole is mercurated with great difficulty. Experience has shown that the ease of mercuration of heterocyclic compounds varies considerably with the nature of the hetero-atom. Generally, those containing oxygen or sulphur are easily mercurated (cf. thiophen). Whereas those containing nitrogen are usually mercurated with greater difficulty.

Pyrrole couples with diazonium salts in the 2-position in weakly acid solution, and in the 2- and 5-positions (to give the bisazo-compound) in alkaline solution. If the 2- and 5-positions are occupied by, *e.g.*, methyl groups, coupling takes place in the 3-position.

When pyrrole is treated with methylmagnesium bromide, pyrrolylmagnesium bromide is formed, and this reacts with a series of alkyl halides to give both 2- and 3-alkylpyrroles in comparable amounts (Skell *et al.*, 1962). Reinecke *et al.* (1964) from their NMR spectra studies of the indole Grignard reagent, have concluded that the structure of the indole and pyrrole reagents are resonance hybrids. This would explain attack at positions 2 and 3.

MgX^+

Pyrrole is oxidised by chromium trioxide in acetic acid to maleinimide:

[O]

When pyrrole is refluxed with an ethanolic solution of hydroxylamine, *the ring is opened* and succinaldehyde dioxime is formed:

$$NH + 2NH_2OH \longrightarrow \begin{matrix} CH_2CH{=}NOH \\ | \\ CH_2CH{=}NOH \end{matrix} + NH_3$$

When potassiopyrrole is heated with chloroform and sodium ethoxide, *ring expansion* takes place, the product being 3-chloropyridine:

$$+ CCl_2 \xrightarrow[EtONa]{CHCl_3} \xrightarrow{-Cl^-}$$

Reduction Products of Pyrrole

Pyrrole is reduced by zinc and acetic acid to pyrroline (1,5-*dihydropyrrole*), b.p. 91°C. This, on heating with hydriodic acid and red phosphorus, gives pyrrolidine (*tetrahydropyrrole*), b.p. 88°C:

$$\xrightarrow[AcOH]{Zn} \xrightarrow[P]{HI}$$

Pyrrolidine may also be prepared by catalytically reducing pyrrole using nickel at 200°C, or by the electrolytic reduction of succinimide.

Pyrroline and pyrrolidine are both strong bases, and do not show any tendency to polymerise. Pyrrolidine is used in the preparation of enamines.

2-Pyrrolidone, m.p. 25°C, may be prepared by the electrolytic reduction of succinimide:

$$\xrightarrow{e;\,H^+}$$

2-Pyrrolidone is the *lactam* (cf. lactones) of γ-aminobutyric acid 3-*Pyrrolidone* is also known.

Table 3.2

Base	*pKa*	*Base*	pK_a
Pyrrole	–0.27	Thiazole	2.53
(acidic)	(15)	Pyridine	5.23
Pyrazole	2.53	Piperidine	11.22
(acidic)	(14)	Quinoline	4.94
Imidazole	7.03	Isoquinoline	5.2
(acidic)	(14.5)	Acridine	5.60

Pyrrole Derivatives

Methods of preparation have already been described; here will be mentioned some of their reactions.

Pyrrole-2-aldehyde does not undergo this benzoin condensation or the Cannizzaro reaction; it is readily reduced to pyrrolylmethanol by lithium aluminium hydride or sodium borohydride.

Pyrrolecarboxylic acids are readily decarboxylated when heated, and so this reaction offers a means of preparing various pyrrole derivatives. This becomes an important method when taken in conjunction with the fact that the 3-ester group is often hydrolysed by sulphuric acid at 40–60°C, whereas the 2-ester group is unaffected. Furthermore, the 2-ester group is hydrolysed by cold dilute sodium hydroxide solution, but not the 3-group.

Fig. 3.9

Complete hydrolysis with hot alkali, followed by decarboxylation of the free acid, leads to alkylpyrroles.

Alkylated pyrroles may also be prepared from pyrrole-Grignard reagents.

Indole (*benzopyrrole*), m.p. 52°C, occurs in coal-tar, jasmine flowers and orange blossoms. Indole is the parent substance of indigotin.

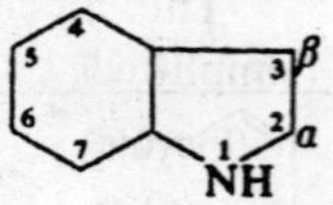

Indoles may be synthesised in various ways:

(i) **Fischer's indole synthesis (1886).** This is the most important method of preparing indole derivatives, and is carried out by heating the phenylhydrazone or substituted phenylhydrazone of an appropriate aldehyde, ketone or ketonic acid with zinc chloride, polyphosphoric acid sulphuric acid in ethanol, etc.

Various mechanisms have been proposed for this reaction: a widely accepted one is that suggested by Robinson (1918) and modified by later workers:

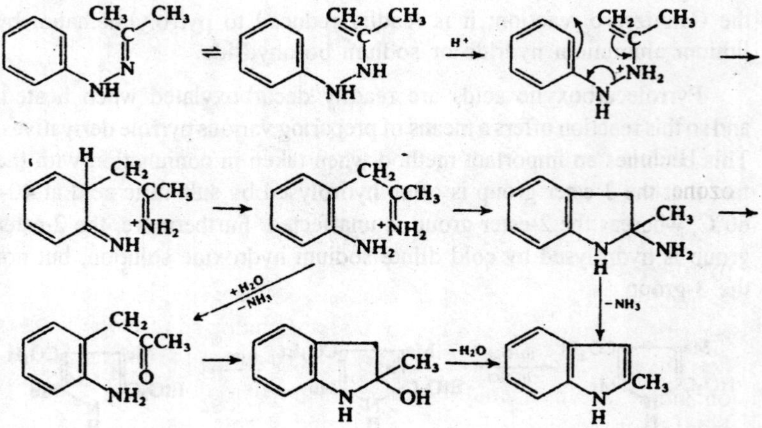

Fig. 3.10

(ii) **Madelung synthesis** (1912). This is the cyclisation of an o-acylamidotoluene by means of a strong base, *e.g.*, indole (from o-formamidotoluene) and 2-methylindole (from o-acetamidotoluene):

(79%) (R = H) ← t-BuOK — o-CH_3-C$_6$H$_4$-NHCOR — $NaNH_2$ → (80%) (R = Me)

The mechanism of this reaction is uncertain; a possibility (simplified) is:

o-CH_3-C$_6$H$_4$-NHCOR —t-BuO$^-$→ o-$\bar{C}H_2$-C$_6$H$_4$-NH-C(=O)R —$+H^+$→ 2-hydroxy-2-R-indoline (OH, R, NH) —$-H_2O$→ 2-R-indole

(iii) **Reissert synthesis** (1897). This is a very good method, and is carried out with o-nitrotoluene (or its substituted derivatives) and ethyl oxalate as follows:

o-CH_3-C$_6$H$_4$-NO_2 + $(CO_2Et)_2$ —EtONa→ o-CH_2COCO_2Et-C$_6$H$_4$-NO_2 —Zn/AcOH→ o-CH_2COCO_2Et-C$_6$H$_4$-NH_2 → 2-hydroxy-2-CO_2Et-indoline —$-H_2O$→ indole-2-CO_2Et

Indole resembles pyrrole in many of its properties. It is oxidised by ozone to indigotin.

Electrophilic substitution normally occurs in the 3-position (cf. pyrrole), but if this is occupied then 2-substitution occurs. If both the 2- and 3-positions are blocked, then the benzene ring is attacked at the 6-position. Thus, indole is brominated and iodinated to the 3-halogeno-derivative; 3-chloroindole is prepared by the action of sulphuryl chloride on indole. 3-Nitroindole has been prepared by treating a mixture of indole and sodium ethoxide with ethyl nitrate. On the other hand, mixed acids convert 2,3-dimethylindole into the corresponding 6-nitro-derivative.

An unusual case of orientation is the formation of the 2-sulphonic acid by the action of sulphur trioxide in pyridine on indole at about 120°C. Indole-3-aldehyde is formed by the Gattermann reaction or by the action of dimethylformamide and phosphoryl chloride on indole. This compound may also be prepared by the Reimer-Tiemann reaction, but in this case 3-chloroquinoline is also produced (cf. pyrrole). Mercuration with mercuric acetate leads to 2,3-diacetoxymercuri-indole. Indole readily forms a Grignard reagent, which may be used to prepare substituted indoles.

Indole also undergoes the Mannich reaction with formaldehyde and dimethylamine to form *gramine* (3-dimethylaminomethylindole):

$$\text{Indole} + CH_2O + Me_2NH \longrightarrow \text{3-}(CH_2NMe_2)\text{indole}$$

The dimethylamino-group is easily displaced by certain reagents, particularly when in the form of the trimethylammonium ion, *e.g.*,

$$\text{3-}(CH_2NMe_2)\text{indole} \xrightarrow{MeI} \text{3-}(CH_2\overset{+}{N}Me_3I^-)\text{indole} \xrightarrow{KCN} \text{3-}(CH_2CN)\text{indole}$$

Indole is reduced electrolytically, by tin and hydrochloric acid, or by zinc dust and phosphoric acid-to 2,3-dihydroindole (*indoline*). This is also produced by catalytic reduction using copper chromite, but if Raney nickel is used, the product is octahydroindole.

Indole behaves in some ways as a tautomeric substance, the tautomer being known as *indolenine*. Indolenine itself has never yet been isolated, but its derivatives are known, and these may be prepared, *e.g.*, by condensation between indoles and aldehydes.

$$\text{2-}R^1\text{-indole} \rightleftharpoons \text{2-}R^1\text{-indolenine} \xrightarrow[HCl]{R^2CHO} \text{3-}(=CHR^2)\text{-2-}R^1\text{-indolenine}$$

Indoxyl is the term usually applied to the keto-form of **3-hydroxyindole** (the enolic form):

$$\text{3-OH-indole} \rightleftharpoons \text{indolin-3-one (C=O)}$$

Derivatives of both are known (for its preparation). Indoxyl is a bright yellow solid, m.p. 85°C, and is readily oxidised in alkaline solution by air to indigotin. It gives a red colour with ferric chloride, and reacts with diazonium salts. These are typical reactions of an enol, but since it also reacts with aldehydes, this has been taken as evidence for the presence of the keto-form (aldehydes condense with an active methylene group).

Oxindole, m.p. 126°C. There are three possible formulae for oxindole:

N H O (I) N OH (II) N H OH (III)

Evidence obtained from u.v. and i.r. spectra studies indicates that oxindole is almost completely the lactam (amide) form (I).

Oxindole is most conveniently from aniline and chloroacetyl chloride as follows:

NH_2 —$ClCH_2COCl$→ NH–CO–CH_2Cl —$AlCl_3$→ N H O

Dioxindole (3-*hydroxyoxindole*), m.p. 180°C, is tautomeric with 2, 3-dihydroxyindole:

OH OH N H ⇌ OH N H O

It may be prepared by reducing isatin with zinc and hydrochloric acid. Sodium amalgam in alkaline solution or with sodium dithionite.

O N H O —e; H⁺→ OH N H O

Isatin exists in two forms, the term ψ-isatin being applied to the *lactim* form (I), and isatin to the *lactim* form (II) Both are derivatives of 2,3-dihydroindole: and the name isatin is often applied to I):

O N H O ⇌ O N OH

(I) (II)

This is an example of the *amido-imidol* tautomeric system:

—NH—CH— ⇌ —N = C(OH)—

It appears to be the first case of tautomerism to be recognised. X-ray analysis of crystalline isatin has shown that it is predominantly in the amide (lactam) form.

In aqueous solution, however, the species present depends on the pH, but in chloroform solution, i.r. spectra studies have shown that only amide form is present.

Isatin was first obtained by the oxidation of indigotin with nitric acid. It is best prepared by heating a solution of concentrated hydrochloric acid containing aniline, chloral hydrate, hydroxylamine and sodium sulphate. Isonitrosoacetanilide' (oximinoacetanilide) crystallises out and this, on treatment with concentrated sulphuric acid, forms isatin by ring closure:

$C_6H_5NH_2 + NH_2OH + CCl_3C(OH)_2 \xrightarrow{HCl}$ $C_6H_5NHCOCH{=}NOH$ (80–91%) $+ 3HCl + H_2O \xrightarrow[(H_2SO_4)]{H_2O}$ isatin (71–78%)

The following synthesis from o-nitrobenzoyl chloride clearly shows the structure of isatin (Claisen *et al.* 1879):

$C_6H_4(COCl)(NO_2) \xrightarrow{NaCN} C_6H_4(COCN)(NO_2) \xrightarrow{H^+} C_6H_4(COCO_2H)(NO_2) \xrightarrow[NH_3]{FeSO_4} C_6H_4(COCO_2H)(NH_2) \xrightarrow[(warm)]{-H_2O}$ isatin (NH)

Isatin is a red solid, m.p. 200°C. With phosphorus pentachloride it forms *isatin chloride*, and with warm sodium hydroxide, isotonic acid.

isatin chloride $\xleftarrow{PCl_5}$ isatin $\xrightarrow{NaOH}$ $C_6H_4(COCO_2Na)(NH_2)$

Carbazole (*dibenzopyrrole*), m.p. 245°C, may be isolated from the anthracene fraction of coal-tar. It may be synthesised by passing diphenylamine through a red-hot tube, or better, by heating 2,2'-diaminobiphenyl at 200°C with concentrated phosphoric acid (the yield is almost quantitative; Leditschke, 1953).

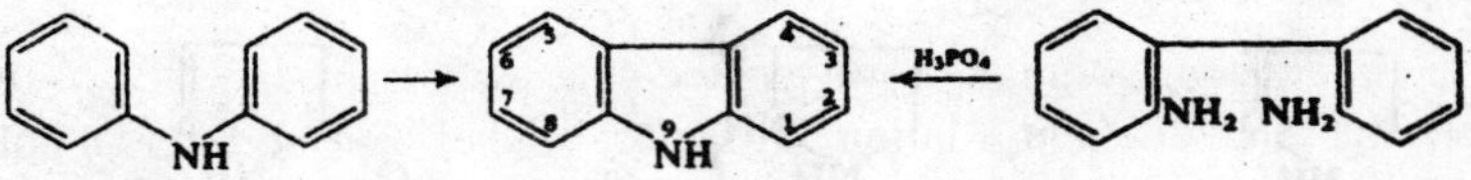

$$\text{Ph-NH-Ph} \longrightarrow \text{carbazole (9-NH)} \xleftarrow{H_3PO_4} \text{2,2'-}(NH_2)_2\text{-biphenyl}$$

A convenient preparation is the reduction of 2-nitrobiphenyl with triethyl phosphite (82.5%; Cadogan *et al.*, 1962).

Silver oxide converts carbazole into N,N'-*dicarbazyl*, which is a colourless compound but gives coloured solutions due to its dissociation into carbazyl free radical. Carbazole is used in the preparation of polyvinylcarbazole plastics.

AZOLEST

Azole is the suffix used for five-membered rings containing two or more hetero-atoms, at least one of which is nitrogen.

Pyrazoles. Pyrazole may be prepared by passing acetylene into a cold ethereal solution of diazomethane; this is a 1,3-dipolar addition:

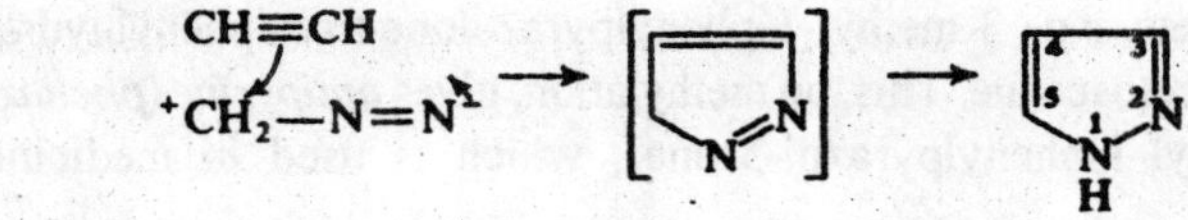

$$CH{\equiv}CH + {}^{+}CH_2{-}N{=}N^{-} \longrightarrow [\text{3H-pyrazole}] \longrightarrow \text{pyrazole (1-NH)}$$

Pyrazole, m.p. 70°C, has aromatic properties, readily undergoing substitution (with the usual electrophilic reagents) in the 4-position.

Pyrazole is a tautomeric substance; this cannot be demonstrated in pyrazole itself, but may be shown as follows. If pyrazole is tautomeric, then positions 3 and 5 are identical; if not tautomeric, these positions are different.

When the phenyl group in 3-methyl-l-phenyl- and 5-methyl-l-phenylpyrazole is removed, both compounds give the same methylpyrazole. Hence, positions 3 and 5 are equivalent, and this can only be explained by assuming that pyrazole is tautomeric.

Pyrazole may be catalytically reduced to pyrazoline (I) and then to pyrazolidine (II). Both are stronger bases than pyrazole. 5-Ketopyrazoline or *pyrazol-5-one* is (III).

(I) (II) (III)

Pyrazole Derivatives

One of the chief methods for preparing these is by reaction between hydrazines and 1,3-dicarbon compounds, *e.g.*, hydrazine and acetylacetone form 3,5-dimethylpyrazole.

5-Pyrazolones are formed by reaction between hydrazines and β-ketonic esters, *e.g.*, 3-methyl-1-phenylpyrazolone from phenylhydrazine anti ethyl acetoacetate. This, on methylation, gives *antipyrine* (*phenazone*, 2,3-dimethyl-1-phenylpyrazol-5-one), which is used in medicine as febrifuge.

$CH_3COCH_2CO_2Et + C_6H_5NHNH_2 \xrightarrow{} \xrightarrow{-EtOH} \xrightarrow[NaOH]{(CH_3)_2SO_4}$

Imidazoles (glyoxalines, iminazoles). Imidazole is isomeric with pyrazole, and occurs in the purine nucleus and in histidine.

A general method for preparing imidazoles is by reaction between (x-dicarbonyl compounds, ammonia and an aldehyde.

$$R^1\!-\!C\!=\!O,\ R^2\!-\!C\!=\!O + 2NH_3 + R^3CHO \longrightarrow \text{(imidazole: } R^1, R^2, R^3\text{, N, NH)} + 3H_2O$$

Glyoxal, ammonia and formaldehyde produce imidazole itself. This, however, is best prepared as follows: tartaric acid is treated with a mixture of concentrated nitric and sulphuric acids, the product, tartaric acid dinitrate ('dinitrotartaric acid'), then treated with formalin and ammonia solution, and the imidazole-4,5-dicarboxylic acid formed is then decarboxylated by heating with copper oxide.

$$CO_2H\!-\!CHOH\!-\!CHOH\!-\!CO_2H \xrightarrow[H_2SO_4]{HNO_3} CO_2H\!-\!CHONO_2\!-\!CHONO_2\!-\!CO_2H \longrightarrow CO_2H\!-\!CO\!-\!CO\!-\!CO_2H$$

$$\xrightarrow[2NH_3]{CH_2O} \text{imidazole-4,5-}(HO_2C)_2 \ (43\text{–}48\%) \xrightarrow{CuO} \text{imidazole} \ (68\text{–}76\%)$$

Imidazole, m.p. 90°C, is tautomeric (positions 4 and 5 are identical).

Oxazoles are formed by reaction between acid amides and a-halogenoketones, *e.g.*, acetamide and bromcacetone form 2,4-dimethyloxazole.

$$CH_3CO\!-\!CH_2Br + NH_2\!-\!COCH_3 \xrightarrow{130^\circ C} \text{2,4-dimethyloxazole} + HBr + H_2O$$

Isoxazoles are formed by warming the monoximes of 1,3-diketones, *e.g.*, 3,5-dimethylisoxazole from acetylacetonemonoxime.

$$CH_3C(=N\!-\!OH)\!-\!CH_2\!-\!COCH_3 \xrightarrow{-H_2O} \text{3,5-dimethylisoxazole}$$

Thiazoles may be prepared by reaction between an a-chlorocarbonyl compound and a thioacid amide.

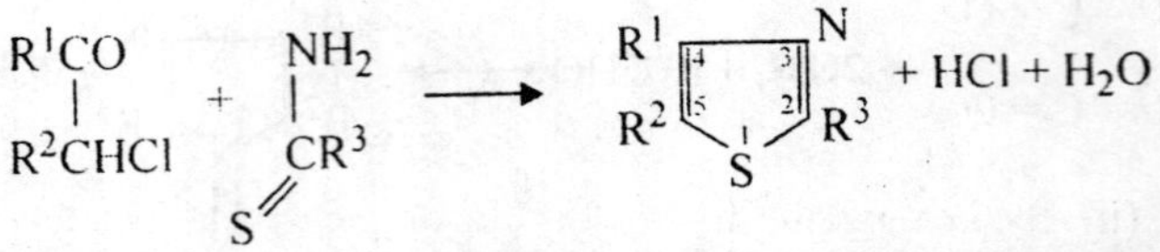

$$R^1CO\text{–}CHClR^2 + NH_2\text{–}C(=S)R^3 \longrightarrow \text{thiazole} + HCl + H_2O$$

Thiazole, b.p. 117°C, may be prepared from chloroacetaldehyde and thioformamide. Vitamin B_1 contains the thiazole nucleus.

Triazoles and **tetrazoles** are also known

osotriazole triazole tetrazole

All are tautomeric substances.

An unusual five-membered ring is that in sydnones. These were prepared by Earl and Mackney (1935), and the peculiar feature of these compounds is that it is not possible to give them a structure which represents covalent bonds by the usual paired electrons, *e.g.*, sydnone (I), (II), (III). Dipole and infra-red studies indicate this compound is a resonance hybrid; it has been named a *meso-ionic structure*.

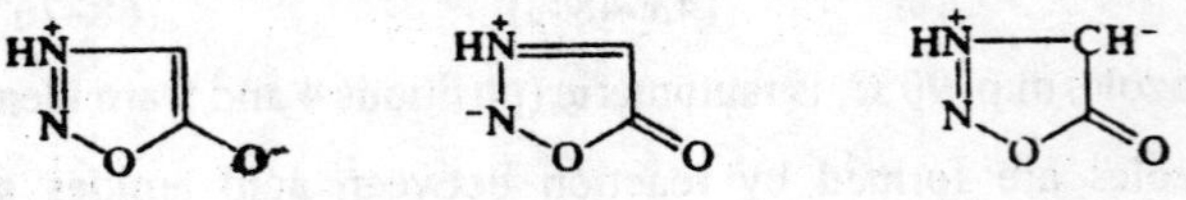

(I) (II) (III)

Six-membered rings

PYRIDINE AND ITS DERIVATIVES

Pyridine, b.p. 115°C, occurs in the light oil fraction of coal-tar and in bone oil, and is a decomposition product of several alkaloids. It is obtained from light oil by treating the latter with dilute sulphuric acid. This dissolves pyridine and other basic substances. The acid layer is neutralised with sodium hydroxide and the liquid repeatedly fractionated. Pyridine is basic, but it resembles benzene in many of its properties and this partly led Korner (1864) to adopt the ring structure for pyridine, a structure which is confirmed by synthesis.

Synthesis of Pyridine

There are many methods available, *e.g.*,

(i) By passing a mixture of acetylene and hydrogen cyanide through a red-hot tube:

$$2C_2H_2 + HCN \longrightarrow \text{pyridine}$$

(ii) By heating the hydrochloride of pentamethylenediamine and oxidising the product *piperidine*, with concentrated sulphuric acid at 300°C, or by catalytic dehydrogenation (Pd—C)

$$CH_2(CH_2CH_2NH_3^+Cl^-)_2 \xrightarrow[-HCl]{-NH_4Cl} \text{piperidine (NH)} \xrightarrow{Pd-C} \text{pyridine}$$

Pyridine derivatives may be prepared by various methods. One important method is the **Hantzsch synthesis** (1882). A β-dicarbonyl compound (2 molecules) is condensed with are aldehyde (1 molecule) and ammonia (1 molecule). The dihydropyridine derivative is obtained and this gives the pyridine derivative on oxidation with nitric acid. The mechanism of the reaction is not fully understood, but intermediates proposed have been shown to give pyridines under the conditions of the experiment (proton transfers are given in one step):

(i) $MeCOCH_2CO_2Et \xrightarrow{NH_3} MeC(NH_2) = CHCO_2Et$

(ii) $MeCHO + MeCOCH_2CO_2CO_2Et \xrightarrow{NH_3} MeCH{=}C(COMe)(CO_2Et)$

(iii)

$EtO_2C{-}CH{=}C(Me)NH_2 + MeHC{=}C(CO_2Et)COMe \longrightarrow$ $EtO_2C{-}C(=C(Me)\overset{+}{N}H_2){-}CH(Me){-}C(CO_2Et){=}C(O^-)Me \longrightarrow$

$EtO_2C{-}C{=}C(Me){-}NH_2 \ldots CH(Me){-}CHCO_2Et{-}C(=O)Me \longrightarrow$ EtO_2C, Me, CO_2Et, OH, Me, NH (cyclic hydroxy intermediate) $\xrightarrow{-H_2O}$

EtO_2C, Me, CO_2Et, Me, Me, NH (dihydropyridine) $\xrightarrow{HNO_3}$ EtO_2C, Me, CO_2Et, Me, Me, N (pyridine) $\xrightarrow[\text{(ii) CaO; heat}]{\text{(i) OH}^-}$ Me, Me, Me, N (trimethylpyridine)

On the other hand, aldehydes (or ketones) alone (*i.e.*, no β-dicarbonyl compound need be present) form pyridines when heated with ammonia, *e.g.*,

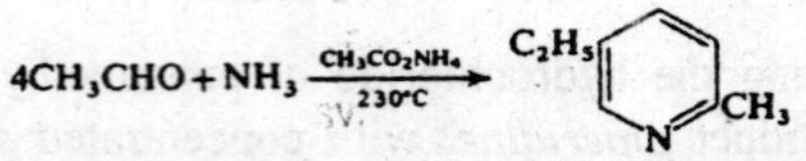

Pyridones may be prepared by the action of ammonia on 2- or 4-pyrones:

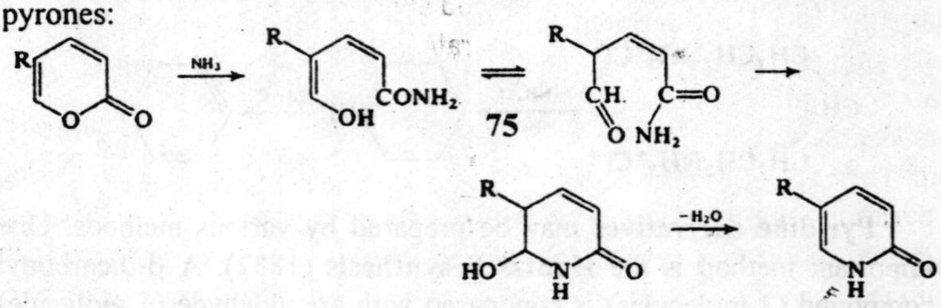

2-Furyl ketones may be used as starting material; when heated with ammonia and ammonium chloride, the product is a 3-hydroxypyridine (Leditschke 1953):

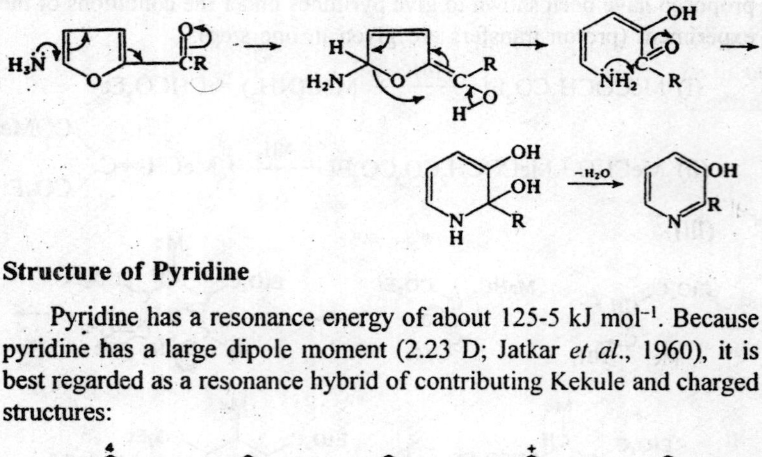

Structure of Pyridine

Pyridine has a resonance energy of about 125-5 kJ mol^{-1}. Because pyridine has a large dipole moment (2.23 D; Jatkar *et al.*, 1960), it is best regarded as a resonance hybrid of contributing Kekule and charged structures:

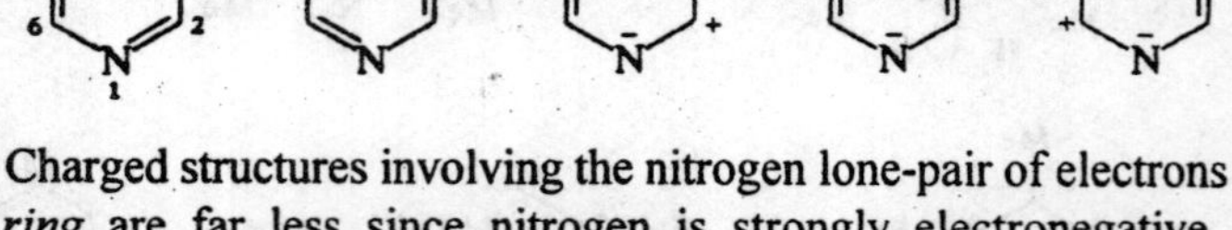

Charged structures involving the nitrogen lone-pair of electrons *into the ring* are far less since nitrogen is strongly electronegative, and consequently a negatively charged nitrogen is more stable than a positively charged one.

Substitution in Pyridine

Examination of the contributing structures of pyridine shows, from the point of view of charge distribution in the molecule, that positions 3 and 5 will be sites for electrophilic attack, whereas positions 2, 4, and 6 will be sites for nucleophilic attack. Furthermore, because of the withdrawal of electrons from the ring-carbon atoms (towards the nitrogen atom), the ring is deactivated towards electrophilic reagents, thereby resembling the benzene ring in nitrobenzene. Also, in strongly acid solution, pyridine is protonated, and so the positively charged nitrogen atom deactivates the ring even more than the unprotonated nitrogen atom. This is clearly shown by the difficulty in nitration and sulphonation of pyridine. If a + 1 group is present, electronphilic reactions are accelerated. When the substituent group is in the 3-position, the entering group takes up the 2- or 6-position; and if in the 2- or 4-position, then the 3- or 5-position is the attack.

The position of substitution may also be considered from the point of view of the stabilities of the intermediate carbonium ions (σ-complexes), *e.g.*, there are three resonating structures for 2-, 3- and 4-substitution (electrophilic), but in the 2- and 4-cations, one contributing structure has a positively charged nitrogen atom. This makes these ions less stable than the 3-cation.

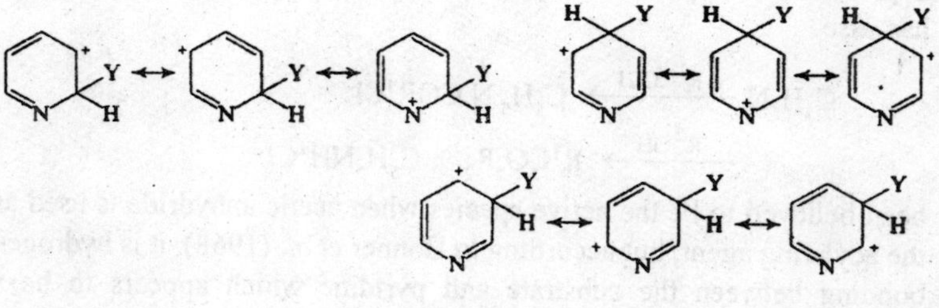

Fig. 3.11

In a similar way, consideration of the anions in nucleophilic substitution shows that the 2-, and 4-anions have a nitrogen carrying a negative charge, and so these ions will be more stable than the 3-anion:

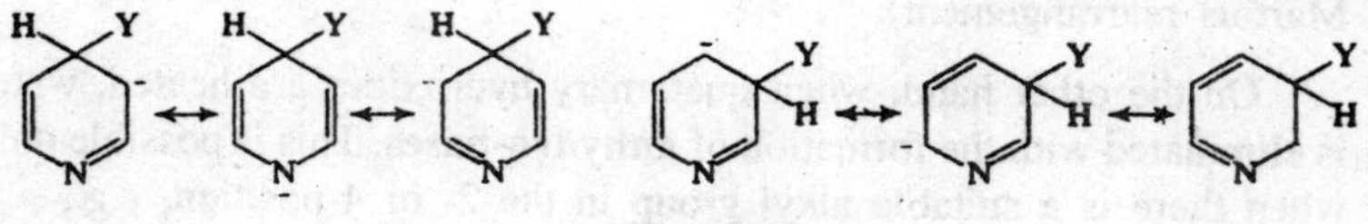

Fig. 3.12

Basicity of Pyridine

Since the nitrogen atom carries a negative charge (and the lone pair is 'completely' available for protonation; cf. pyrrole), pyridine would be expected to be a very strong base. In actual fact, pyridine is far less basic than expected (see Table 3.2), being less basic than the alkylamines. It is difficult to explain this, but a possibility is as follows. Since the nitrogen atom in pyridine is sp^2 hybridised, it is more electron-attracting than nitrogen that is sp^3 hybridised (as in RNH_2). Consequently, the lone pair is more tightly held and so is less available for protonation.

Reactions of Pyridine

A number of reactions of pyridine occur at the nitrogen atom. The basic property of pyridine has already been discussed, and it is this property that is used in many reactions where it is desired to carry out dehydrohalogenation, *e.g.*, removal of HBr from bromosuccinic (KOH/EtOH would hydrolyse the ester at the same time).

Pyridine also reacts with halogen in the cold to form 1-halogeno-pyridinium halides, *e.g.*, $C_5H_5\overset{+}{N}BrBr^-$, and with acyl chlorides to form 1-acylpyridinium chlorides. This type of compound is believed to be the active species in the acylation of hydroxy- and amino-compound with acyl halides in the presence of pyridine. This type of quaternary compound has also

$$C_5H_5N \xrightarrow{R^1COCl} C_5H_5\overset{+}{N}COR^1Cl^-$$

$$\xrightarrow{R^2OH} R^1CO_2R_2 + C_5H_5NH^+Cl^-$$

been believed to be the active species when acetic anhydride is used as the acylating agent, but according to Bonner *et al.* (1968), it is hydrogen bonding between the substrate and pyridine which appears to be a prerequisite for reaction.

Pyridine forms quaternary salts when heated with alkyl halides, *e.g.*, pyridine methiodide or 1-methylpyridinium iodide, $C_5H_5\overset{+}{N}MeI^-$. This, when heated at 300°C, gives 2- and 4-methyl-pyridine (cf. the Hofmann-Martius rearrangement).

On the other hand, when quaternary hydroxides are heated, water is eliminated with the formation of **anhydro-bases**. This is possible only when there is a suitable alkyl group in the 2- or 4-position, *e.g.*,

$$\text{1,2-dimethylpyridinium hydroxide } (\overset{+}{N}\text{-Me, 2-Me, } OH^-) \xrightarrow{-H_2O} \text{1-methyl-2-methylene-1,2-dihydropyridine } (N\text{-Me, } =CH_2)$$

Anhydro-bases, on treatment with aqueous inorganic acids, regenerate the quaternary hydroxide.

Pyridine is readily oxidised by per-acids (see later), and complexes with many metal ions and Lewis acids. It is this complex-formation that inhibits the Friedel-Crafts reaction with pyridine.

When pyridine and bromine are passed over a catalyst of charcoal at 300°C, a mixture of 3-bromopyridine and 3,5-dibromopyridine is obtained; at 500°C a mixture of the 2- and 2,6-bromo-derivatives is obtained. At 300°C the reaction is electrophilic in character, whereas at 500°C the mechanism is via free radicals. 3-Chloropyridine may be obtained by heating potassiopyrrole with chloroform and sodium ethoxide. Chlorination of pyridine in the presence of aluminium chloride (2 molecules) gives 3-chloropyridine. 2- and 4-Chloropyridines may be obtained by diazotising the corresponding amino-compounds in concentrated hydrochloric acid and treating with cuprous chloride.

Halogen in the 2- or 4-position is reactive, being fairly readily replaced by OH, CN, NH_2, etc. The bromopyridines form Grignard reagents in the presence of ethyl bromide.

Pyridine reacts with nitric acid fairly readily only if a hydroxyl or an amino-group is present in the ring. Pyridine itself is nitrated to 3-nitropyridine by heating with concentrated sulphuric acid and fuming nitric acid at 300°C. 2- and 4-Nitropyridines may be obtained by oxidising the corresponding amino-compounds with hydrogen peroxide in sulphuric acid.

Sulphonation of pyridine is difficult, but when heated with, 20 per cent oleum (containing a little mercuric sulphate) at 220°C, pyridine gives pyridine-3-sulphonic acid (70%). 2- and 4-Pyridinesulphonic acids may be prepared by oxidation of the corresponding thiols (prepared by the action of potassium hydrogen sulphide on the chloropyridine).

The sulphonic acid group can be replaced by hydroxyl or by the cyano-group.

Pyridine may be fairly easily mercurated in the 3-position with aqueous mercuric acetate.

Sodium and ethanol, electrolytic reduction, or catalytic reduction using nickel convert pyridine into piperidine, but reduction with sodium in liquid ammonia in the presence of ethanol (Birch reduction) gives 1,4-dihydropyridine and with LAH, 1,2-dihydropyridine. On the other hand, when pyridine is heated with hydriodic acid at 300°C, the ring is opened with the formation of n-pentane and ammonia.

In addition to electrophilic reactions, pyridine undergoes nucleophilic reactions, but it does so only with strong nucleophiles. When heated with sodamide in toluene solution, pyridine forms 2-aminopyridine; excess of sodamide produces 2,6-diaminopyridine (**Tschitschibabin (Chichibabin) reaction**, 1914). Actually, the sodium salts are formed, but on treatment with water, are hydrolysed to the amine.

$+ NaNH_2 \longrightarrow$ (2-aminopyridine, NH_2) $+ NaH \longrightarrow$ ($\bar{N}HNa^+$) $+ H_2$

The three monoaminopyridines can be obtained by means of the Hofmann amine preparation the amides of the pyridine monocarboxylic acids.

3-Aminopyridine can be diazotised easily; 2- and 4-aminopyridines are difficult to diazotise. Angyal *et al.* (1952), from an examination of the infra-red absorption spectra, have concluded that 2- and 4-aminopyridine (and aminoquinolines) are mainly in the amino form (cf. the corresponding hydroxy- compounds).

Pyridine forms 2-hydroxypyridine when heated with potassium hydroxide at 320°C.

Pyridine $\xrightarrow[\text{(ii) } H_3O^+]{\text{(i) KOH}}$ [2-hydroxypyridine (OH)] $\longrightarrow$ 2-pyridone (N–H, =O)

However, the properties of 2- and 4-hydroxypyridine are not phenolic. X-ray and infra-red studies have shown the hydroxy-form (*pyridol*) is present only in small amount, the amide form (*pyridone*) being the predominant structure. On the other hand, 3-hydroxypyridine exhibit typical phenolic properties.

When heated with n-butyl-lithium, pyridine forms 2-n-butylpyridine (Ziegler *et al.*, 1930).

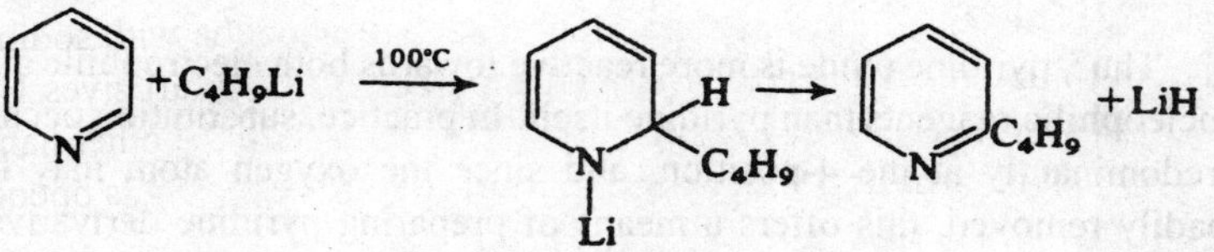

$$C_5H_5N + C_4H_9Li \xrightarrow{100°C} \text{(N-Li-2-butyl-2-H-dihydropyridine)} \longrightarrow \text{2-}C_4H_9\text{-pyridine} + LiH$$

Pyridine-1-oxide. Pyridine is oxidised by per-acids (peracetic, perbenzoic, etc.) to *pyridine oxide* which, on the basis of dipole moment studies, is formulated as a resonance hybrid.

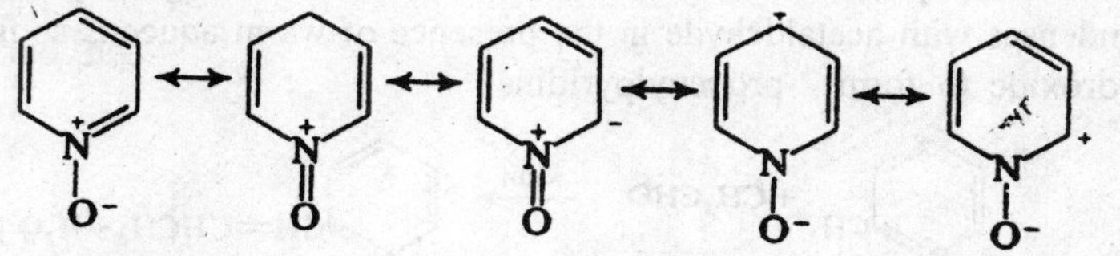

Examination of these canonical forms shows that in pyridine oxide, high or low charge densities are produced at positions 2 and 4.

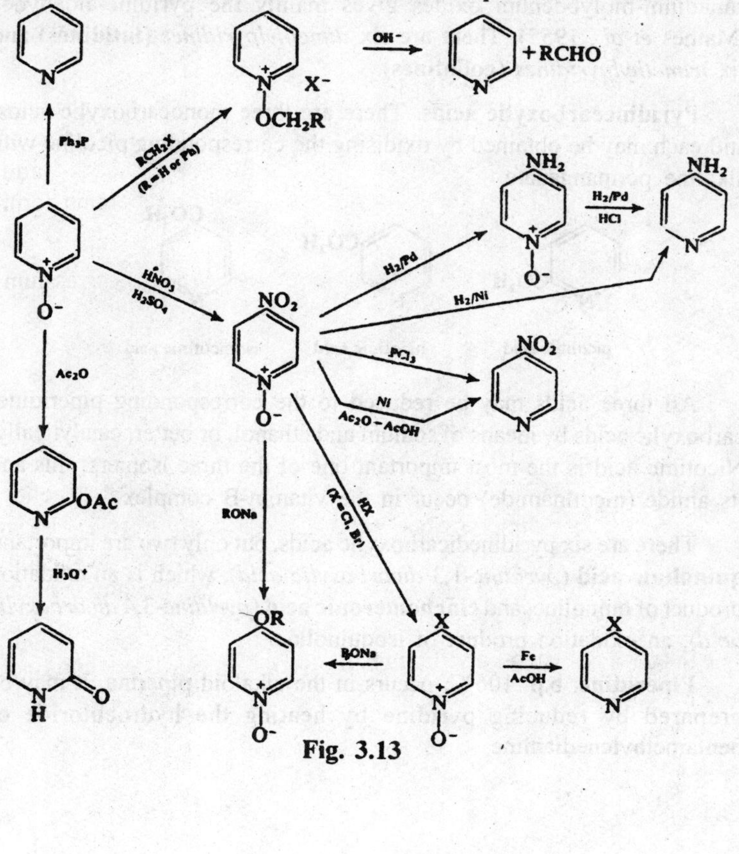

Fig. 3.13

Thus, pyridine oxide is more reactive towards both electrophilic and nucleophilic reagents than pyridine itself. In practice, substitution occurs predominantly at the 4-position, and since the oxygen atom may be readily removed, this offers a means of preparing pyridine derivatives that are difficult to prepare by other routes.

Pyridine homologues occur in coal-tar and bone oil. There arc' three methylpyridines and these a known as **picolines**. The methyl group in 2- and 4-picolines is reactive (cf. nitrotoluenes, *e.g.*, 2-picoline condenses with acetaldehyde in the presence of warm aqueous sodium hydroxide to form 2-propenylpyridine:

$$C_5H_4N\text{-}CH_3 + CH_3CHO \xrightarrow{NaOH} C_5H_4N\text{-}CH{=}CHCH_3 + H_2O$$

Oxidation of gaseous picolines with air at 380°C over mixed vanadium-molybdenum oxides gives mainly the pyridine-aldehydes (Mathes *et al.*, 1955). There are six *dimethylpyridines* (**lutidines**) and six *trimethylpyridines* (**collidines**).

Pyridinecarboxylic acids. There are three monocarboxylic acids, and each may be obtained by oxidising the corresponding picoline with alkaline permanganate.

$C_5H_4N\text{-}CO_2H$ (2-position) — picolinic acid; $C_5H_4N\text{-}CO_2H$ (3-position) — nicotinic acid; $C_5H_4N\text{-}CO_2H$ (4-position) — isonicotinic acid

picolinic acid nicotinic acid isonicotinic acid

All three acids may be reduced to the corresponding piperidine-carboxylic acids by means of sodium and ethanol, or better, catalytically. Nicotinic acid is the most important one of the three isomers; this and its amide (nicotinamide) occur in the vitamin-B complex.

There are six pyridinedicarboxylic acids, but only two are important: **quinolinic acid** (*pyridine*-1,3 *dicarboxylic acid*), which is an oxidation product of quinoline, and **cinchomeronic acid** (*pyridine*-3,4 *dicarboxylic acid*), an oxidation product of isoquinoline.

Piperidine, b.p. 106°C, occurs in the alkaloid piperine. It may be prepared by reducing pyridine by heating the hydrochloride of pentamethylenediamine:

$$CH_2\begin{cases}CH_2CH_2NH_2\cdot HCl\\CH_2CH_2NH_2\cdot HCl\end{cases} \longrightarrow \text{piperidine (N–H)} \xleftarrow{Na/C_2H_5OH} \text{pyridine}$$

Piperidine is manufactured by the catalytic hydrogenation of pyridine.

Piperidine behaves as a secondary aliphatic amine, and concentrated sulphuric acid at 300°C or catalytic dehydrogenation convert it into pyridine.

Methods of Ring Fission

Many heterocyclic compounds (containing nitrogen) occur naturally, and an extremely important step in the determination of their structure is to ascertain the disposition of the carbon atoms.

A common procedure is to reduce the heterocyclic compounds and then open the ring of the product by the following methods (in which piperidine is used as the example):

(i) *Secondary* cyclic amines may be opened by treatment with 3 per cent hydrogen peroxide, *e.g.*, piperidine gives δ-aminovaleraldehyde

$$\text{C}_5\text{H}_{10}\text{NH} \xrightarrow{H_2O_2} NH_2CH_2CH_2CH_2CH_2CHO$$

(ii) *Von Braun's method* (1910) may be used to open *secondary* cyclic amines:

$$\text{C}_5\text{H}_{10}\text{NH} \xrightarrow[NaOH]{PhCOCl} \text{C}_5\text{H}_{10}\text{NCOPh} \xrightarrow[0°C]{PBr_3/Br_2} [\text{C}_5\text{H}_{10}\text{NCBr}_2\text{Ph}] \longrightarrow Br(CH_2)_5Br + PhCN$$

It might be noted that this provides a good method of preparing 1,5-dibromopentane.

(iii) *von Braun cyanogen bromide reaction* (1900). This is the reaction of a tertiary amine with cyanogen bromide to form an alkyl bromide and a disubstituted cyanamide.

$$R_3N + BrCN \longrightarrow RBr + R_2NCN$$

This reaction has been used in alkaloid chemistry for opening tertiary *cyclic* amines.

$$\text{(piperidine ring)NR} \xrightarrow{BrCN} \text{(ring)}\overset{+}{N}(R)(CN)\ Br^- \xrightarrow[\text{heat}]{\text{aq. HBr}} \left[CH_2 \begin{matrix} CH_2-CH_2Br \\ CH_2-CH_2NRCN \end{matrix} \right] \longrightarrow Br(CH_2)_5NHR$$

(iv) Piperidine may be converted into n-pentane by heating with hydriodic acid at 300°C:

$$\text{(piperidine)NH} \xrightarrow[300°C]{HI} CH_3(CH_2)_3CH_3 + NH_3$$

(v) **Hofmann exhaustive methylation or degradation** (1881) is the most important method of opening heterocyclic rings, but it fails with unhydrogenated pyridine, quinoline and isoquinoline derivatives, and with hydrogenated quinolines.

Consider piperidine as our example. The process is repeated twice, the first step opening the ring, and the second step then eliminating trimethyl-amine to form penta-l,4-diene (for mechanism). This, however, isomerises to *piperylene* (penta-l,3-diene). This isomerisation is general, an isolated double bond system rearranging, if possible, to form a conjugated system.

$$\text{(ring)NH} \xrightarrow{MeI} \text{(ring)}\overset{+}{N}Me_2I^- \xrightarrow{\text{'AgOH'}} \text{(ring)}\overset{+}{N}Me_2OH^- \xrightarrow[(-H_2O)]{\text{heat}}$$

$$Me_2NCH_2CH_2CH_2CH{=}CH_2 \xrightarrow[\text{(ii) AgOH}^+]{\text{(i) MeI}}$$

$$HO^-Me_3\overset{+}{N}\ CH_2CH_2CH_2HC = CH_2$$

$$\xrightarrow[(-H_2O)]{\text{heat}} Me_3N + [CH_2 = CHCH_2CH = CH_2]$$

$$\longrightarrow CH_3CH = CHCH = CH_2$$

Mass Spectra

Heterocyclic compounds undergo fragmentation in the same ways that resemble the benzene analogues.

However, the fragmentation patterns also depend on the nature of the hetero-atom, *e.g.*,

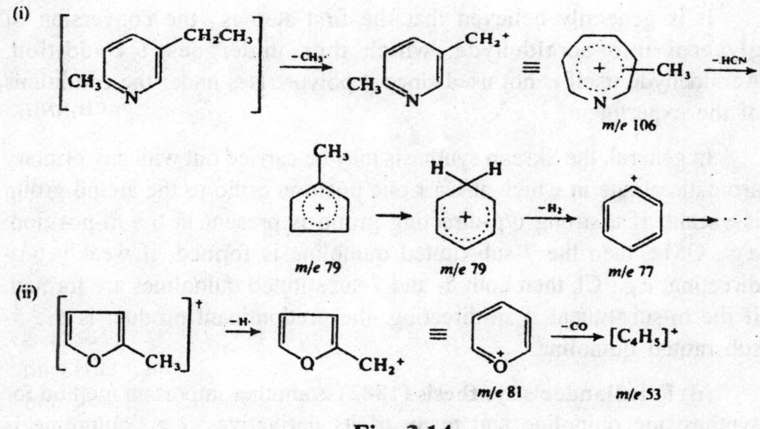

(i)

m/e 106

m/e 79 m/e 79 m/e 77

(ii)

m/e 81 m/e 53

Fig. 3.14

Quinoline is present in coal-tar and bone oil, and was first obtained from the alkaloid *quinine* by alkaline decomposition. It is obtained commercially from coal-tar, or prepared synthetically.

(i) **Skraup synthesis** (1880) is a very important method, and may be carried out by heating a mixture of aniline, nitrobenzene, glycerol, concentrated sulphuric acid and ferrous sulphate. Nitro-benzene acts as an oxidising agent, and ferrous sulphate makes the reaction less violent. Arsenic acid may be used instead of nitrobenzene and the former is better since the reaction is less violent. The mechanism of the Skraup synthesis is not certain.

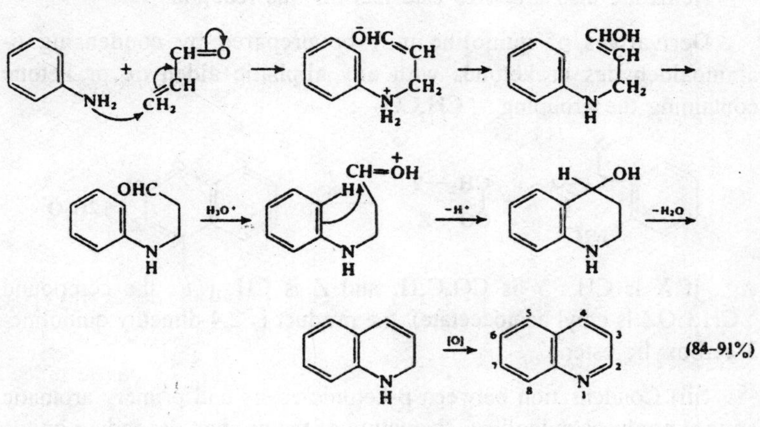

(84–91%)

Fig. 3.15

It is generally believed that the first step is the conversion of glycerol into acraldehyde, which then undergoes 1,4-addition. Acraldehyde itself is not used since it polymerises under the conditions of the experiment.

In general, the Skraup synthesis may be carried out with any primary aromatic amine in which at least one position ortho to the amino-group is vacant. If a strong o/p-directing group is present in the m-position. *e.g.*, OMe, then the 7-substituted quinoline is formed; if weakly o/p-directing, *e.g.*, Cl, then both 5- and 7-substituted quinolines are formed. If the m-substituent is m-directing, the predominant product is the 5-substituted quinoline.

(ii) **Friedlander's synthesis** (1882) is another important method for synthesising quinoline and many of its derivatives: *e.g.*, quinoline is formed when o-aminobenzaldehyde is condensed with acetaldehyde in aqueous sodium hydroxide:

CHO, NH₂ + CH₃–CHO →(OH⁻) CHO, CH₃, N=CH →(−H⁺ (OH⁻)) CHO, CH₂⁻, N=CH →(+H⁺) H, OH (N) →(−H₂O) (N)

Fig. 3.16

Acids are also effective catalysts for this reaction.

Derivatives of quinoline may be prepared by condensing o-aminoaldehydes or ketones with any aliphatic aldehyde or ketone containing the grouping —CH_2CO—:

X–C=O, NH_2 + CH_2—Y, CO—Z →(NaOH) X, Y, Z, N + $2H_2O$

e.g., if X is CH_3, Y is $CO_2C_2H_5$ and Z is CH_3 (*i.e.*, the compound YCH_2COZ is ethyl acetoacetate), the product is 2,4-dimethy quinoline-3-carboxylic ester.

(iii) Condensation between β-ketonic esters and primary aromatic amines produces quinolines, the nature of the product depending on the conditions, *e.g.*, aniline and E.A.A.

(a) **Conrad-Limoach synthesis** (1887).

$$C_6H_5NH_2 + CH_3COCH_2CO_2C_2H_5 \xrightarrow{\text{room temp.}} C_6H_5NH{-}C(CH_3){=}CHCO_2C_2H_5 \xrightarrow[-C_2H_5OH]{250^\circ C}$$

(4-quinolone, CH_3 at 2) ⇌ (4-hydroxy-2-methylquinoline, OH, CH_3)

(b) **Knorr quinoline synthesis** (1886, 1888)

$$C_6H_5NH_2 + CH_3COCH_2CO_2C_2H_5 \xrightarrow{110^\circ C} C_6H_5NHCOCH_2COCH_3 \xrightarrow[-H_2O]{H_2SO_4}$$

(4-methyl-2-quinolone, CH_3) ⇌ (2-hydroxy-4-methylquinoline, CH_3, OH)

(iv) Homologues of quinoline may be prepared by the **Doebner-Miller synthesis** (1881); *e.g.*, aniline and paraldehyde heated with sulphuric acid form *quinaldine* (2-methylquinoline).

Since, in general, the aldehyde may be any α, β-unsaturated aldehyde, the mechanism may be formulated (cf. the Skraup synthesis):

(i) $2CH_3CHO \xrightarrow{H_3O^+} CH_3CH{=}CHCHO$

(ii) $C_6H_5NH_2 + CHCHO{=}CHCH_3 \longrightarrow$ (OHC, CH_3, NH tetrahydro adduct) $\xrightarrow[\text{(ii) } -H_2O]{\text{(i) } H_3O^+}$

(1,2-dihydro-2-methylquinoline, CH_3, NH) $\xrightarrow{PhN{=}CHCH_3}$ (2-methylquinoline, CH_3) $+ PhNHCH_2CH_3$

In this synthesis, no oxidising agent is added, and the final dehydrogenation is believed to occur by the action of the Schiff base produced from aniline and acetaldehyde.

(v) **Pfitzinger reaction** (1886). This is carried out by heating isatin with alkali in the presence of a ketone: *e.g.*,

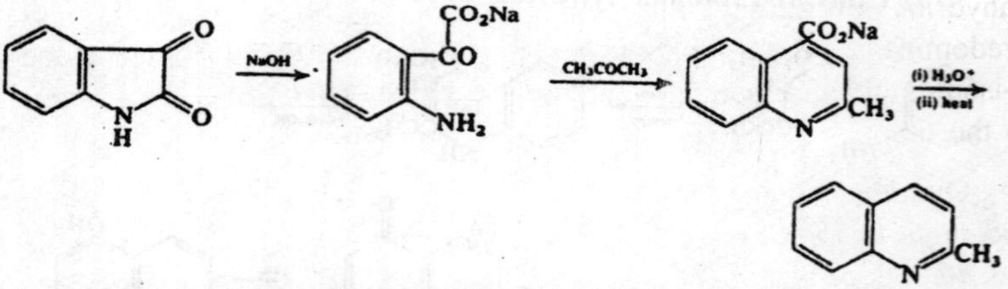

(vi) When methyl-lithium is added to indole in methylene dichloride solution, quinoline is produced by ring expansion via the addition of chloromethylene (Closs *et al.*, 1961):

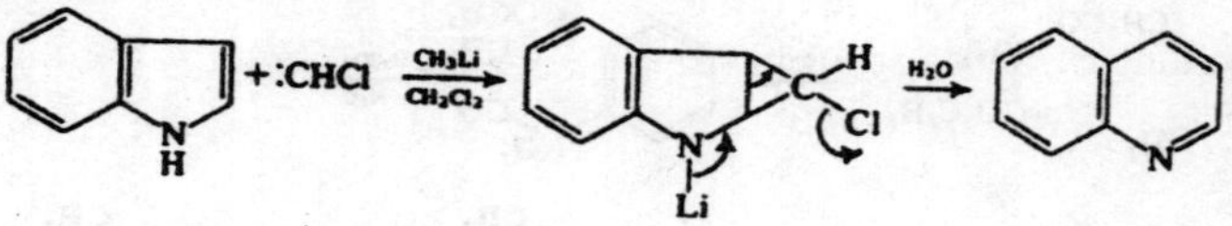

Quinoline, b.p. 238°C, is a tertiary base and forms salts with inorganic acids; with alkyl halides it forms *quinolinium salts* (quaternary salts), *e.g.*, with methyl iodide it forms *quinoline methiodide* or 1-*methylquinolmium iodide* (I); with methyl sulphate it forms *quinoline methylmethosulphate* or 1-*methylquinolinium methyl sulphate* (II):

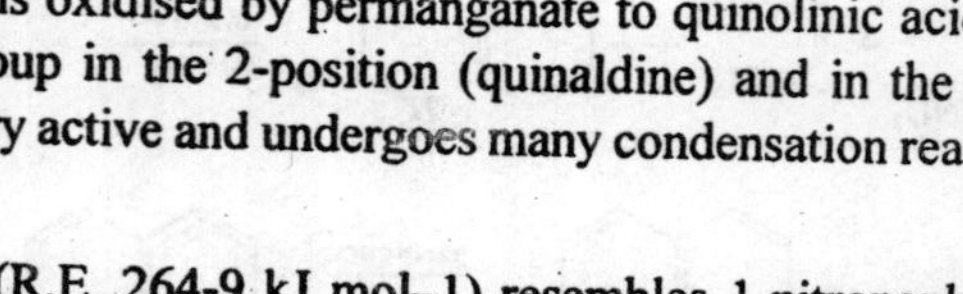

(I) (II)

Quinoline is oxidised by permanganate to quinolinic acid (p. 855). The methyl group in the 2-position (quinaldine) and in the 4-position (lepidine) is very active and undergoes many condensation reactions (see cyanine dyes).

Quinoline (R.E. 264-9 kJ mol–1) resembles 1-nitronaphthalene in many ways. It is attacked preferentially at position 8 by electrophilic reagents, and positions 2 and 4 by nucleophilic reagents. It should also be remembered that in acid solution, it is the quinolinium cation that is present (cf. pyridine).

Vapour-phase bromination of quinoline produces 3-bromoquinoline, but at 500°C the product is 2-bromoquinoline (cf. pyridine). However, under conditions of high acidity, brominations occur in the 5- and 8-positions (de la Mare *et al.*, 1958). Nitration with mixed acid gives a mixture of 5- and 8-nitroquinolines, but nitration with nitric acid-acetic anhydride forms the 3-nitrocompound. Sulphonation at 220°C gives predominantly the 8-sulphonic acid, together with a small amount of the 5-isomer. When the 8-sulphonic acid is heated to 300°C, it rearranges to the 6-acid.

Quinoline undergoes the Tschitschibabin reaction when heated with sodamide to form 2-aminoquinoline, and with n-butyl-lithium it forms the 2-butyl derivative.

Quinoline is oxidised to the 1-oxide by perbenzoic acid and is reduced by lithium aluminium hydride or sodium in liquid ammonia to 1,2-dihydroquinoline. On the other hand, it is reduced to 1,2,3,4-tetrahydroquinoline by tin and hydrochloric acid or by catalytic reduction (nickel). Decahydroquinoline is obtained by catalytic reduction using platinum as catalyst in acetic acid solution.

Hydroxyquinolines have been prepared and all, except the 2- and 4-isomers, exist in the hydroxy-form (cf. hydroxypyridines) 2-Quinolone (III), is known as *carbostyril*, and 8-hydroxyquinoline (IV), is known as *oxine*. The latter compound is used in gravimetric analysis of many metals, with which it forms chelated complexes. Oxine is *intramolecularly* hydrogen-bonded.

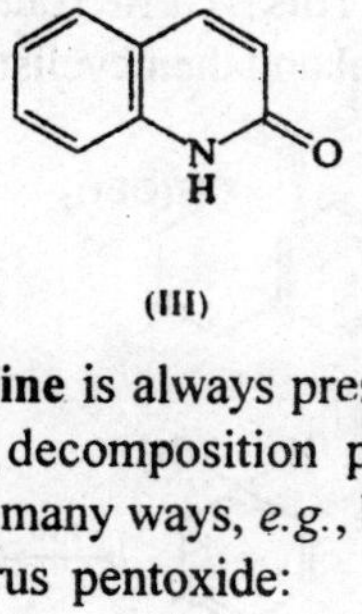

N H O

(III)

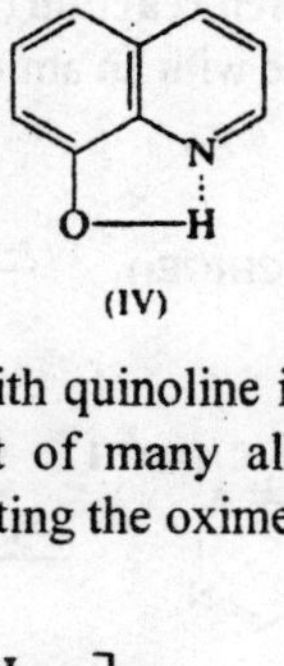

N H O

(IV)

Isoquinoline is always present with quinoline in coal-tar and bone oil, and is a decomposition product of many alkaloids. It may be synthesised in many ways, *e.g.*, by heating the oxime of cinnamaldehyde with phosphorus pentoxide:

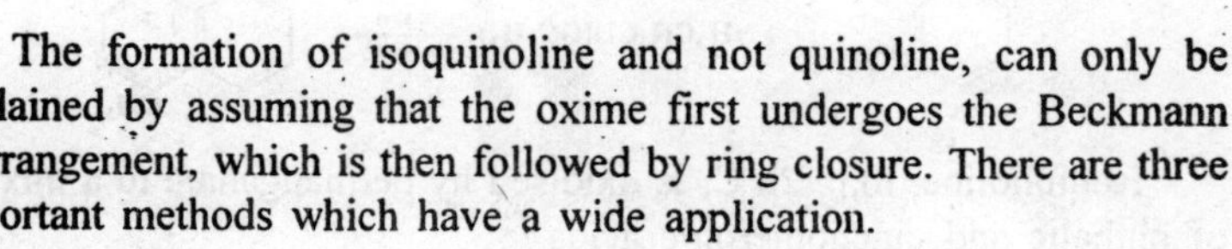

CH CH CH HON → [CH CH N CHOH] $\xrightarrow{-H_2O}$ 5 4 6 3 7 N 2 8 1

The formation of isoquinoline and not quinoline, can only be explained by assuming that the oxime first undergoes the Beckmann rearrangement, which is then followed by ring closure. There are three important methods which have a wide application.

1. Bischler-Napieralski reaction (1893). In this method a β-phenylethylamide is made to undergo cyclodehydration to a 3,4-dihydroisoquinoline by heating with phosphoryl chloride.

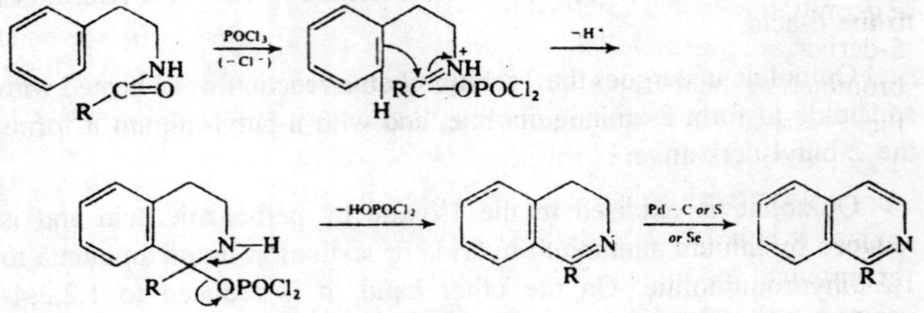

Fig. 3.17

2. Pictet-Spengler reaction (1911). Condensation between a β-arylethylamine and an aldehyde in the presence of a large excess of hydrochloric acid at 100 °C produces a 1,2,3,4-tetrahydro-isoquinoline.

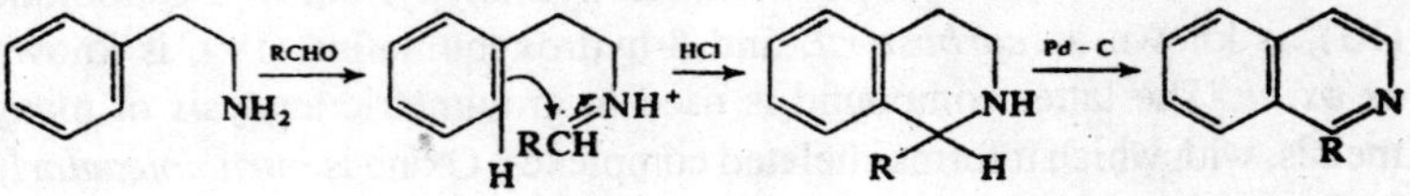

Pomeranz-Fritsch reaction (1893). This is carried out by condensing an aromatic aldehyde with an aminoacetal and then cyclising the product with sulphuric acid *e.g.*,

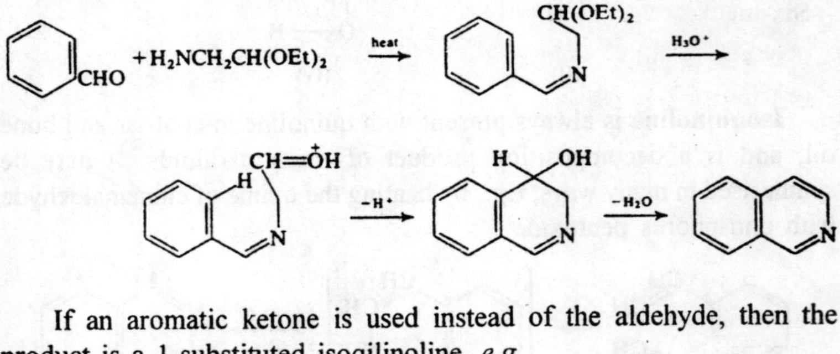

If an aromatic ketone is used instead of the aldehyde, then the product is a 1-substituted isoqilinoline, *e.g.*,

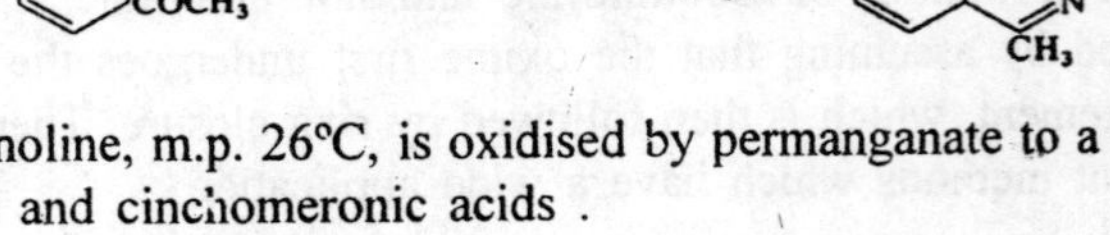

Isoquinoline, m.p. 26°C, is oxidised by permanganate to a mixture of phthalic and cinchomeronic acids .

Isoquinoline resembles 2-nitronaphthalene in many of its properties. Electrophilic attack occurs at the 5- (predominantly) and 8-positions, *e.g.*, nitration and sulphonation (cation present) produce mainly the 5-derivative, together with a small amount of the 8-. On the other hand, bromination and mercuration (with mercuric acetate) give predominantly the 4-derivative. Nucleophilic substitution occurs at position 1, *e.g.*, 1-aminoisoquinoline is formed by the action of sodamide. Isoquinoline forms quaternary salts, and is oxidised by perbenzoic acid to the N-oxide. It is reduced by sodium in liquid ammonia to the 1,2-dihydro-compound, by tin and hydrochloric acid to the 1,2,3,4-tetrahydro-compound, and catalytic hydrogenation give' octahydroisoquinoline.

Acridine occurs in the anthracene fraction of coal-tar. It may be synthesised by passing the vapour of benzylaniline or o-aminodiphenylmethane through a red-hot tube (the outside numbering is to be used now):

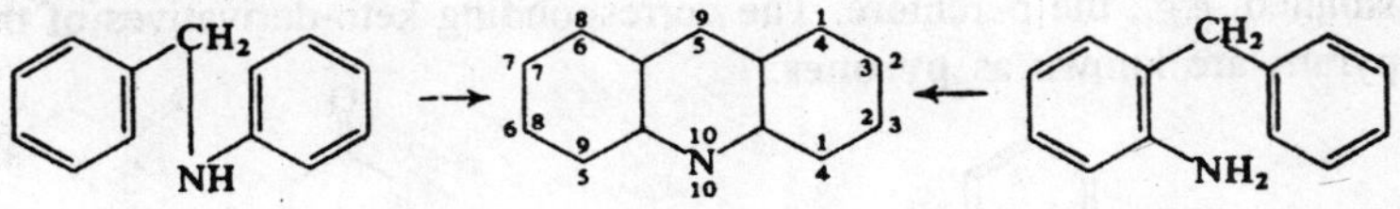

CH_2 NH → (acridine, numbered 1–10) ← CH_2 NH_2

Acridine, m.p. 110°C, is a tertiary base, but weaker than quinoline. It is the parent substance of a number of dyes and antiseptics.

Six-membered Rings with One Oxygen Atom

The simplest six-membered rings with one oxygen are 2-H- (or α-) **pyran** (I), and 4-H- (or γ-) **pyran** (II)

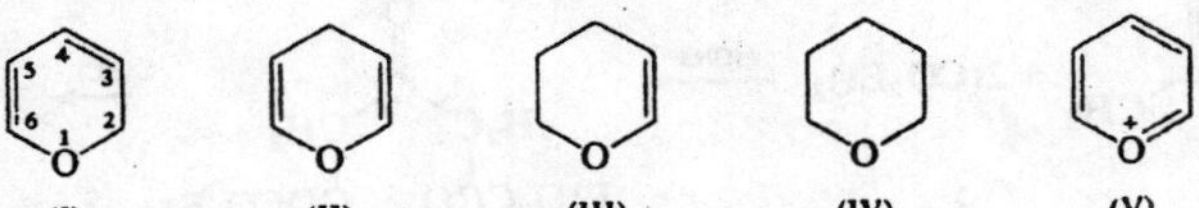

(I) (II) (III) (IV) (V)

2,3-Dihydro-4-pyran, (III), is prepared by heating tetrahydrofurfuryl alcohol with alumina

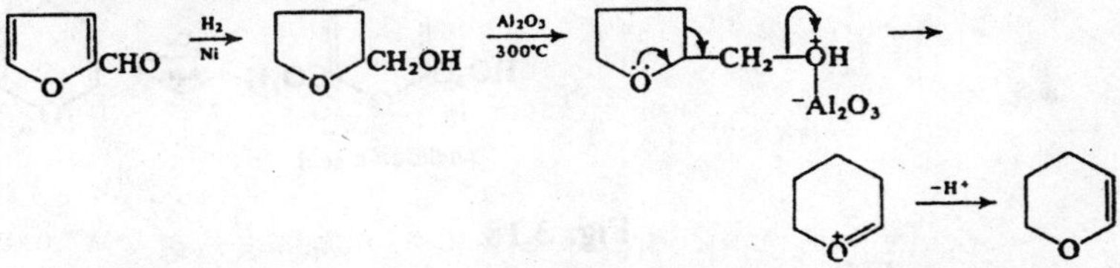

CHO $\xrightarrow[Ni]{H_2}$ CH_2OH $\xrightarrow[300°C]{Al_2O_3}$ CH_2—OH, $^{-}Al_2O_3$ → $\xrightarrow{-H^+}$

This, on catalytic reduction (Raney nickel), gives tetrahydropyran (IV). This compound is oxidised by nitric acid to glutaric acid, and is converted into pimelic acid by means of carbon monoxide and steam in the presence of a catalyst. Also, tetrahydropyran is readily converted into 1,5-dibromopentane by hydrobromic acid in the presence of sulphuric acid. The pyranose sugars are derivatives of tetrahydropyran.

The double bond in dihydropyran is very reactive, readily reacting with hydroxylic compounds such as alcohols and phenols, in the presence of p-toluenesulphonic acid, to form ethers. These ethers are stable to alkali, but are easily split by dilute acids. Thus, this reaction offers a very good means of protecting a hydroxyl group in reactions involving alkaline media:

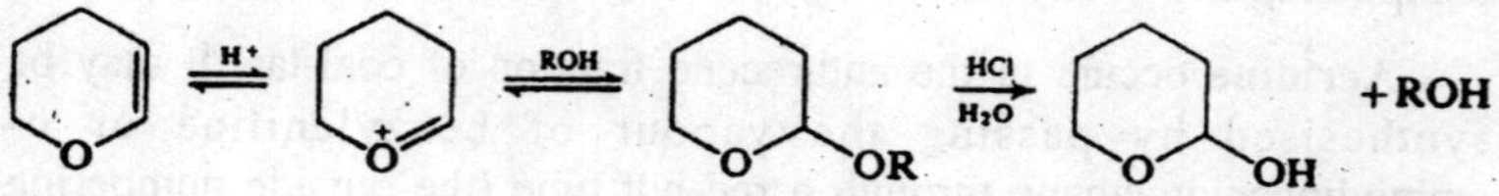

H^+ ⇌ ROH ⇌ OR $\xrightarrow[H_2O]{HCl}$ OH + ROH

(V) is the aromatic **pyrylium cation,** and some salts have been isolated, *e.g.*, the perchlore. The corresponding keto-derivatives of the pyrans are known as **pyrones**:

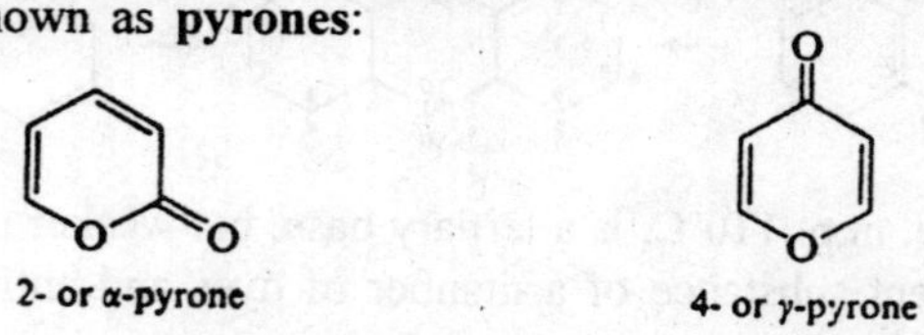

2- or α-pyrone

4- or γ-pyrone

4-Pyrone may be prepared by heating *chelidonic acid* just above its m.p. (262°C). Chelidonic acid naturally occurring substance) may be prepared from acetone and ethyl oxalate.

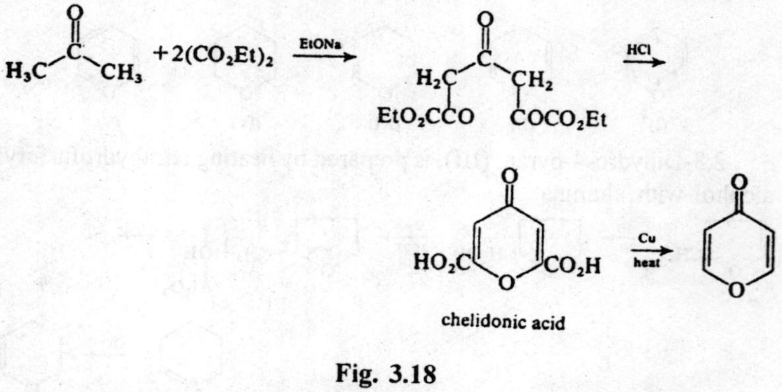

$H_3C{-}CO{-}CH_3 + 2(CO_2Et)_2 \xrightarrow{EtONa} EtO_2CCO{-}H_2C{-}CO{-}CH_2{-}COCO_2Et \xrightarrow{HCl}$

HO₂C, CO₂H $\xrightarrow[heat]{Cu}$

chelidonic acid

Fig. 3.18

4-Pyrone, m.p. 32.5°C, is basic and shows some aromatic properties, and so is a resonance hybrid with contributing structures such as (III).

4-Pyrone does *not* form an oxime or phenylhydrazone. Similarly, the structure of 4-pyrone salts may be (VI) rather than (IV). According to Brown (1951), however, calculation of general charge distribution suggests (V) as the structure of the oxonium salts (this corresponds to (II) for 4-pyrone itself).

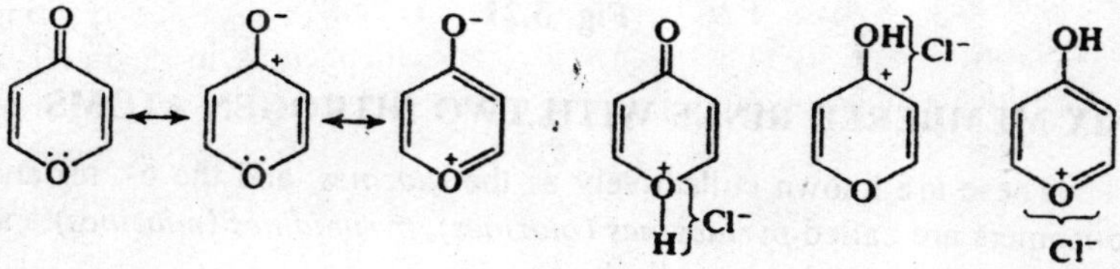

(I) (II) (III) (IV) (V) (VI)

Fig. 3.19

2,6-Dimethyl-4-pyrone is a very important derivative of 4-pyrone from the theoretical point of view, since its salt with hydrochloric acid was the first *oxonium* salt to be prepared (Collie and Tickle, 1890); the structure of this salt corresponds to that of (V) or (VI).

The dimethyl-4-pyrone may be prepared from the copper salt of ethyl acetoacetate as follows:

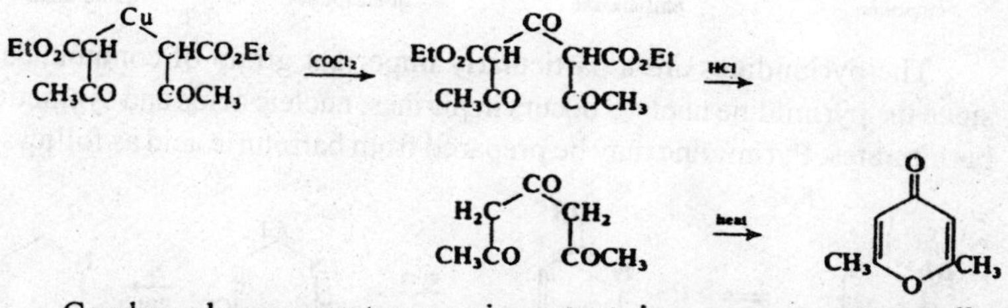

Condensed pyrone systems are important since many occur naturally:

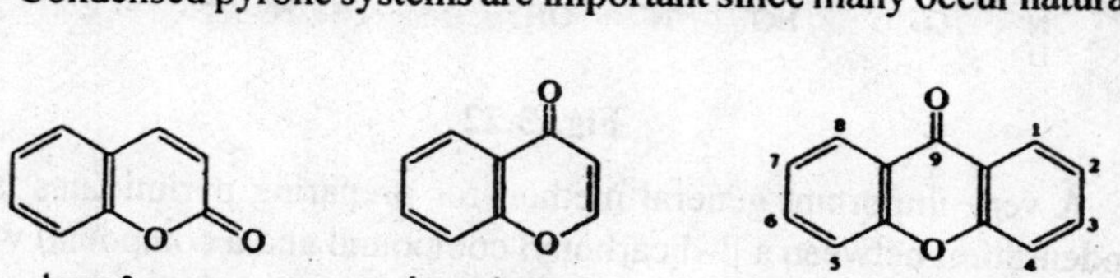

benzo-2-pyrone *or* coumarin (α-chromone)

benzo-4-pyrone (γ-chromone)

xanthone

Fig. 3.20

α-Flavone is 2-phenyl-4-chromone, and many of its derivatives are the colouring matter of flowers

Fig. 3.21

SIX-MEMBERED RINGS WITH TWO NITROGEN ATOMS

These are known collectively as the *diazines*, and the o-, m- and p-isomers are called *pyridazines* (*oiazines*), *pyrimidines* (*miazines*) and *pyrazines* (*piazines*), respectively.

The corresponding *benzodiazines* are:

The pyrimidines are a particularly important group of compounds, since the pyrimidine nucleus occurs in purines, nucleic acids and synthetic barbiturates. Pyrimidine may be prepared from barbituric acid as follows:

Fig. 3.22

A very important general method for preparing pyrimidines is the condensation between a β-dicarbonyl compound and a compound which has the amidine structure, *e.g.*, 6-hydroxy-2,4-dimethylpyrimic from ethyl acetoacetate and acetamidine.

$$CH_3C(NH_2)=NH + C_2H_5O_2C\text{-}CH_2\text{-}COCH_3 \longrightarrow \text{(6-hydroxy-2,4-dimethylpyrimidine: OH, } CH_3, CH_3) + H_2O + C_2H_5OH$$

Some pyrimidines found in nucleic acids are:

cytosine 5-methylcytosine thymine uracil

AROMATIC AMINO-COMPOUNDS

Aromatic amines are of two types, one in which the amino-group is attached directly to the ring, and the other in which the amino-group is in the side-chain. The latter behave very much like the aliphatic amines, and so may be regarded as aryl-substituted aliphatic amines.

Nuclear amino-compounds fall into the same three classes as the aliphatic amines, *viz., primary amines,* $ArNH_2$; *secondary amines,* Ar_2NH and ArNHR; and *tertiary amines,* Ar_3N, Ar_2NR and $ArNR_2$. Quaternary aromatic compounds containing 4, 3, or 2 aryl groups attached to nitrogen have not yet been prepared. This is possibly due partly to decreased basicity and partly to steric effects.

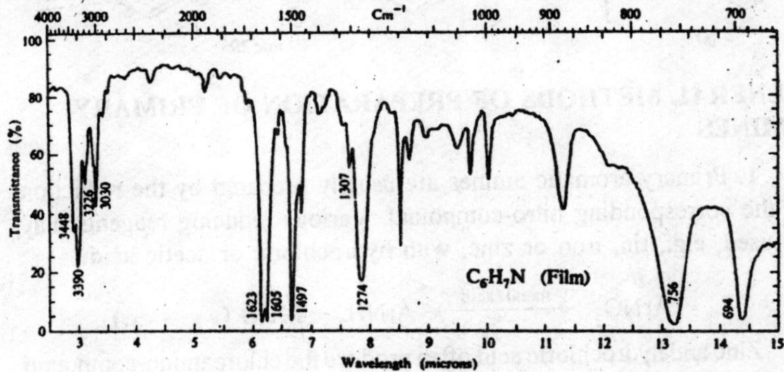

Fig. 3.23

(1623), (1605), and 1497 cm^{-1}: C = C (str.) in aromatic nuclei. 3030 cm^{-1}: = C—H (str.) in aromatics. 3448 and 3390 cm^{-1}: N—H (str.) in primary amines. 3226 cm^{-1}: N—H (str.); association. (1623) and (1605) cm^{-1}: N—H (def.) in primary amines. 1307 and 1 274 cm^{-1}: C — N (str.) in primary aromatic amines. 756 and 694 cm^{-1}: Monosubstituted benzenes.

The N—H (stretch) infra-red absorption regions are the same as those for aliphatic amines, but the C—N (stretch) region is different.

Primary aromatic amines absorb in the region 1 340 – 1250 cm^{-1} (s); secondary, 1350–1280 cm^{-1} (s); and tertiary, 1360–1310 cm^{-1} (s). At the same time, the absorption regions of the aromatic nucleus will show its presence, with additional peaks according to the pattern of substitution.

Fig. 3.23 is the infra-red spectrum of aniline (as a film),

MASS SPECTRA

The peak for the molecular ion is strong and has an odd mass number. Primary amines lose a molecule of HCN, *e.g.*-,

$$[C_6H_5NH_2]^{+} \xrightarrow{-HCN} [C_5H_6]^{+} \xrightarrow{-H\cdot} C_5H_5^{+}$$

M^{+} 93   *m/e* 66   *m/e* 65

Alkyl groups attached to the nucleus or to the nitrogen atom readily undergo β-cleavage, e.g.,

$$[C_6H_5NHCH_2R]^{+} \xrightarrow{-R} C_6H_5\dot{N}HCH_2^{+} \longleftrightarrow C_6H_5\overset{+}{N}H{=}CH_2$$

M^{+} *m/e* 106

GENERAL METHODS OF PREPARATION OF PRIMARY AMINES

1. Primary aromatic amines are usually prepared by the reduction of the corresponding nitro-compound. Various reducing reagents may be used, e.g., tin, iron or zinc, with hydrochloric or acetic acid:

$$ArNO_2 \xrightarrow{\text{metal / acid}} ArNH_2 + 2H_2O\ (v.g - ex.)$$

Zinc and hydrochloric acid often produce the chloroamino-compound as well as the desired amino-compound. Acetic acid is very useful for compounds which contain a group that is attacked by inorganic acid; *e.g.*, reduction of p-nitroacetanilide in hydrochloric acid solution leads to some hydrolysis of the acetyl group; use of acetic acid prevents this deacetylation.

Titanous chloride in hydrochloric acid reduces nitro-compounds to amines, and may be used for the quantitative estimation of nitro-groups in a compound:

$$ArNO_2 + 6TiCl_3 + 6HCl \longrightarrow ArNH_2 + 6TiCl_4 + 2H_2O$$

By means of aqueous ethanolic ammonium hydrogen sulphide, aqueous sodium sulphide, methanolic sodium hydrogen sulphide, or stannous chloride in hydrochloric acid, nitro-groups in a polynitro-compound can be reduced one at a time, *e.g.*, m-nitroaniline from m-dinitro-benzene:

$$C_6H_4(NO_2)_2 + 3NH_4HS \longrightarrow C_6H_4(NH_2)(NO_2) + 3NH_3 + 3S + 2H_2O \quad (70\text{–}80\%)$$

It appears that ammonium hydrogen sulphide and sodium sulphide will not reduce a nitro-group in the o-position to an alkyl group, whereas stannous chloride in hydrochloric acid reduces this nitro-group preferentially:

$$\text{3-nitro-}p\text{-toluidine} \xleftarrow{NH_4HS} CH_3C_6H_3(NO_2)_2 \xrightarrow{SnCl_2/HCl} \text{5-nitro-}o\text{-toluidine}$$

3-nitro-*p*-toluidine **5-nitro-*o*-toluidine**

When the ring contains a group other than alkyl, then the course of reduction is less certain, *e.g.*, it is the o-nitro-group that is reduced in 2,4-dinitroaniline by ammonium hydrogen sulphide (52–58%), and the o-nitro-group in 2,4-dinitrophenol by sodium sulphide (58–61%).

Nitro-compounds may be reduced catalytically to the corresponding amine at 200 – 400°C; if Raney nickel or platinum is used as catalyst, the reduction may be carried out at room temperature.

Catalytic reduction is particularly useful for compounds which contain a group that is hydrolysable in acid solution, *e.g.*, p-nitroacetanilide. Under milder conditions of catalytic reduction, nitro-compounds produce azo-compounds.

Nitro-compounds are also reduced to amines by hydrazine hydrate or sodium borohydride in the presence of palladised charcoal, and by cyclohexene and palladium. Sodium borohydride alone produces the corresponding radical anion, $ArNO_2^{-}$. (Waters *et al.*, 1970).

If a nitro-compound contains an unsaturated chain or an aldehyde group in the nucleus, alkaline ferrous sulphate may be used as the reducing agent to avoid reduction of these other groups. If only an aldehyde group is present, then stannous chloride may be used, *e.g.*, m-nitro-benzaldehyde may be reduced to m-aminobenzaldehyde:

$$\text{CHO}\ldots\text{NO}_2 \xrightarrow{SnCl_2/HCl} \text{CHO}\ldots\text{NH}_2$$

2. Amino-compounds may be prepared by reduction of aromatic nitroso-, azo-, azoxy-, and hydrazo-compounds, but since these are obtained from nitro-compounds, their use is limited to special cases.
3. Aromatic amines may be prepared by methods used for aliphatic amines, *e.g.*, amination of aryl halides, Hofmann amine preparation from amides.

   Schmidt reaction with acids. Curtius reaction, etc. The Beckmann rearrangement of oximes is also useful in certain cases because of the ease of preparation of alkaryl ketones by the Friedel-Crafts reaction.

Aniline *(aminobenzene).* In the laboratory, aniline is prepared by the reduction of nitrobenzene with metal and acid, *e.g.*, tin and hydrochloric acid:

$$C_6H_5NO_2 \xrightarrow{e;\,H^+} C_6H_5NH_2 + 2H_2O \ (80\%)$$

The aniline remains in solution as aniline stannichloride $(C_6H_5NH_2)_2.H_2SrCl_6$; when the solution is made alkaline and steam distilled, the aniline comes over in the distillate.

Commercially, aniline is prepared by the reduction of nitrobenzene with iron, water and 30 per cent hydrochloric acid, or by catalytic reduction (copper catalyst).

Aniline (whose properties are characteristic of primary aromatic amines in general) is, when freshly prepared, a colourless liquid, b.p. 184°C; it has an unpleasant odour and is poisonous. When exposed to air, it rapidly darkens, since it is very sensitive to oxidation.

Aniline forms crystalline salts with strong inorganic acids, and these salts are considerably hydrolysed in solution. Aniline is a weaker base than the primary aliphatic amines, and this may be explained by resonance (which is not possible in aliphatic primary amines).

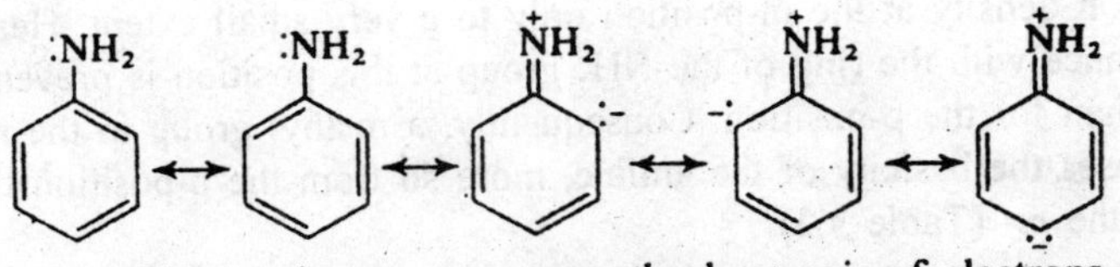

In aniline, owing to resonance, the lone pair of electrons on the nitrogen atom is less available for co-ordinating with a proton; at the same time, the small positive charge on the nitrogen atom would tend to repel a proton. Alternatively, since there are more resonating structures possible for aniline itself than for the cation $C_6H_5\overset{+}{N}H_3$, the former will be stabilised with respect to the latter.

The effect of a ring substituent on basicity depends on whether the substituent is electron-attracting or releasing, its ability to enter into resonance with the amino-group, and its position. All the nitroanilines are weaker bases than aniline (see Table 3.3). The nitro-group has a strong – R effect, and *o*- and p-nitroaniline are therefore more resonance-stabilised than aniline itself.

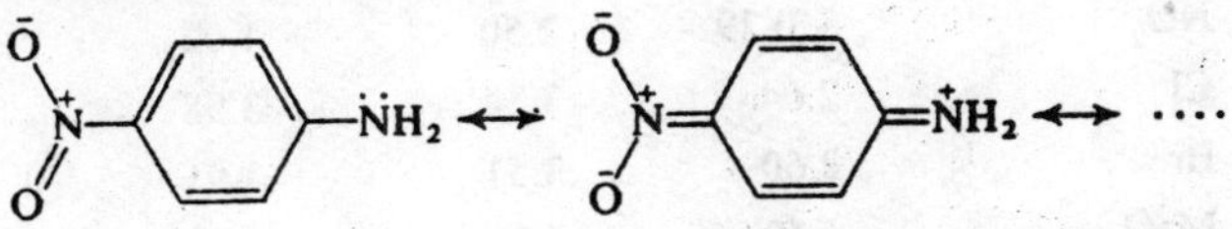

A m-nitro-group cannot enter into resonance with the amino-group, but nevertheless m-nitro-aniline is a much weaker base than aniline. In this case, the explanation is that, in addition to exhibiting a – R effect when possible, the nitro-group always has a strong – I effect (due to the positive charge on the nitrogen atom). This – I effect tends to draw into the ring the lone pair on the N atom of the NH_2 group, with consequent decrease in basicity. In the same way may be explained the decreased basicity due to halogen atoms in the ring, Table 3.3. The + R effect of halogens is very small compared with their – I effect. It can therefore be expected that increasing the number of nitro-groups or halogen atoms in the ring will decrease basicity. When the ring contains an electron-releasing group, this reduces resonance of the NH_2 group with the ring, and consequently should increase basicity. The methyl group raises the electron density more at the *o*- and p-positions than at the *m*-. Thus, the carbon *atom para* to the methyl group has a high electron density, and the lone pair on the nitrogen atom is therefore prevented, to some extent, from entering into resonance with the ring. A methyl group raises the electron density at the m-position only to a very small extent. Hence, resonance with the ring of the NH_2 group at this position is prevented less than for the p-position. Consequently, a methyl group in the ring increases the basicity of the aniline, more so from the p-position than from the *m*- (Table 3.3).

Table 3.3

Base	pK_a	*Base*	pK_a
$PhNH_2$	4.58	PhNHEt	5.11
PhNHMe	4.85	$PhNEt_2$	6.56
$PhNMe_2$	5.06	Ph_2NH	0.9
		PhNHAc	0.4

Substituent in $PhNH_2$	*o*-	*m*-	PK_a *P*-
Me	4.39	4.69	5.12
NO_2	– 0.29	2.50	1.02
Cl	2.64	3.34	3.98
Br	2.60	3.51	3.91
MeO	4.49	4.20	5.29
NH_2	4.47	4.88	6.08

The p-OMe group has a + R effect, and its presence increases basicity. In the m-position, it can exert only a – I effect, and this decreases the basicity (see also Resonance and the Hammett equation).

It can be seen from the foregoing account that when the substituent group can enter into resonance with the rest of the molecule, it has a marked effect on the basicity of the compound. It therefore follows that this effect will be diminished (or be completely absent) if this resonance can be diminished (or completely prevented). An essential requirement for resonance is that the resonating structures-must be planar (or nearly planar). Thus, if planarity is partially reduced or completely prevented by the steric effect of a group, resonance is diminished or prevented. In these circumstances, the phenomenon is known as **steric inhibition of resonance.** This phenomenon may be used to explain the fact that 3,5-dimethyl-4-nitroaniline is a stronger base than the corresponding 2,6-dimethyl isomer. In the former, the nitro-group *cannot* enter into resonance with the amino-group, whereas it can in the latter. It might also be noted that a steric effect is operating between the amino group and the two o-methyl groups in the 2,6-dimethyl isomer, but in this case, since only hydrogen atoms (which are very small) are involved, the steric effect will be very much smaller than that due to oxygen atoms.

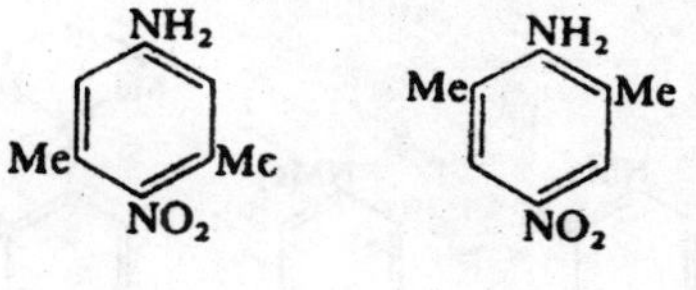

NH_2 / Me, Me / NO_2 NH_2 / Me, Me / NO_2

pK_a 2·49   0·95

N-Alkylated anilines are stronger bases than aniline. At first sight one might have expected the opposite to be true, since alkyl groups have a +1 effect and consequently increased resonance of the lone pair with the ring would be anticipated. Also, since an N-Et substituent increases the basic strength more than an N-Me, the observed results cannot be explained on the basis of polar effects. The explanation offered is that a steric effect operates. Since the ethyl group is larger than the methyl, the steric effect is greater for the former, and hence there is greater steric inhibition of resonance in the former.

Thus, in the former, the lone pair on the nitrogen atom is more available for protonation, and consequently the basicity of N-ethylamine is greater than that of N-methylaniline. Steric inhibition of resonance will be further increased by the introduction of a second alkyl group on the nitrogen atom, and so basicity should increase. This has been found to be the case (Table 3.3). This argument for the steric effect is supported by other evidence, *e.g.*, (i) N-t-butylaniline has a pK_a of 7.1 (the t-butyl group is very large), (ii) Quinuclidine ((I); pK_a of 10.58) is a very strong base; the bridge is perpendicular to the other ring, and a flat nitrogen atom is therefore not possible and so resonance is inhibited.

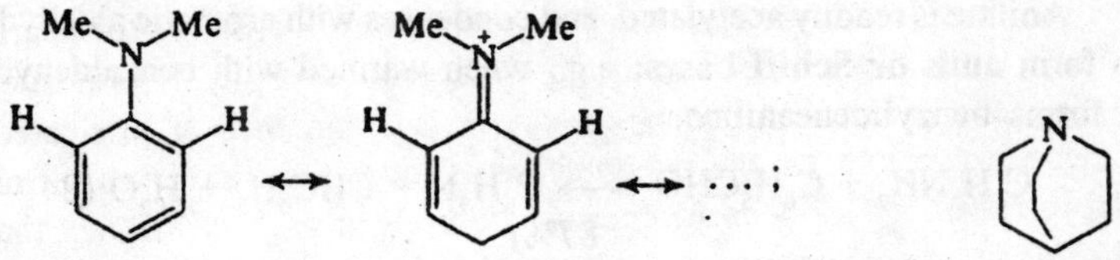

Me, Me, N, H, H $\longleftrightarrow$ Me, Me, $\overset{+}{N}$, H, H $\longleftrightarrow$ ... ;   N (I)

Fig. 3.24

If we proceed with the above argument, then we can expect that o-methylaniline would be a stronger base than the *m*- and p-isomers, in which steric hindrance is absent. In practice, the reverse is true (Table 23.1). Also, there are the basicities of the following compounds (in 50 per cent aqueous ethanol):

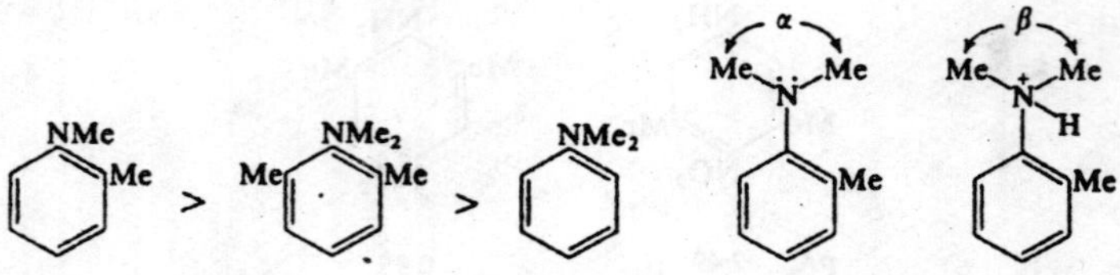

NMe, Me > Me, NMe₂, Me > NMe₂ (II) (III)

pK 5·07 4·69 4·09.

Fig. 3.25

The reason for these orders is not certain. A possible explanation is that angle a in the free base (II) is smaller than angle β in the protonated base (III). Thus protonation relieves crowding on the nitrogen atom but increases steric interaction with an o-methyl group. With one o-methyl group, nitrogen relief overcomes the steric effect of this methyl group, but with two o-methyl groups, their steric effect overcomes nitrogen relief.

Hydrogen atoms of the amino-group are replaceable by halogen when aniline is treated with hypohalous acid and by sodium or potassium on heating:

$$C_6H_5NH_2 + K \longrightarrow [C_6H_5NH]^-K^+ + \frac{1}{2}H_2$$

Aniline combines with alkyl halides to give finally a quaternary ammonium compound, and gives the *isocyanide reaction* when heated with chloroform and ethanolic potassium hydroxide:

$$C_6H_5NH_2 + CHCl_3 + 3KOH \longrightarrow C_6H_5NC + 3KCl + 3H_2O$$

Aniline is readily acetylated, and condenses with aromatic aldehydes to form **anils or Schiff bases;** *e.g.,* when warmed with benzaldehyde, it forms benzylideneaniline:

$$C_6H_5NH_2 + C_6H_5CHO \longrightarrow C_6H_5N = CHC_6H_5 + H_2O \text{ (84 – 87\%)}$$

These Schiff bases are easily hydrolysed to the free amine, and so their formation offers a means of 'protecting' an amino-group, *e.g.,* during nitration. Schiff bases may also be used to prepare secondary amines:

$$C_6H_5N = CHAr \xrightarrow[Ni]{H_2} C_6H_5NHCH_2Ar$$

When refluxed with ethanolic carbon disulphide and solid potassium hydroxide, aniline forms **N,N'-diphenylthiourea** *(thiocarbanilide),* m.p. 154°C, which is used as a rubber accelerator:

$$CS_2 + 2C_6H_5NH_2 + 2KOH \longrightarrow S = C(NHC_6H_5)_2 + K_2S + 2H_2O \ (70\%)$$

When treated with concentrated hydrochloric- acid, diphenylthiourea is converted into phenyl isothiocyanate:

$$S = C(NHC_6H_5)_2 + HCl \longrightarrow C_6H_5NCS + C_6H_5NH_2.HCl$$

Phenyl isothiocyanate may be conveniently prepared by running aniline into a cooled mixture of carbon disulphide and concentrated aqueous ammonium hydroxide, and decomposing the precipitated ammonium phenyldithiocarbamate with lead nitrate solution:

$$C_6H_5NH_2 + CS_2 + NH_4OH \longrightarrow C_6H_5NHCS_2NH_4 + H_2O$$

$$\xrightarrow{Pb(NO_3)_2} C_6H_5NCS + NH_4NO_3 + HNO_3 + PbS \ (74 - 78\%)$$

The oxidation products of aniline are far more complex than those obtained from primary aliphatic amines. Nitrosobenzene, nitrobenzene, phenylhydroxylamine, *p-benzo*-quinone, azobenzene, azoxybenzene, aniline black (a dye), etc., may be isolated, the actual substance obtained depending on the nature of the oxidising agent used, *e.g.*

Caro's acid gives nitrosobenzene; peroxytrifluoroacetic acid, nitrobenzene (89%; Emmons, 1954); chromic acid, p-benzoquinone; sodium hypochlorite gives a purple coloration — this reaction is characteristic of aniline.

Aniline may be easily mercurated with an aqueous ethanolic solution of mercuric acetate in the cold or on warming.

The most important difference between aniline and primary aliphatic amines is their behaviour towards nitrous acid; aliphatic amines liberate nitrogen, whereas aniline (and all other primary aromatic monoamines) forms *diazonium salts,* which are stable in cold aqueous solution:

$$C_6H_5NH_2 + HNO_2 + HCl \longrightarrow \underset{\text{benzenediazonium chloride}}{C_6H_5H_2^+Cl^-} + 2H_2O$$

Acetanilide (*N-phenylacetamide*), m.p. 114°C, may be readily prepared, in very good yield, by heating aniline with acetyl chloride, acetic anhydride, or glacial acetic acid. It is hydrolysed by strong acids or alkalis to aniline. It is used in medicine, under the name of *antifebrin,* as a febrifuge. It is a useful intermediate in various reactions of aniline in which it is desirable to 'protect' the amino-group, *e.g.,* in nitration, halogenation, etc.; the acetamido-group is predominantly p-orienting. The N-phenyl derivative of the amide of any acid is known as an *anilide.*

The amino-group may also be protected by the formation of urethans with ethyl chloroformate. Urethans are more readily hydrolysed than acetamido-compounds.

Halogenated anilines. The amino-group in aniline activates the o- and p-positions to a very large degree; *e.g.,* when aniline is treated with excess of chlorine- or bromine-water, an immediate precipitate of the 2,4,6-*trihalogeno-derivative* is obtained:

$$C_6H_5NH_2 + 3Br_2 \longrightarrow 2,4,6\text{-}Br_3C_6H_2NH_2 + 3HBr$$

The yield is quantitative, and so this reaction is used to estimate aniline.

In order to introduce, only one chlorine or bromine atom, the activating effect of the amino-group must be lowered; this may be done by acetylation:

$$C_6H_5NHCOCH_3 \xrightarrow{Br_2/CH_3CO_2H} p\text{-}BrC_6H_4NHCOCH_3\ (80\%) + o\text{-}BrC_6H_4NHCOCH_3\ (\text{small amount})$$

These, on hydrolysis, yield the corresponding bromoanilines, which may be separated by steam distillation (the o-compound being steam volatile).

The deactivating effect of the acetyl group may be explained as follows. The lone pair of the NH_2 group enters into resonance with the carbonyl group, and consequently resonance with the ring is strongly diminished. At the same time, this also explains the very weak basic character of acetanilide (see Table 3.3).

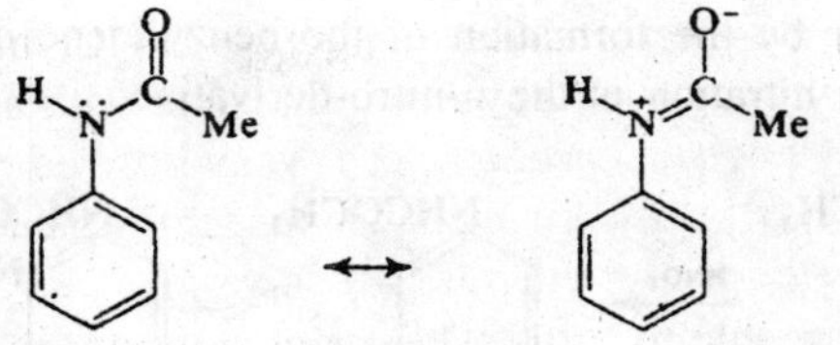

o-Bromoaniline may conveniently be prepared by first protecting the p-position by sulphonation.

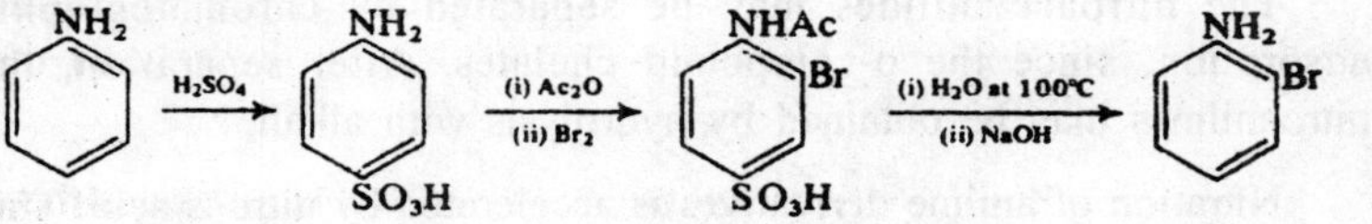

p-Iodoaniline may be prepared by the action of iodine on aniline in the presence of aqueous sodium hydrogen carbonate. The ring is highly activated by the amino-group and elemental iodine is now capable of direct iodination.

m-Chloro- and bromo-anilines may be prepared by reducing the corresponding halogenonitrobenzenes. They are also formed by the halogenation of aniline in concentrated sulphuric acid, the m-orienting effect being due to the presence of the—NH_3 group.

2,6-Dichloroaniline may be prepared by first blocking the p-position of aniline with a sul-sulphonamide group, refluxing the sulphanilamide with hydrochloric acid and hydrogen peroxide, and subsequently refluxing the product, 3,5-dichlorosulphanilamide, with 70 per cent sulphuric acid, thereby removing the sulphonamide group:

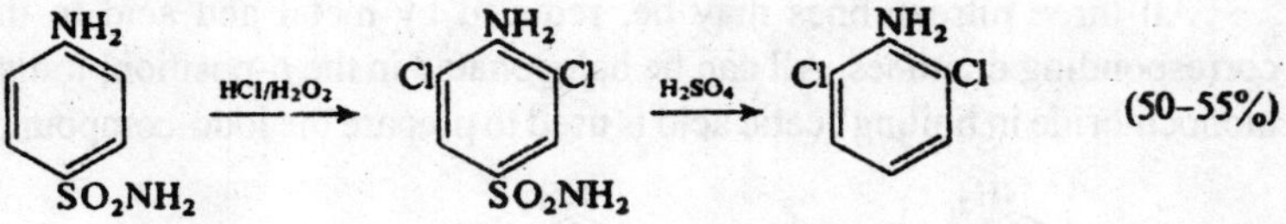

(50–55%)

Using HBr instead of HCl gives 2,6-dibromoaniline (57–75%). The presence of a halogen atom decreases the basic properties of aniline (see Table 23.1).

Nitroanilnes. Direct treatment of aniline with nitric acid gives a complex mixture of mono-, di- and trinitro-compounds, and oxidation products. If, however, the amino-group is protected (and deactivated) by acetylation or by the formation of the benzylidene derivative, the main product on nitration is the p-nitro-derivative:

$$C_6H_5NHCOCH_3 \xrightarrow{HNO_3} p\text{-}NO_2C_6H_4NHCOCH_3\ (90\%) + o\text{-}NO_2C_6H_4NHCOCH_3\ (1\text{–}2\%)$$

The nitroacetanilides may be separated by chromatographic adsorption, since the o-compound chelates. After separation, the nitroanilines may be obtained by hydrolysis with alkali.

Nitration of aniline derivatives is accelerated by nitrous acid. This is believed to be due to the formation of the nitroso-compound, followed by its oxidation to the nitro-compound.

***o*- and p-Nitroanilines** are manufactured by heating *o*- and p-chloronitrobenzenes with ammonia under pressure.

m-Nitroaniline (m.p. 114°C) may be prepared by the partial reduction of m-dinitrobenzene with ammonium hydrogen sulphide (yield: 70–80%), sodium sulphide (yield: 70%), or stannous chloride in hydrochloric acid. m-Nitroaniline may also be prepared by the direct nitration of aniline *in the presence of concentrated sulphuric acid (cf.* m-chloroaniline).

Industrially, it is prepared by reducing m-dinitrobenzene with the calculated quantity of iron (for one nitro-group) and a small amount of hydrochloric acid.

All three nitroanilines may be, reduced by metal and acid to the corresponding diamines. All can be halogenated in the o-position; iodine monochloride in boiling acetic acid is used to prepare the iodo-compound:

$$p\text{-}NH_2C_6H_4NO_2 + 2ICl \longrightarrow 2,6\text{-}I_2\text{-}4\text{-}NO_2C_6H_2NH_2 + 2HCl \quad (56\text{–}64\%)$$

o- and p-Nitroanilines, but not *m*-, undergo activated nucleophilic substitution with boiling aqueous sodium hydroxide to give the corresponding nitrophenol. All the nitroanilines are weaker bases than aniline (see Table 3.3).

2,4-Dinitroaniline, m.p. 180°C, may be prepared by refluxing an ethanolic solution of 1-chloro-2,4-dinitrobenzene with concentrated ammonia solution. This is also an example of activated nucleophilic substitution.

2,6-Dinitroaniline may be prepared as follows (activated nucleophilic substitution):

$$C_6H_5Cl \xrightarrow[H_2SO_4]{KNO_3} \text{(Cl; } O_2N, NO_2; SO_3K) \xrightarrow{NH_4OH} \text{(}NH_2\text{; } O_2N, NO_2; SO_3K) \xrightarrow[\text{boil}]{\text{dil. } H_2SO_4} \text{(}NH_2\text{; } O_2N, NO_2) \; (30\text{–}36\%)$$

Aminobenzenesulphonic Acids

Toluidines (*tolylamines*). These may be obtained by reduction of the corresponding nitro-toluenes. *o*- and m-Toluidines are oils, b.p. 201° and 200°C, respectively; p-toluidine is a solid, m.p. 45°, b.p. 200°C.

N-Alkylanilines. The commonest N-alkylanilines are mono- and dimethylaniline. These may be prepared by heating aniline with methyl iodide:

$$C_6H_5NH_2 + CH_3I \longrightarrow C_6H_5NHCH_3.HI$$

$$C_6H_5NHCH_3 + CH_3I \longrightarrow C_6H_5N(CH_3)_2.HI$$

Commercially, these methylanilines are prepared by heating a mixture of aniline, methanol and sulphuric acid under pressure at 230°C. No quaternary compound is formed under these conditions. Monomethylaniline is the main product (50%) when about 1-2 molecules of methanol are used; dimethyl-aniline and unchanged aniline are also present. Dimethylaniline is obtained as the main product (95%) by using a large excess of methanol; monomethylaniline and unchanged aniline are also present, but these are removed by acetylating the mixture and distilling off the dimethylaniline.

To prepare higher N-alkyl derivatives of aniline, hydrochloric acid is used instead of sulphuric, since the latter tends to dehydrate the alcohol to alkene.

An interesting example of *reductive methylation* is the conversion of aniline into dimethylaniline; aniline and formaldehyde, in aqueous ethanolic sulphuric acid, is hydrogenated in the presence of platinum as catalyst.

$$C_6H_5NH_2 + 2CH2O + 2H_2 \xrightarrow{Pt} C_6H_5N(CH_3)_2 + 2H_2O\ (74\%)$$

Mono- and dimethylaniline are liquids, b.p. 196° and 193°C, respectively, and cannot be readily separated by fractional distillation. They can be separated, however, by acetylating the monomethyl derivative.

Pure monomethylamline may be prepared by treating the sodium derivative of acetanilide with methyl iodide in toluene solution:

$$[C_6H_5NCOCH_3]^-Na^+ + CH_3I \longrightarrow C_6H_5N(CH_3)COCH_3 + NaI$$
$$\xrightarrow{KOH} C_6H_5NHCH_3\ (90\%)$$

N-Alkylated-anilines are stronger bases than aniline.

Mono- and dimethylaniline resemble secondary and tertiary aliphatic amines in many ways, but differ in that they can also undergo substitution reactions (in the *o*- and p-positions) due to the presence of the benzene ring. Monoalkylanilines form pale yellow N-nitrosoamines with nitrous acid, and these give Liebermann's nitroso-reaction:

$$C_6H_5NHCH_3 + HNO_2 \longrightarrow C_6H_5N(NO)CH_3 + H_2O\ (87\text{–}93\%).$$

This may be converted into the original compound by heating with hydrochloric acid:

$$C_6H_5N(NO)CH_3 + H_2O \longrightarrow C_6H_5NHCH_3 + HNO_2.$$

The nitroso-compound may also be reduced to 1,1-methylphenylhydrazine:

$$C_6H_5N(NO)CH_3 \xrightarrow{Zn/CH_3CO_2H} C_6H_5N(CH_3)NH_2 + H_2O \quad (52\text{–}56\%)$$

N-Methylformanilide, b.p. 253°C, may be prepared by distilling a mixture of methylaniline, formic acid and toluene:

$$C_6H_5NHCH_3 + HCO_2H \longrightarrow C_6H_5N(CH_3)CHO + H_2O\ (93 - 97\%)$$

It may also be prepared (in 80 per cent yield) by oxidising N,N-dimethylaniline with hydrated manganese dioxide (Henbest *et al.*, 1957).

It is used as a formylating agent in the preparation of aldehydes.

Dialkylanilines form p-nitroso-compounds with nitrous acid, *e.g.*, dimethylaniline forms p-nitrosodimethylaniline (green flakes):

$$C_6H_5N(CH_3)_2 + HNO_2 \longrightarrow p\text{-}ON\cdot C_6H_4N(CH_3)_2 + H_2O \quad (80\text{–}90\%)$$

p-Nitrosodimethylaniline readily undergoes activated nucleophilic substitution, *e.g.*, it forms p-nitrosophenol with boiling aqueous sodium hydroxide. It is also readily oxidised by permanganate to p-nitrodimethylaniline, and may be reduced to p-aminodimethylaniline.

The dimethylamino-group activates the *p-hydrogen* atom so much that dimethylaniline will condense with formaldehyde and carbonyl chloride to form diphenylmethane derivatives, and with aromatic aldehydes to form triphenylmethane derivatives; *e.g.*, with carbonyl chloride, it forms *Michler's ketone*:

$$COCl_2 + 2C_6H_5N(CH_3)_2 \longrightarrow CO[C_6H_4N(CH_3)_2]_2 + 2HCl$$

Mono- and dialkylanilines (and the quaternary compounds) as their hydrochlorides (or hydrobromides) undergo rearrangement on strong heating, an alkyl group migrating from the nitrogen atom and entering preferentially the p-position or, if this is occupied, the *o*-; *e.g.*, when phenyltrimethylammonium chloride is heated under pressure, the following rearrangement takes place:

$$C_6H_5\overset{+}{N}(CH_3)_3Cl^- \xrightarrow{300°C} p\text{-}CH_3C_6H_4N(CH_3)_2\cdot HCl \longrightarrow 2,4\text{-}(CH_3)_2C_6H_3NHCH_3\cdot HCl \longrightarrow 2,4,6\text{-}(CH_3)_3C_6H_2NH_2\cdot HCl$$

This reaction is known as the **Hofmann-Martius rearrangement** (1871); it may be used to prepare aniline homologues.

Many mechanisms have been proposed for the Hofmann-Martius rearrangement. Hughes and Ingold (1952) have proposed the following, based largely on the suggestion of Hickinbottom (1934) that the rearrangement occurs via the formation of an alkyl carbonium ion:

$$C_6H_5NH_2C_2H_5Cl^- \rightleftharpoons C_6H_5NH_2 + C_2H_5Cl$$

$$C_2H_5Cl \rightleftharpoons C_2H_5^+ + Cl^-$$

$$C_6H_5NH_2 + C_2H_5^+ \rightleftharpoons C_6H_5NH_3^+ + C_2H_4$$

$$2C_6H_5NH_2 + C_2H_5^+ \rightleftharpoons p\text{-}C_2H_5C_6H_4NH_2 + C_6H_5NH_3^+$$

It should be noted that the heterolytic fission of ethyl chloride is facilitated in the highly polar molten salt.

Rearrangements of this kind have been observed to take place with aniline derivatives of the type $C_6H_5\underset{|}{N}—Z$, where Z is R, X, NH_2, NO or NO_2, *e.g.*,

$$\text{C}_6\text{H}_5\text{NHNH}_2\cdot\text{HCl} \xrightarrow{250^\circ\text{C}} p\text{-H}_2\text{NC}_6\text{H}_4\text{NH}_2\cdot\text{HCl} \text{ (main product)} + o\text{-H}_2\text{NC}_6\text{H}_4\text{NH}_2\cdot\text{HCl}$$

When Z is NO, *i.e.*, the compound is the N-nitroso-derivative of a secondary aromatic amine, the rearrangement is known as the **Fischer-Hepp rearrangement** (1886).

The main product in the Fischer-Hepp rearrangement is the p-isomer. The mechanism of this reaction has long been *assumed* to be intermolecular (*cf.* the Orton rearrangement below), *i.e.*,

$$PhNRNO + HCl \rightleftharpoons PhNHR + NOCl \longrightarrow p\text{-}RNHC_6H_4NO + HCl$$

However, Williams *et al.* (1969), who carried, out this rearrangement in the presence of ^{15}N-labelled sodium nitrite, showed that no ^{15}N was incorporated in the product.

This suggests that the rearrangement is *intramolecular*, possibly by a mechanism similar to that for N-nitroanilines (see below).

Hughes *et al*, (1956, 1964) have studied the rearrangement of N-nitroaniline in sulphuric acid in the presence of enriched ^{15}N tracer.

The products were *o*- and p-nitroaniline and they contained the normal amount of ^{15}N, thus showing that the rearrangement is *intramolecular.*

This is consistent with the following six-centre mechanism for the nitramine rearrangement:

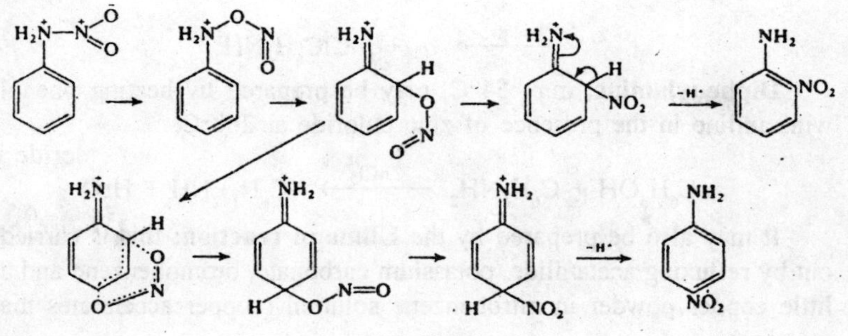

Fig. 3.26

The Orton rearrangement (1901). This is also known as the **chloramine rearrangement,** and involves the rearrangement of aromatic N-halogenoamines. The most widely studied example is the rearrangement of N-chloroacetanilide in aqueous hydrochloric acid; the products obtained are p-chloroacetanilide together with a small amount of the o-isomer.

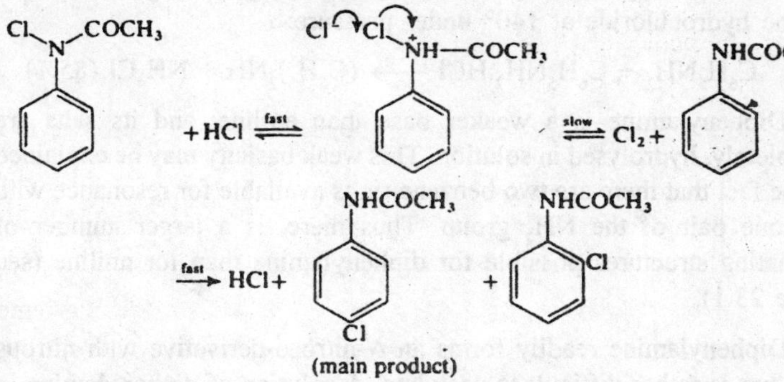

(main product)

Fig. 3.27

One strong piece of evidence for the intermolecular nature of the rearrangement is that when hydrochloric acid labelled with radioactive chlorine is used, the p-chloroacetanilide produced contains some radioactive chlorine (Olson *et al.,* 1936 – 1938):

Although the reduction of nitrobenzene with zinc and hydrochloric acid gives predominantly aniline, fair amounts of *o*- and p-chloroaniline are also obtained. These by-products may possibly be formed via a chloroamine rearrangement:

$$PhNO_2 \xrightarrow[HCl]{Zn} PhNHOH \xrightarrow{HCl} [phNHCl]$$

$$\xrightarrow{\text{rearr.}} o\text{-} + \text{p-ClC}_6\text{H}_4\text{NH}_2$$

Diphenylamine, m.p. 54°C, may be prepared by heating phenol with aniline in the presence of zinc chloride at 260°C:

$$C_6H_5OH + C_6H_5NH_2 \xrightarrow{ZnCl_2} (C_6H_5)_2NH + H_2O$$

It may also be prepared by the **Ullmann reaction;** this is carried out by refluxing acetanilide, potassium carbonate, bromobenzene and a little copper powder in nitrobenzene solution (copper accelerates the reaction)

$$(C_6H_5)_2NHCOCH_3 + CH_6H_5Br + K_2CO_3 \xrightarrow{Cu} (C_6H_5)_2\,NH_2 + CO_2 + CH_3CO_2K + KBr\ (60\%).$$

This is a useful general method for preparing substituted diphenylamines.

Commercially, diphenylamine is prepared by heating aniline with aniline hydrochloride at 140° under pressure:

$$C_6H_5NH_2 + C_6H_5NH_2.HCl \longrightarrow (C_6H_5)_2NH + NH_4Cl\ (85\%)$$

Diphenylamine is a weaker base than aniline, and its salts are completely, hydrolysed in solution. This weak basicity may be explained by the fact that there are two benzene rings available for resonance with the lone pair of the NH_2 group. Thus there, is a larger number of resonating structures possible for diphenylamine than for aniline (see Table 23.1).

Diphenylamine readily forms an *N*-nitroso-derivative with nitrous acid, but is rather difficult to acetylate. A solution of diphenylamine in phosphoric acid gives a blue colour with oxidising agents; this reaction is used as a test for nitric acid.

Diphenylamine is alkylated very slowly, but its sodium salt reacts readily with alkyl halides. The sodium salt is prepared by heating diphenylamine with sodium or sodamide:

$$(C_6H_5)_2NH + Na \longrightarrow [(C_6H_5)_2N]^-Na^+ + \frac{1}{2}H_2$$

When treated with iodine, sodiodiphenylamine forms *tetraphenylhydrazine*:

$$2(C_6H_5)_2N^-Na^+ + I_2 \longrightarrow (C_6H_5)_2NN(C_6R_5)_2 + 2NaI$$

Tetraphenylhydrazine is best prepared by oxidising diphenylamine with potassium permanganate in acetone solution. It is a colourless crystalline solid, burin benzene solution (or any other non-ionising solvent) it produces a green colour due to the formation of the free *diphenylnitrogen* radical, in which the nitrogen is bivalent:

$$(C_6H_5)_2N - N(C_6H_5)_2 \rightleftharpoons 2(C_6H_5)_2N.$$

When the solution is diluted, the colour deepens, thereby showing increased dissociation; 'the colour also deepens on warming and fades on cooling. That we are dealing with a free radical is shown, apart from the colour phenomena (the colour intensity does not obey Beer's law), by the immediate reaction (in solution) with *e.g.*, nitric oxide, to form *N*-nitrososodiphenylamine.

$$(C_6H_5)_2N. + NO \longrightarrow (C_6H_5)_2N - NO$$

The stability of this free radical is attributed to its being a resonance hybrid *(cf.* triphenylmethyl,

Triphenylamine, m.p. 127°C, may be prepared by the Ullmann reaction (in nitrobenzene solution):

$$2(C_6H_5)_2NH + 2C_6H_5I + K_2CO_3 \xrightarrow{Cu} 2(C_6H_5)_3N + CO_2 + 2KI$$

$$+ H_2O \ (82 - 85\%)$$

Triphenylamine dissolves in concentrated sulphuric acid to give a blue solution. It is an extremely weak base, but its salts with HF, HBF4, and $HClO_4$ have been prepared under special conditions, *e.g*:, triphenylammonium borofluorate, $Ph_3NH^+BF_4^-$, is precipitated when boron trifluoride is passed into a solution of triphenylamine in moist benzene (Sharp *et al.*, 1960).

Benzylamine (*α-aminotoluene*), b.p. 185°C, may be prepared by the reduction of phenyl cyanide or benzaldoxime:

$$C_6H_5CN \xrightarrow{Na/C_2H_5OH} C_6H_5CH_2NH_2$$

$$\xleftarrow{Na/C_2H_5OH} C_6H_5CH = NOH$$

It may also be prepared by heating benzyl chloride with ammonia under pressure, or by the Hofmann amine preparation on phenylacetamide:

$$C_6H_5CH_2CONH_2 \xrightarrow{Br_2/KOH} C_6H_5CH_2NH_2 + CO_2$$

Benzylamine resembles the aliphatic amines, and is best regarded (as far as the amino-group is concerned) as a phenyl-substituted methylamine; *e.g.*, it differs from the primary aromatic amines in that it is soluble in water, basic (more so than aniline), and with nitrous acid does not form a diazonium salt but gives benzyl alcohol:

$$C_6H_5CH_2NH_2 + HNO_2 \longrightarrow C_6H_5CH_2OH + N_2 + H_2O$$

Diamines

The aromatic diamines may be prepared by methods used for the monoamines, *e.g.*, by reduction of dinitro-compounds, nitroanilines, and in certain cases, aminoazo-compounds, *e.g.*, reduction of aminoazobenzene gives aniline and p-phenylenediamine:

$$C_6H_5N{=}N\text{-}C_6H_4\text{-}NH_2 \xrightarrow{Na_2S_2O_4} C_6H_5NH_2 + NH_2\text{-}C_6H_4\text{-}NH_2$$

The diamines are colourless or white crystalline solids which turn brown when exposed to the air. They are 'di-acid' bases, and their reactions are characterised by those of the three phenylenediamines.

o-Phenylenediamine (*benzene*-1,2-*diamine*, 1,2-*diaminobenzene*), m.p. 102°C, is best prepared by reducing o-nitroaniline with zinc dust and aqueous ethanolic sodium hydroxide (74 – 85%). The most characteristic property of o-diamines is the ease with which they form heterocyclic compounds.

(i) When o-phenylenediamine is treated with ferric chloride solution, a dark red colour is produced due to the formation of 2,3-*diaminophenazine*:

$$C_6H_4(NH_2)_2 + C_6H_4(NH_2)_2 \xrightarrow{FeCl_3} \text{2,3-diaminophenazine}$$

(ii) *Benzimidazoles* are formed when o-phenylenediamine is heated with organic acids, *e.g.*,

$$C_6H_4(NH_2)_2 + HCO_2H \xrightarrow{-H_2O} \left[C_6H_4(NHCHO)(NH_2) \right] \xrightarrow{-H_2O} \text{benzimidazole} \quad (83\text{–}85\%)$$

(iii) When *o*-phenylenediamine is treated with nitrous acid (a solution of the diamino-compound in acetic acid is treated with aqueous sodium nitrite), *benztriazole* is formed:

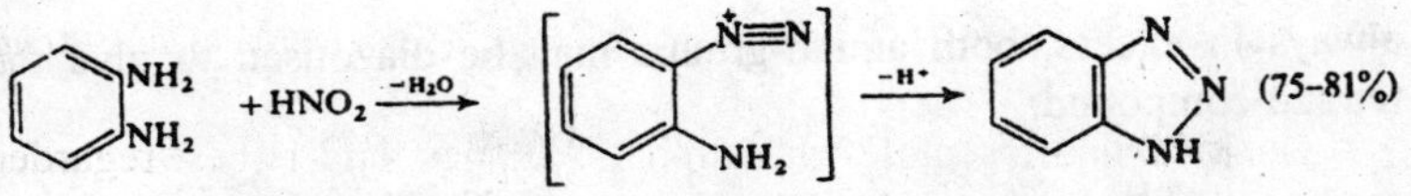

$$\text{o-C}_6\text{H}_4(\text{NH}_2)_2 + \text{HNO}_2 \xrightarrow{-\text{H}_2\text{O}} [\text{o-NH}_2\text{C}_6\text{H}_4\overset{+}{\text{N}}\equiv\text{N}] \xrightarrow{-\text{H}^+} \text{benzotriazole (75–81\%)}$$

(iv) *o*-Phenylenediamine condenses with α-dicarbonyl compounds to form *quinoxalines; e.g.*, with glyoxal, quinoxaline is formed:

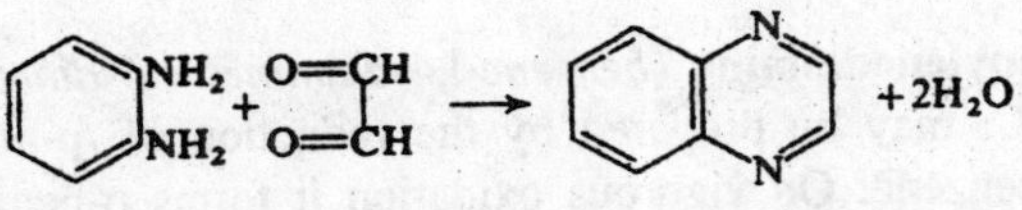

$$\text{o-C}_6\text{H}_4(\text{NH}_2)_2 + \text{O=CH–CH=O} \longrightarrow \text{quinoxaline} + 2\text{H}_2\text{O}$$

This reaction is used to identify o-diamines; the a-dicarbonyl compound employed for this purpose is phenanthraquinone, resulting in the formation of a sparingly soluble phenazine derivative:

$$\text{o-C}_6\text{H}_4(\text{NH}_2)_2 + \text{phenanthraquinone} \longrightarrow \text{phenazine derivative} + 2\text{H}_2\text{O}$$

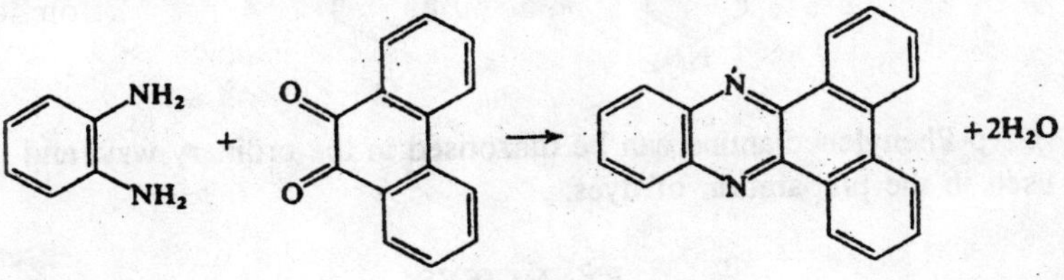

Fig. 3.28

m-Phenylenediamine (*benzene*-1,3-*diamine,* 1,3-*diaminobenzene*), m.p. 63°C, is best prepared by the reduction of m-dinitrobenzene with iron and hydrochloric acid (yield: 90%). Its most characteristic reaction is the formation of brown dyes – *Bismarck Brown* – when it is treated with nitrous acid; a monazo- and a bisazo-compound are formed:

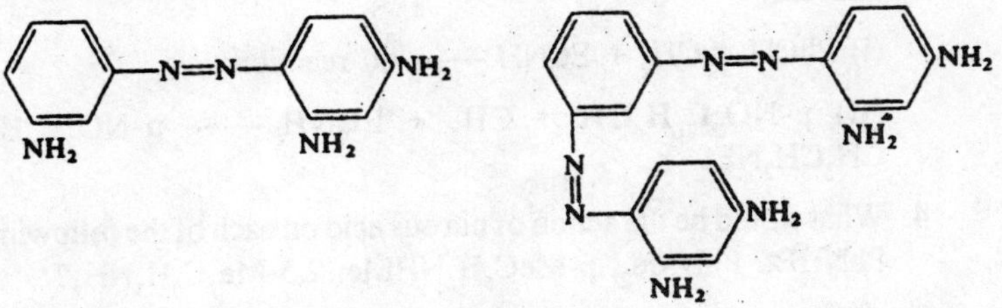

(monazo-compound: $\text{NH}_2\text{C}_6\text{H}_4\text{–N=N–C}_6\text{H}_3(\text{NH}_2)_2$; bisazo-compound: $\text{C}_6\text{H}_4[\text{–N=N–C}_6\text{H}_3(\text{NH}_2)_2]_2$)

Fig. 3.29

This reaction is used as a colorimetric method for the determination of nitrites in water; even when nitrites are present in traces, a yellow colour is produced. On the other hand, by dissolving m-phenylenediamine in concentrated hydrochloric acid, and by keeping the nitrous acid always in excess, both amino-groups may be diazotised to give the tetrazo-compound:

$$C_6H_4(NH_2)_2 + 2HNO_2 + 2HCl \longrightarrow C_6H_4(N_2^+Cl^-)_2 + 4H_2O$$

p-Phenylenediamine (*benzene-*1,4-*diamine,* 1,4-*diaminobenzene),* m.p. 147°C, may be prepared by the reduction of /p-nitroaniline or aminoazobenzene. On vigorous oxidation it forms p-benzoquinone:

$$p\text{-}C_6H_4(NH_2)_2 \xrightarrow[H_2SO_4]{K_2Cr_2O} p\text{-benzoquinone}$$

p-Phenylenediamine can be diazotised in the ordinary way, and is used in the preparation of dyes.

EXERCISES

1. Draw the energy profiles for the reactions:

   (i) $PhNH_2 + Z^+ \longrightarrow$ o- + m- + p-$ZC_6H_4NH_2$

   (ii) $PhCl + Z^+ \longrightarrow$ o- + m- + p-ZC_6H_4Cl

2. Compare the polar effects in $PhNMe_3$ and $PhPMe_3$.

3. Discuss:

   (i) $PhCH = CH_2 + Et_2NH \longrightarrow$ no reaction

   (ii) p-$NO_2C_6H_4CH = CH_2 + Et_2NH \longrightarrow$ p-$NO_2C_6H_4$-$CH_2CH_2NEt_2$

4. What would be the action of nitrous acid on each of the following: PhNHEt, $PhNMe_2$, p-MeC_6H_4NHMe, 2,5-$Me_2C_6H_3NH_2$?

5. What are the effects of the following groups on the basicity of aniline when in: (*a*) the p-position; (*b*) the m-position:

   (i) NO_2; (ii) OR;

   (iii) CO_2R; (iv) NR_2?

6. Complete the equations:

(i) Br–C_6H_4–NH_2 + NaNCO $\xrightarrow{AcOH}$?

(ii) $PhNH_2$ + ? $\xrightarrow{Et_2O}$ PhNHCOCH = $CHCO_2H$

(iii) Me–C_6H_4–NHAc + Br_2 ⟶ ?

(iv) O_2N–C_6H_4–NH_2 + 2ICl $\xrightarrow{AcOH}$?

(v) $PhNH_2$ + $PhSO_2Cl$ ⟶ ? $\xrightarrow[NaOH]{Me_2SO_4}$? $\xrightarrow{H_2O^+}$?

(vi) $PhNH_2CH_2CHMe_2{}^+Br^-$ $\xrightarrow{heat}$?

(vii) $PhNO_2$ $\xrightarrow[Pd-C]{NaBH_4}$

7. Pyrrole is more reactive than furan. Suggest a reason.
8. Complete the following equations:

(i) Furan + $CCO_2Me \equiv CCO_2Me$ ⟶ ?

(ii) Furan-2-CHO + Ac_2O $\xrightarrow{AcONa}$?

(iii) Furan + O_2N–C_6H_4–$N_2{}^+Cl^-$ $\xrightarrow{OH^-}$?

(iv) 2-Bromofuran + maleic anhydride (CO–O–CO ring) ⟶ ?

9. Suggests a mechanism for the reacticn

N-Et-2-(CH_2Cl)pyrrolidine $\xrightarrow{heat}$ N-Et-3-chloropiperidine (Cl, N, Et)

10. Suggest a mechanism for the reaction

$C_6H_4(Cl)CH_2CH_2NHMe$ $\xrightarrow[Et_2O]{PhLi}$ N-Me indoline (N, Me)

11. The nitration of furan with acetyl nitrate proceeds via a 2, 5-addition intermediate with the reagent. Formulate the complete reaction.

12. Pyrrole is much more acidic than s-alkylamines. Suggest a reason.

13. No action $\xleftarrow{\text{furan}}$ $CH_2O + MeNH$ $\xrightarrow{\text{2-Me-furan}}$ A; $\xrightarrow{\text{pyrrole}}$ B

What are A and B? Explain.

14. Complete the equations:

(i) 3-Bromo-2-aminopyridine (Br, NH_2, N) $\xrightarrow[\text{heat}]{\text{M.A.}}$?

(ii) 2-Methylpyridine (CH_3, N) $\xrightarrow{\text{PhLi}}$? $\xrightarrow[\text{(ii) } H^+]{\text{(i) } CO_2}$?

(iii) 3-Methylpyridine N-oxide (Me, $\overset{+}{N}$–O^-) $\xrightarrow[\text{heat}]{\text{M.A.}}$?

(iv) 4-Ethyl-1-methylpyridinium hydroxide (Et, $\overset{+}{N}$ Me, OH^-) $\xrightarrow{\text{heat}}$?

(v) 2-Vinylpyridine ($CH{=}CH_2$, N) $+ HCN \longrightarrow$? (cyanide)

(vi) Aniline (NH_2) $+ MeCOCH{=}CH_2$ $\xrightarrow{H_2SO_4}$?

15. Discuss the behaviour of the three picolines in aqueous NaOH containing D_2O.

Quinoline-2-carboxylic acid (CO_2H, N) $+ PhCHO \xrightarrow{\text{heat}} CO_2 +$ quinoline (N) $+$ 2-quinolyl-CHOHPh (N, CHOHPh)

4

AROMATIC ACIDS

INTRODUCTION

Aromatic acids are compounds containing one or more carboxyl groups which are directly attached to the nucleus. Those acids in which the carboxyl group occurs in the side-chain may be regarded as aryl-substituted aliphatic acids, but they are also classified as aromatic acids, since they exhibit aromatic properties due to the presence of a benzene ring.

MONOCARBOXYLIC ACIDS WITH THE CARBOXYL GROUP ATTACHED TO THE RING

General Methods of Preparation

1. By the oxidation of the corresponding alcohol or aldehyde; *e.g.*, benzaldehyde gives benzoic acid:

$$C_6H_5CHO \xrightarrow{KMnO_4} C_6H_5CO_2H \text{ (v.g.)}$$

2. By the hydrolysis of the corresponding cyanide, *e.g.*, o-tolyl cyanide gives o-toluic acid:

$$o\text{-}CH_3C_6H_4CN + 2H_2O \xrightarrow{H_2SO_4} o\text{-}CH_3C_6H_4CO_2H + NH_3 \text{ (g.-v.g.)}$$

3. By means of a Grignard reagent, *e.g.*,

$$C_6H_5MgBr + CO_2 \longrightarrow C_6H_5CO_2MgBr \xrightarrow{HCl} C_6H_5CO_2H$$

4. By means of the Friedel-Crafts reaction; *e.g.*, treatment of benzene with carbonyl chloride in the presence of anhydrous aluminium chloride gives benzoyl chloride which, on hydrolysis, forms benzoic acid:

$$C_6H_5 + COCl_2 \xrightarrow{AlCl_3} C_6H_5COCl \longrightarrow C_6H_5CO_2H \text{ (55–88\%)}$$

Excess of carbonyl chloride must be used in this method; otherwise benzophenone will be the main product.

5. By the oxidation of benzene homologues with dilute nitric acid, dichromate and sulphuric acid, alkaline permanganate, or chromium trioxide in glacial acetic acid. In some cases it may be more convenient to chlorinate the hydrocarbon, and then oxidise the chloro-derivative. *e.g.*,

$$C_6H_5CH_3 \xrightarrow{Cl_2} C_6H_5CH_2Cl \xrightarrow[KMnO_4]{Na_2CO_3} [C_6H_5CH_2OH]$$

$$\xrightarrow{[O]} C_6H_5CO_2H \text{ (80\%)}$$

In this way the oxidation of the side-chain is much easier, since the intermediate alcohol is much more readily oxidised than the hydrocarbon itself.

General Properties

In general, the aromatic acids are slightly stronger acids than the aliphatic, less soluble in water, and less volatile. The reactions of the carboxyl group are very similar to those in the aliphatic acids, but apparently they differ in that they do not undergo coupling by the Kolbe electrolytic synthesis. However, according to Fichter (1939), the electrolysis of a solution of benzoic acid in pyridine gave a number of products, among which was biphenyl.

The infra-red absorption region of the carbonyl group (C=O stretch) in aromatic acids is 1700-1800 cm^{-1} (s), which is a lower frequency than for saturated aliphatic acids, and so can be distinguished from the latter. This region, however, lies within that of the α-β-unsaturated acids, and so the identification of a carboxyl group by infra-red studies will afford no certain distinction between aryl and the α, β-unsaturated acids (the molecular formula will help here). However, by identifying the presence of the benzene ring by its own characteristic frequencies, these two types of acids may then be distinguished. Also, the nature of the orientation may be identified.

Benzoic acid (*benzenecarboxylic acid*), m.p. 122°C, is present in certain resins, particularly gum-benzoin; it is also present in balsams. Benzoic acid is also found as *hippuric acid* (*benzoylglycine*) in the urine of horses.

Benzoic acid may be prepared in the laboratory by any of the general methods. Commercially, it is prepared as follows:

(i) By the hydrolysis of benzotrichloride with aqueous calcium hydroxide in the presence of iron powder as catalyst (this is an example of indirect oxidation):

$$C_6H_5CCl_3 + 2H_2O \xrightarrow[Fe]{Ca(OH)_2} C_6H_5CO_2H + 3HCl$$

(ii) By the catalytic oxidation of toluene with air and Co—Mn acetates as catalyst:

$$2C_6H_5CH_3 + 3O_2 \xrightarrow{170\,^{\circ}C} 2C_6H_5CO_2H + 2H_2O$$

(iii) When phthalic anhydride and steam are passed over a metal phthalate catalyst, *e.g.*, zinc chromium or nickel salt at 200-300°C, some of the resulting phthalic acid is decarboxylated to benzoic acid:

$$C_6H_4\begin{matrix}CO\\CO\end{matrix}\!\!>\!O + H_2O \longrightarrow C_6H_4\begin{matrix}CO_2H\\CO_2H\end{matrix} \longrightarrow C_6H_5CO_2H + CO_2$$

Benzoic acid readily forms esters when it is refluxed with an alcohol in the presence of a small amount of concentrated sulphuric acid or hydrogen chloride. Benzoic esters are pleasant-smelling liquids which are denser than water. Phenyl benzoate which may be prepared by the action of benzoyl chloride on an alkaline solution of phenol, is a solid, m.p. 71°C. in general, it has been found that benzoic acid containing a substituent in the o-position does not esterify as easily as the m- or p-isomer. If both o-positions (with respect to the carboxyl group) are occupied, then esterification occurs with the greatest difficulty, if at all. Furthermore, once the ester is formed from these di-ortho-substituted benzoic acids, it is very difficult to hydrolyse them. The explanation for these abnormal reactions is not yet complete (see the *ortho*-effect, for further details).

Benzoyl chloride (*benzenecarbonyl chloride*), b.p. 197°C, was the first acid chloride to be discovered. It may be readily prepared by distilling benzoic acid with phosphorus pentachloride or with thionyl chloride:

$$C_6H_5CO_2H + SOCl_2 \longrightarrow C_6H_5COCl + HCl + SO_2 \ (75–80\%)$$

It is prepared commercially by chlorinating benzaldehyde in the cold (benzaldehyde contains no α-hydrogen atom):

$$C_6H_5CHO + Cl_2 \longrightarrow C_6H_5COCl + HCl \text{ (ex)}$$

Another commercial method is to heat benzoic acid with benzotrichloride:

$$C_6H_5CO_2H + C_6H_5CCl_3 \longrightarrow 2C_6H_5COCl + HCl$$

Benzoyl chloride is only very slowly decomposed by water or by dilute sodium hydroxide (cf. acetyl chloride), and because of this, compounds containing an active hydrogen atom can be benzoylated in the presence of dilute aqueous sodium hydroxide. This method of benzoylation is known as the Schotten-Baumann reaction, *e.g.*,

$$C_6H_5COCl + C_6H_5NH_2 + NaOH \longrightarrow C_6H_5CONHC_6H_5 + NaCl + H_2O$$

p-Nitrobenzoyl and 3,5-dinitrobenzoyl derivatives of the alcohols are usually well-defined crystalline substances, and so are used to characterise alcohols.

Hippuric acid (*benzoylglycine*), m.p. 188°C, may be readily prepared by the action of benzoyl chloride on glycine:

$$C_6H_5COCl + NH_2CH_2CO_2H \longrightarrow C_6H_5CONHCH_2CO_2H + HCl$$

Its most important reaction is its condensation with aromatic aldehydes to form acetones: *e.g.*, when heated with benzaldehyde, acetic anhydride and sodium acetate, hippuric acid forms benzoyl-α-aminocinnamic aziactone (*3-phenyl-4-benzylideneoxazol-5-one*):

$$C_6H_5CHO + \underset{\displaystyle NHCOC_6H_5}{\underset{|}{CH_2CO_2H}} \longrightarrow C_6H_5CH{=}\overset{4}{C}{-}\overset{5}{C}O \text{ (ring: } {}^4C{-}N^3{=}{}^2C(C_6H_5){-}{}^1O{-}{}^5CO)$$

This reaction is usually referred to as the **Erienmeyer aziactone synthesis** (1893). Aziactones are very important as intermediates in the preparation of amino- and keto-acids; *e.g.*, phenylalanine may be prepared from the above aziactone as follows:

$$C_6H_5CH{=}C{-}CO \text{ (ring: } C{-}N{=}C(C_6H_5){-}O{-}CO) \xrightarrow{NaOH} C_6H_5CH{=}\underset{\displaystyle NHCOC_6H_5}{\underset{|}{C}}CO_2H \xrightarrow{Na/Hg}$$

$$C_6H_5CH_2CH(NHCOC_6H_5)CO_2H \xrightarrow{HCl} C_6H_5CH_2CH(NH_2)CO_2H + C_6H_5CO_2H$$

Benzamide (*benzenecarbonamide*), m.p. 130°C, may be prepared by the action of concentrated aqueous ammonia on benzoyl chloride or by hydrolysing phenyl cyanide with warm alkaline 3 per cent hydrogen peroxide:

$$C_6H_5COCl + 2NH_3 \longrightarrow C_6H_5CONH_2 + NH_4Cl \text{ (v.g.)}$$

$$C_6H_5CN \xrightarrow{H_2O_2} C_6H_5CONH_2 \text{ (ex.)}$$

Benzamide undergoes most of the usual reactions of an aliphatic acid amide; *e.g.*, it is readily hydrolysed by dilute acids or alkalis to benzoic acid and ammonia; it forms mercury benzamide, $(C_6H_5CONH)_2Hg$, with mercuric oxide.

Benzoic anhydride, m.p. 42°C, may be prepared by heating a mixture of sodium benzoate and benzoyl chloride:

$$C_6H_5CO_2Na + C_6H_5COCl \longrightarrow (C_6H_5CO)_2O + NaCl \text{ (g.)}$$

Better yields are obtained by shaking an aqueous solution of sodium benzoate with benzoyl chloride at room temperature in the presence of pyridine (97.5%;).

An alternative method is to slowly distil a mixture of benzoic acid and acetic anhydride:

$$2C_6H_5CO_2H + (CH_3CO)_2O \longrightarrow (C_6H_5CO)_2O + 2CH_3CO_2H \quad (72\text{–}74\%)$$

McGookin *et al.* (1951) have prepared benzoic anhydride by the action of dry silver oxide or yellow mercuric oxide on benzoyl chloride in benzene:

$$2C_6H_5COCl + Ag_2O \longrightarrow (C_6H_5CO)_2O + 2AgCl$$

Benzoic anhydride is only very slowly decomposed by water (cf. acetic anhydride).

Benzoyl peroxide, m.p. 104°C, may be prepared by adding a mixture of benzoyl chloride and aqueous sodium hydroxide to cool hydrogen peroxide with vigorous shaking:

$$(C_6H_5CO)_2O_2 + CH_3ONa \longrightarrow C_6H_5COO_2Na + C_6H_5CO_2CH_3 \text{ (82\%)}$$

It is widely used as a catalyst in polymerisation processes (it readily generates the phenyl free-radical).

Perbenzoic acid, m.p. 41°C, may be prepared as its sodium salt (from which the acid is obtained by acidification):

$$(C_6H_5CO)_2O_2 + CH_3ONa \longrightarrow C_6H_5COO_2Na + C_6H_5CO_2CH \ (82\%)$$

The most satisfactory preparation appears to be that of Swem *et al.* (see peracetic acid).

Perbenzoic acid is a fairly active oxidising agent; *e.g.*, it converts the ethylenic bond quantitatively into the epoxide:

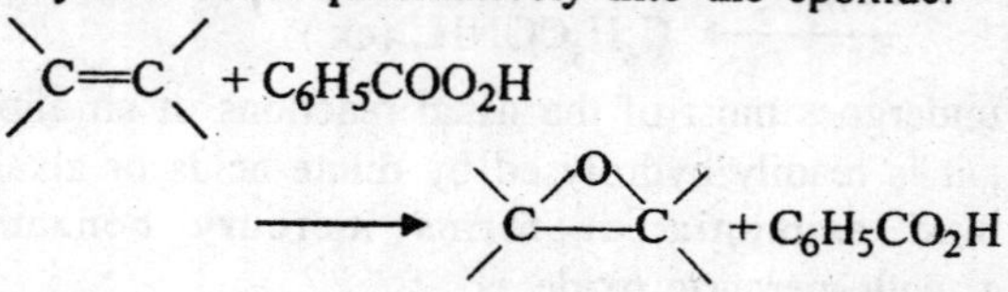

$$\gt C{=}C\lt \ + C_6H_5COO_2H \longrightarrow \gt C\overset{O}{—}C\lt \ + C_6H_5CO_2H$$

Perbenzoic acid is therefore used for the detection and estimation of ethylenic bonds (Prileschaiev reaction). It is far less troublesome to handle than peracetic acid, and so is used preferably (often as its sodium salt).

p-Nitroperbenzoic acid is better to use than perbenzoic acid, and has been prepared by treating p-nitrobenzoyl chloride with sodium peroxide in aqueous tetrahydrofuran (Vikas, 1959):

$$2p\text{-}NO_2C_6H_4COCl + Na_2O_2 \longrightarrow (p\text{-}NO_2C_6H_4CO)_2O_2 + 2NaCl$$

Phenyl cyanide (*benzonitrile, benzenecarbonitrile*) may be prepared by heating benzamide with phosphorus pentoxide, by means of the Sandmeyer reaction with benzenediazonium chloride, by fusing sodium benzenesulphonate with sodium cyanide, or by refluxing benzaldehyde in formic acid containing hydroxylamine hydrochloride and sodium acetate (van Es, 1965); the yield with aromatic aldehydes is 81-97 per cent (cf. aliphatic aldehydes). It is a colourless oil, b.p. 191°C, and behaves similarly to the aliphatic cyanides.

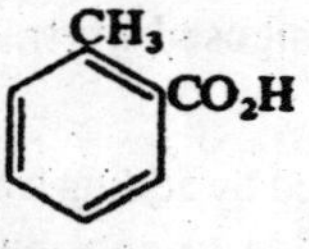

CH_3, CO_2H	CH_3, CO_2H	CH_3, CO_2H
o-toluic acid, m.p. 105°C	***m*-toluic acid, m.p. 111°C**	***p*-toluic acid, m.p. 180°C**

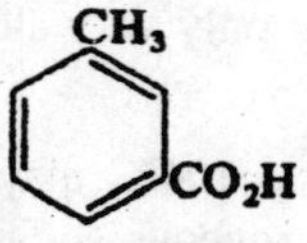

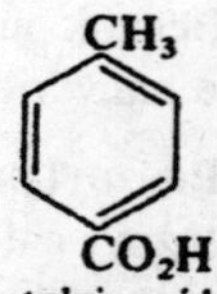

There are many homologues of benzoic acid, but only the toluic acids need be mentioned:

Each may be prepared by oxidising the corresponding xylene with dilute nitric acid, or from the corresponding toluidine (via the cyanide).

Substituted derivatives of benzoic acid. Benzoic acid is attacked by the usual electrophilic reagents chlorine, bromine, nitric and sulphuric acids, to give mainly the m-derivatives; *e.g.*, m-bromobenzoic acid may be obtained by heating benzoic acid, bromine and water under pressure:

$$C_6H_5CO_2H + Br_2 \longrightarrow m\text{-}BrC_6H_4CO_2H + HBr \quad (80\%)$$

Waters el at. (1950) have prepared m-iodobenzoic acid (75% yield) by adding iodine to a solution of benzoic acid in concentrated sulphuric acid containing silver sulphate. The-active species in this case is the iodinium ion :

$$I_2 + H_2O \rightleftharpoons HOI + H^+ + I^-$$

$$Ag_2SO_4 + 2I^- \longrightarrow 2AgI + SO_4^{2-}$$

$$HOI + H^+ \rightleftharpoons H_2OI^+ \rightleftharpoons H_2O + I^+$$

o- and p-Substituted benzoic acids may be obtained by oxidising the corresponding toluene derivatives, *e.g.*,

(i) $$o\text{-}ClC_6H_4CH_3 \xrightarrow{KMnO_4} o\text{-}ClC_6H_4CO_2H \quad (76\text{–}78\%)$$

(ii) $$p\text{-}CH_3C_6H_4HgCl \xrightarrow{KMnO_4} p\text{-}HO_2CC_6H_4HgCl \; (61\text{–}74\%) \xrightarrow{I_2/C_2H_5OH} p\text{-}IC_6H_4CO_2H \; (72\text{–}81\%)$$

Many substituted benzoic acids may also be prepared from the corresponding aminobenzoic acids (via the diazonium salts), o- and p-Substituted benzoic acids may be obtained directly by using, *e.g.*, the sodium salt of benzoic acid.

2,4,6-Tribromobenzoic acid may be prepared as follows:

$$\text{m-}H_2N\text{-}C_6H_4CO_2H \xrightarrow[Br_2]{\text{in HCl;}} \text{2,4,6-}Br_3\text{-3-}H_2N\text{-}C_6HCO_2H \xrightarrow[\text{(ii) } H_3PO_2]{\text{(i) } NaNO_2-H_2SO_4} \text{2,4,6-}Br_3C_6H_2CO_2H \quad (70\text{–}80\%)$$

Nitrobenzoic acids. o- and p-Nitrobenzoic acids may be prepared by oxidising the corresponding nitrotoluenes with acid dichromate:

$$\text{p-}O_2N\text{-}C_6H_4CH_3 \xrightarrow{[O]} \text{p-}O_2N\text{-}C_6H_4CO_2H \quad (82\text{–}86\%)$$

m-Nitrobeozoic acid may be prepared by direct nitration of methyl benzoate:

$$C_6H_5CO_2CH_3 \xrightarrow{HNO_3/H_2SO_4} \text{m-}O_2N\text{-}C_6H_4CO_2CH_3 \xrightarrow[\text{(ii) HCl}]{\text{(i) NaOH}} \text{m-}O_2N\text{-}C_6H_4CO_2H \quad (81\text{–}85\%)$$

This method is more satisfactory than that by the direct nitration of benzoic acid.

3,5-Dinitrobenzoic acid, m.p. 204°C, is formed when benzoic acid is nitrated with a mixture of fuming nitric and sulphuric acids:

$$C_6H_5CO_2H \xrightarrow{HNO_3/H_2SO_4} \text{3,5-}(NO_2)_2C_6H_3CO_2H \quad (60\%)$$

It is used to characterise alcohols.

2,5-Dinitrobenzoic acid may be prepared by treating 2-amino-5-nitrotoluene with potassium persulphate and sulphuric acid (*i.e.*, Caro's acid) and further oxidising the product, 5-nitro-2-nitrosotoluene, with potassium dichromate and sulphuric acid:

$$\text{2-}H_2N\text{-5-}O_2N\text{-}C_6H_3CH_3 \longrightarrow \text{2-ON-5-}O_2N\text{-}C_6H_3CH_3 \longrightarrow \text{2,5-}(NO_2)_2C_6H_3CO_2H \quad (55\text{–}66\%)$$

Aminobenzoic acids. All three aminobenzoic acids may be obtained by reduction of the corresponding nitrobenzoic acids.

Anthranilic acid (o-aminohenzoic acid) may be prepared by reducing o-nitrobenzoic acid.

Commercially, anthranilic acid is prepared by oxidising phthalimide with aqueous sodium hydroxide and sodium hypochlorite (the Hofmann amine preparation):

$$C_6H_4(CO)_2NH \xrightarrow{NaOH} C_6H_4(CO_2Na)(CONH_2) \xrightarrow{NaOCl} C_6H_4(CO_2Na)(NH_2) \xrightarrow{HCl} C_6H_4(CO_2H)(NH_2)$$

Anthranilic acid, m.p. 145°C, behaves as an acid and as an amine; it does not exist as a dipolar ion in the solid state (cf. sulphanilic acid). When distilled, anthranilic acid is decarboxylated to aniline.

Methyl anthranilate is used in perfumery, since it is the characteristic constituent of jasmine and orange blossoms. Anthranilic acid is used in the industrial preparation of indicating.

m-Aminobenzoic acid, m.p. 174°C, is used in the preparation of azo-dyes.

p-Aminobenzoic acid, m.p. 186°C, is one of the substances comprising the vitamin-B complex? Certain derivatives of p-aminobenzoic acid are used as local anaesthetics, *e.g.*, *novocaine*,

$$NH_2C_6H_4CO_2CH_2CH_2N(C_2H_5)_2$$

Sulphobenzoic acids. Sulphonation of benzoic acid gives o-sulphobenzoic acid. The o- and p-isomers may be prepared by oxidising the corresponding toluenesulphonic acids.

Saccharin (o-*sulphobenzoic imide*) may be prepared from o-toluenesulphonyl chloride as shown. If the solution is kept neutral during the permanganate oxidation, the product saccharin. With *alkaline* permanganate, the salt of o-sulphamidobenzoic acid is obtained and this, on acidification, gives the free acid which, on heating, forms saccharin.

$$C_6H_4(CH_3)(SO_2Cl) \xrightarrow{NH_3} C_6H_4(CH_3)(SO_2NH_2) \xrightarrow{\text{neutral } KMnO_4} C_6H_4(CO)(SO_2)NH$$

$$C_6H_4(CH_3)(SO_2NH_2) \xrightarrow{KMnO_4,\ NaOH} C_6H_4(CO_2Na)(SO_2NH_2) \xrightarrow{(i)\ H_3O^+\ (ii)\ heat} C_6H_4(CO)(SO_2)NH$$

This was one of the first industrial methods, but others are also used, *e.g.*, starting with anthranilic acid:

$$o\text{-}C_6H_4(CO_2H)(NH_2) \xrightarrow[\text{(ii) } SO_2-Cu]{\text{(i) } NaNO_2-H_2SO_4} o\text{-}C_6H_4(CO_2H)(SO_2H) \xrightarrow[NaOH]{Cl_2} o\text{-}C_6H_4(CO_2H)(SO_2Cl) \xrightarrow[\text{(ii) heat}]{\text{(i) } NH_3} C_6H_4\langle{}^{CO}_{SO_2}\rangle NH$$

Saccharin, m.p. 224°C. is about 550 times sweeter than sugar. It is almost insoluble in water and hence is sold as its sodium salt, which is very soluble. Saccharin is very sweet in dilute but is bitter in concentrated solutions. It is used instead of sugar for many purposes, *e.g.*, sweetening preserves, drinks, etc. It is also used by diabetics and obese persons.

Phenolic acids. There are three hydroxybenzoic acids; only the o-isomer, salicylic acid, is important.

Salicylic acid (o-*hydfoxybenzoic acid*) occurs as its methyl ester in many essential oils. It may be obtained by the oxidation of salicylaldehyde or salicyl alcohol; or fusing o-sulphobenzoic acid with sodium hydroxide. Salicylic acid may be prepared by replacing the amino-group in anthranilic acid by hydroxyl (via the diazonium salt), and also by the Reimer-Tiemann reaction using an alkaline solution of phenol and carbon tetrachloride:

$$C_6H_5OH \xrightarrow[CCl_4]{NaOH} o\text{-}C_6H_4(OH)(CO_2H)$$

The industrial preparation of salicylic acid is carried out by the *Kolbe-Schmitt reaction* (1885); sodium phenoxide and carbon dioxide are heated at 120-140°C under pressure.

$$C_6H_5ONa + CO_2 \longrightarrow o\text{-}C_6H_4(OH)(CO_2Na) \xrightarrow{HCl} o\text{-}C_6H_4(OH)(CO_2H)$$

A small amount of the p-derivative is formed at the same time, and if the temperature rises above 140°C, the p-isomer is the main product.

The mechanism of the Kolbe-Schmitt reaction is uncertain; a possibility is:

$$C_6H_5O^- + O{=}C{=}O \longrightarrow \left[\text{cyclohexadienone with } \overset{}{C}(H)(CO_2^-) \text{ adjacent to } C{=}O \right] \longrightarrow o\text{-}C_6H_4(OH)(CO_2^-).$$

Salicylic acid, m.p. 159°C, is used as an antiseptic, in medicine, and in the preparation of azo-dyes.

Salicylic acid behaves as a phenol and as an acid. Its aqueous solutions give a violet colour with ferric chloride. When heated quickly, salicylic acid sublimes; but when heated slowly, it undergoes decarboxylation:

OH, CO_2H → OH + CO_2

When reduced with sodium and isopentanol, salicylic acid is converted into pimelic acid. The mechanism of this reaction is uncertain; one that has been proposed is:

CO_2H, OH $\xrightarrow{e;\ H^+}$ CO_2H, OH ⇌ H, CO_2H, O (19) $\xrightarrow[\text{(acid)}]{H_2O}$ CH_2CO_2H, CH_2, CO_2H, CH_2, CH_2, CH_2

The final step is 'acid hydrolysis' of a β-keto-acid (cf. acetoacetic ester syntheses).

Bromination and nitration of salicylic acid cause displacement of the carboxyl group with formation of the corresponding 2,4,6-trisubstituted phenol.

Br, OH, Br, Br $\xleftarrow[H_2O]{Br_2}$ OH, CO_2H $\xrightarrow[HNO_3]{\text{fuming}}$ O_2N, OH, NO_2, NO_2

The reason for this ready displacement of the carboxyl group by bromine or by the nitro-group is not certain, but it may possibly occur by the following mechanism:

HO—C(=O)O⁻ + Br—Br → $\bar{B}r$ + HO=, C(=O)O⁻, Br → HO—Br + CO_2

Decarboxylation in alkaline solution may be explained similarly (a proton from water is the attacking agent instead of Br^+).

These displacements of the carboxyl group occur when the reactions are carried out in aqueous solution; they do not occur in acetic acid solution, *e.g.*,

When treated with carbonates, only the carboxyl group in salicylic acid forms salts; with alkali, both the carboxyl and hydroxyl groups form salts. Sodium salicylate is used in the treatment of rheumatism. Salicylic acid is a stronger acid than its m- and p-isomers (see the *ortho*-effect, below).

Methyl salicylate, o-$HOC_6H_4CO_2CH_3$, b.p. 224°C, is the principal constituent of oil of wintergreen. It may be prepared by direct esterification, and is used in perfumery and as a flavouring material. *Salol (phenyl salicylate)*, o-$HOC_6H_4CO_2C_6H_5$; m.p. 43°C, may be prepared by heating salicylic acid with phenol in the presence of phosphoryl chloride. It is used as an internal antiseptic.

Aspirin (acetylsalicylic acid), o-$CH_3COOC_6H_4CO_2H$, m.p. 135°C, may be prepared by acetylating salicylic acid with a mixture of acetic anhydride and glacial acetic acid.

An interesting point about the alkaline hydrolysis of aspirin is that it occurs by the following intra-molecular mechanism (St. Pierre et at., 1968; Kirby *et al.*, 1968):

m-Hydroxybenzoic acid, m.p. 201°C, may be prepared from m-aminobenzoic acid:

It does *not* give a coloration with ferric chloride.

p-Hydroxybenzoic acid, m.p. 214°C, may be prepared by heating potassium salicylate at 230°C:

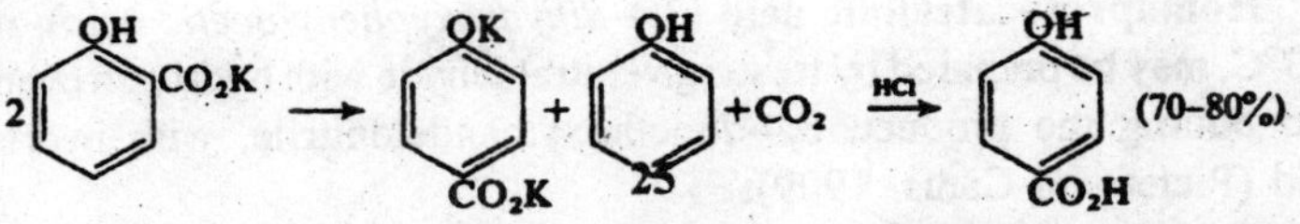

It may also be prepared by oxidising the methyl group in **petrosal** (with protection of the hydroxyl group during the oxidation):

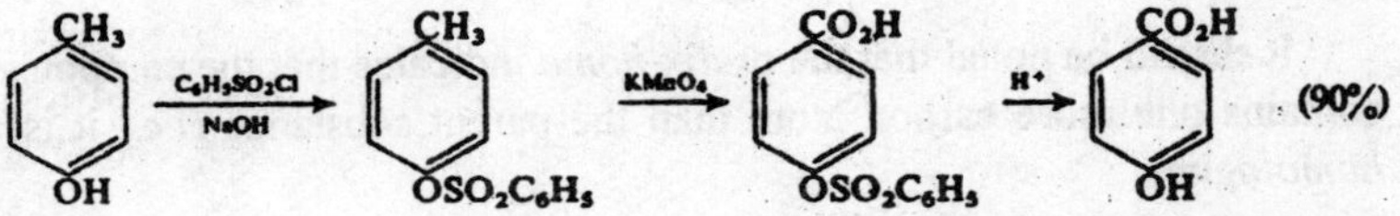

Many esters of p-hydroxybenzoic acid have antiseptic properties. The acid gives a red colour with ferric chloride.

Anisic acid *(p-methoxybenzoic acid)*, m.p. 184°C, may be prepared by the oxidation of anethole. When heated with concentrated hydriodic acid, anisic acid forms p-hydroxybenzoic acid and methyl iodide, and on distillation with calcium oxide it forms anisole.

Protocatechuic acid (3,4-*dihydroxybenzoic acid*) occurs in various plants, and is formed when certain resins (catechin, gum-benzoin, etc.) are fused with alkali. It may be prepared by heating catechol with aqueous ammonium carbonate at 140°C under pressure, or from vanillin as follows (note the demethylation):

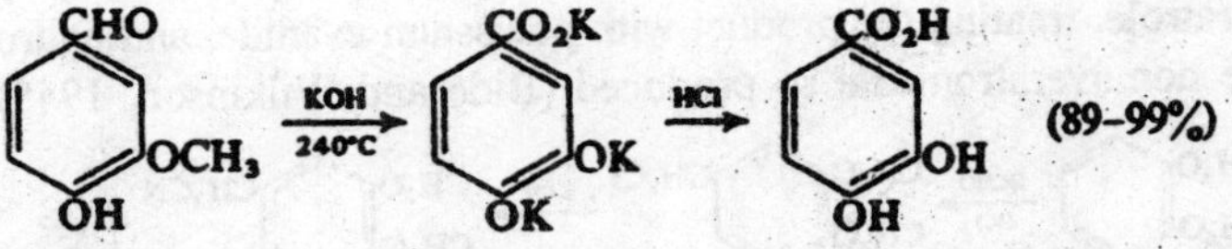

Protocatechuic acid crystallises with one molecule of water of crystallisation; the anhydrous acid melts at 199°C with decomposition into catechol and carbon dioxide.

Veratric acid (3,4-*dimethoxybenzoic acid*), m.p. 181°C, may be prepared by methylating proto-catechuic acid with methyl sulphate in the presence of aqueous sodium hydroxide:

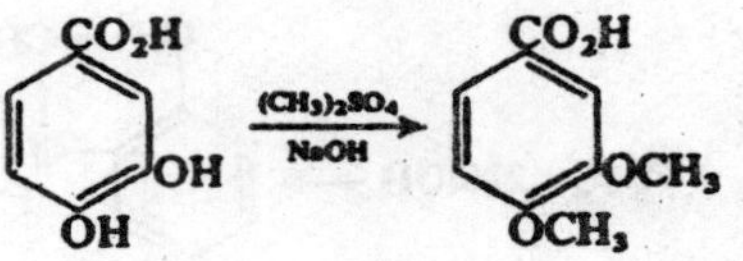

Homoprotocatechuic acid (3,4-*dihydroxyphenylacetic acid*), m.p. 127°C, may be prepared by treating veratraldehyde with hydrogen cyanide and boiling the product, 3,4-dimethoxymandelonitrile, with hydrodic acid (Pictet and Cams. 1909):

$(CH_3O)_2C_6H_3CHO \xrightarrow{HCN} (CH_3O)_2C_6H_3CH(OH)CN \xrightarrow{HI} (HO)_2C_6H_3CH_2CO_2H$

It should be noted that the prefix *homo* indicates that the compound contains one more carbon atom than the parent substance (*i.e.*, it is a *homologue*).

Homoveratric acid (3, 4-*dimethoxyphenylucetic acid*), m.p. 99°C, may be prepared by methylating homoprotocatechuic acid with methyl sulphate in the presence of aqueous sodium hydroxide. It may be prepared via the aziactone synthesis as follows:

$(CH_3O)_2C_6H_3CHO + CH_2(CO_2H)NHCOC_6H_5 \xrightarrow[CH_3CO_2Na]{(CH_3CO)_2O} (CH_3O)_2C_6H_3CH{=}C\text{—}CO\text{—}O\text{—}C(C_6H_5){=}N \xrightarrow[reflux]{NaOH}$

$(CH_3O)_2C_6H_3CH_2COCO_2Na \xrightarrow[NaOH]{H_2O_2} (CH_3O)_2C_6H_3CH_2CO_2Na$

Homoveratric acid may also be prepared by chloromethylating veratrole, treating the product with potassium cyanide, and hydrolysine the homoveratronitrile so produced (Bide and Wilkinson, 1945);

$(CH_3O)_2C_6H_4 \xrightarrow[HCl]{HCHO} (CH_3O)_2C_6H_3CH_2Cl \xrightarrow{KCN} (CH_3O)_2C_6H_3CH_2CN \xrightarrow{H^+} (CH_3O)_2C_6H_3CH_2CO_2$

Piperonylic acid (3,4-*methylenedioxybenzoic acid*), m.p. 229°C, may be prepared by the oxidation of piperonal, or by heating protocatechuic acid with methylene di-iodide and aqueous sodium hydroxide:

$(OH)_2C_6H_3CO_2H + CH_2I_2 + 2NaOH \longrightarrow (CH_2O_2)C_6H_3CO_2H + 2NaI + 2H_2O$

Piperonal and piperonylic acid are oxidation products of piperic acid, which occurs in the alkaloid piperine. Veratric acid is one of the products of oxidation of the alkaloid papaverine. The orientation of the groups in these acids (and in the corresponding homo-acids) may be shown as follows. Sulphonation of p-hydroxybenzoic acid (the orientation of which is known) produces a sulphonic acid derivative in which, according to the rules of direction power. The sulphonic acid group is o- to the hydroxyl group.

This derivative, on fusion with potassium hydroxide, produces the dihydroxybenzoic acid in which the two hydroxyl groups are o- to each other. This is confirmed by the fact that the acid, viz., protocatechuic acid, gives catechol (orientation known) when heated to its melting point:

$$\text{(CO}_2\text{H, OH)} \xrightarrow{H_2SO_4} \text{(CO}_2\text{H, SO}_3\text{H, OH)} \xrightarrow{KOH} \text{(CO}_2\text{H, OH, OH)} \xrightarrow{\text{heat}} \text{(OH, OH)}$$

Thus, protocatechuic acid must be 3,4-dihydroxybenzoic acid. Hence the dimethoxy-derivative, veratric acid, must be 3,4-dimethoxybenzoic acid. At the same time, the orientations of the corresponding aldehydes, protocatechualdehyde and veratraldehyde, are established. Similarly, since piperonylic acid is the methylene ether of protocatechuic acid (shown by analytical and synthetic evidence), its orientation is also established. Finally, since homoprotocatechuic acid (and its methylated derivative, homoveratric acid) can be synthesised from veratraldehyde (see above), its orientation is thus ascertained.

Gallic acid (3,4,5-*trihydroxlbenzoic acid*), m.p. 253°C, occurs in the free state in tea and many plants. It is best prepared by boiling tannin (tannic acid) with dilute acids. When heated at its melting point, it forms pyrogallol and carbon dioxide. It is a powerful reducing agent and hence is used as a photographic developer. It is readily soluble in water, and its aqueous solution gives a bluish-black precipitate with ferric chloride. This property is used in the manufacture of ink.

Strengths of Aromatic Acids

It has been pointed out that replacement of the hydrogen atom in formic acid by an alkyl group weakens the strength of the acid, and the greater the + I effect of the R group, the weaker is the acid. Phenylacetic acid, $PhCH_2CO_2H$ (pK_a 4.31) is stronger than acetic acid (pK_a 4.76), and therefore the phenyl group has a –I effect. On the other hand, benzoic acid (pK_a 4.17) is weaker than formic acid (pK_a 3.75). In this case, the phenyl group has a + I effect (which is smaller than that of methyl group). These apparently contradictory results may be explained as follows. When the carboxyl group is *directly* attached to the nucleus, the resonance effect (+ R) overcomes the – I effect (in phenylacetic acid the phenyl group is insulated from the carboxyl group by a CH_2 group, and so the +R effect is not possible):

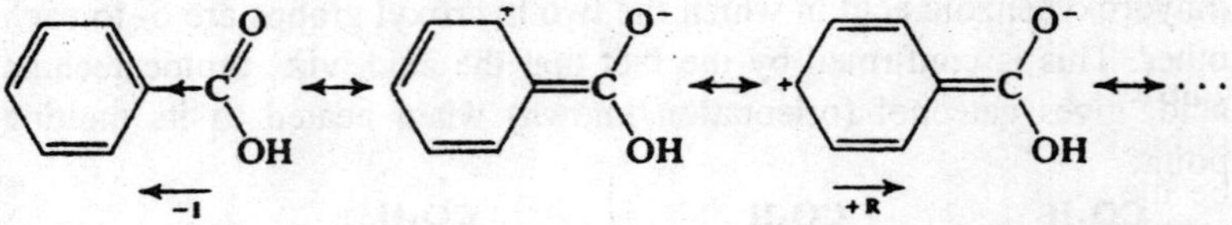

This prevents, to a large extent, the lone pair on the O atom of the OH group from entering into resonance with the CO group. The result is a smaller positive charge on the O atom of the OH group, and so proton release is more difficult than in formic acid. The fact that benzoic acid is stronger than acetic acid means that [–I + (+R)] < + I of the methyl group.

The same arguments may be applied to ionised benzoic acid.

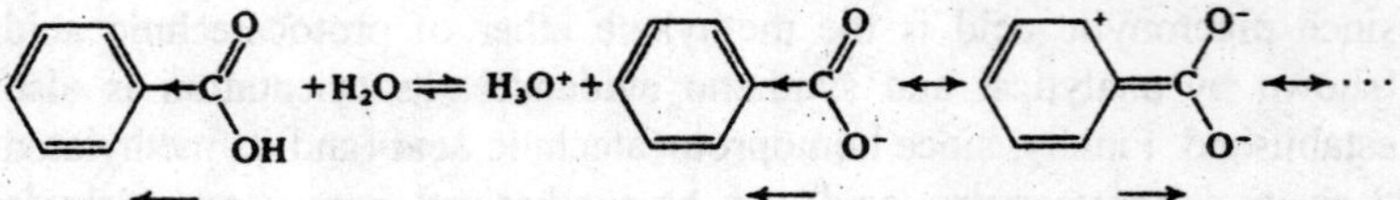

Now let us consider substituted benzoic acids (Table 4.1). At this point we shall discuss effects from the m- and p-positions. The methyl group decreases the strength of benzoic acid from these positions. The methyl group has a +1 effect, and so the net result is to increase the + R effect in the m- or p-tolyl group. Thus the two acids are weaker than benzoic acid. Resonance of the carboxyl group with the ring is from 'ring to the group' and not the other way round.

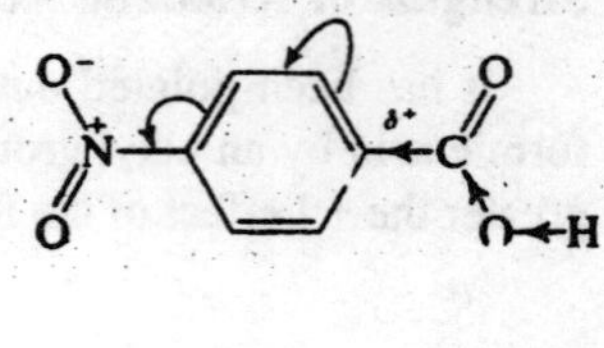

Thus, any substituent group that reduces the electron density at the carbon atom attached to the carboxyl group will cause a withdrawal of electrons from the O atom of the OH group, and consequently facilitate proton release, thereby increasing the strength of the acid. Such a substituent is the nitro-group. This has a –I effect, and this is greater from the p-position than from the m-. Therefore p-nitrobenzoic acid is stronger than the m-isomer.

Groups such as MeO can enter into resonance with the carboxyl group in the p-position, but not from the m-, in which case it is the –I effect that operates. Thus the p-substituted acid will be expected to be weaker and the m-substituted acid stronger than benzoic acid (Table 4.1; see also the Hammett equation). With chlorine as substituent, since the + R effect is very small, the –I effect controls acidity, and so m- and p-chlorobenzoic acid are stronger than benzoic acid.

The alternative way of explaining the strengths of these acids is to consider the stabilising effect of the substituent on the anion. The more stable the anion is with respect to the un-ionised acid, the stronger will be the acid +I and/or + R groups will oppose, whereas –I and/or –R groups will assist (by dispersing the negative charge), the +I effect of the *carboxylate ion group*. Hence, the former substituents weaken and the latter strengthen the acid.

Table 4.1

Substituent	*pKa* *o-*	*m-*	*p-*
None: 4.17			
Me	3.91	4.27	4.37
NO_2	2.17	3.49	3.43
F	3.27	3.87	4.14
Cl	2.94	3.83	3.98
OMe	4.09	4.09	4.47
OH	2.98	4.08	4.58
NH_2	4.98	4.79	4.92
2, 4, 6-tri-NO_2: 0.65			

The *ortho*-effect. This is a special effect that is shown by o-substituents, but is not necessarily just a steric effect, *e.g.*, the basicities of some o-substituted anilines were explained in terms of steric effects and differences in crowding round the nitrogen atom. This *ortho*-effect also operates with the benzoic acids. Examination of Table 4.1 shows irrespective of the polar type, nearly all o-substituted benzoic acids are stronger than benzoic acid. As we have seen, benzoic acid is a resonance hybrid, and so the carboxyl group is coplanar with the ring. An o-substituent tends to prevent this coplanarity. Thus resonance is diminished (or prevented), and so the O atom of the OH group has a greater positive charge, resulting in increased acid strength. It follows from this that the greater the steric inhibition of resonance, the stronger is the acid. Support for this is the following order of strengths of substituted benzoic acids:

2, 6-di-Me (pK_a 3.21) > 2-t-Bu (pK_a 3.46) > 2-Me (pK_a 3.91)

Here again, if we consider the stability of the anion, steric inhibition of resonance prevents the +R effect of the *ring* coming into operation (see above), and since this weakens acid strength, its absence results in increased acid strength.

o-Hydroxybenzoic acid (salicylic acid) is far stronger than the corresponding m- and -isomers. Steric inhibition of resonance cannot explain this very large increase, since the corresponding methoxybenzoic acids all have similar strengths. The explanation offered is hydrogen bonding; H of the o-OH group can form a hydrogen bond with the carboxyl group, whereas the Me of OMe cannot. The carboxylate ions of o-hydroxybenzoic acids are stabilised by intramolecular hydrogen bonding, and support for this is given by the following order of acid strength:

2,6-di-OH (pKa 2.30) > 2-OH (pK_a 2.98) > benzoic acid (pK_a 4.17)

It can be seen that two hydrogen bonds would be expected to bring about more stabilisation than one hydrogen bond:

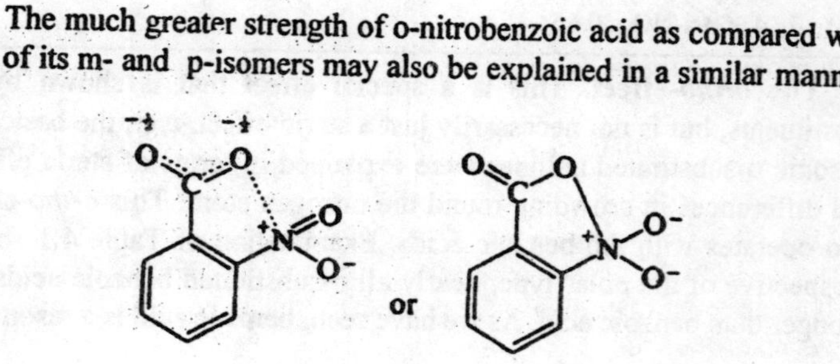

The much greater strength of o-nitrobenzoic acid as compared with that of its m- and p-isomers may also be explained in a similar manner:

or

The actual geometry of these highly substituted benzenes is an interesting problem. In benzene the ring is flat and the hydrogen atoms are coplanar with this ring. Electron-diffraction investigations polyhalogenobenzenes suggest that such molecules are non-planar (Hassel *et al.*, 1947), whereas X-ray studies indicate that in the solid state such molecules are very closely or even exactly planar (Tulinsky *et al.*, 1958; Gafner *et.al.*, 1960). Ferguson *et al.* (1959, 1961) have examined, by X-ray analysis, various substituted benzoic acids, *e.g.*, o-chloro- and bromo-benzoic acid, and 2-chloro-5-nitrobenzoic acid. In all three molecules the steric strain is relieved by small out-of-plane displacements of the exocylic valency bonds in addition to the larger in-plane displacement of these bonds away from one another This type of deformation is referred to as molecular overcrowding.

MONOCARBOXYLIC ACIDS WITH THE CARBOXYL GROUP IN THE SIDE-CHAIN

Phenylacetic acid (α-*toluic acid*), m.p. 77°C, may be prepared from benzyl chloride as follows:

$$C_6H_5CH_2Cl \xrightarrow{KCN} C_6H_5CH_2CN \xrightarrow[(H_2SO_4)]{H_2O} C_6H_5CH_2H\ (60\text{–}70\%)$$

It may also be prepared by reducing mandelonitrile with hydrodic acid and red phosphorus:

$$C_6H_5CH(OH)CN \xrightarrow{HI/P} C_6H_5CH_2CO_2H \quad (90\%)$$

Phenylacetic acid contains two (x-hydrogen atoms and so is reactive in the side-chain as well as in the ring, *e.g.*, chlorination in the cold in the presence of a halogen carrier, gives o- and p-nuclear substitution; chlorination at the boiling point of the acid, in the absence of a halogen carrier, gives replacement of the a-hydrogen (cf. toluene). When oxidised with chromic acid, phenylacetic acid is converted into benzoic acid.

Mandelic acid (*phenylglycollic acid*) may be obtained from the glucoside amygdalin by regulated hydrolysis (cf. benzaldehyde). It may be prepared as follows:

$$C_6H_5CHO \xrightarrow{NaHSO_3} C_6H_5CH(OH)SO_3Na \xrightarrow{NaCN}$$

$$C_6H_5CH(OH)CN \xrightarrow[(HCl)]{H_2O} C_6H_5CH(OH)CO_2H \quad (50\text{-}52\%)$$

Mandelic acid may also be prepared by the hydrolysis of phenacylidene chloride.

Mandelic acid behaves as a hydroxy acid, and is optically active. The acid obtained from amygdalin is laevorotatory, m.p. 133°C; the m.p. of the acid is 118°C. On vigorous oxidation, mandelic acid gives benzoic acid.

Hydratropic acid (*α-phenylpropionic acid*), $C_6H_5CH(CH_3)CO_2H$, b.p. 265°C, may be prepared by the reduction of *atropic acid,* $C_6H_5C(=CH_2)CO_2H$, obtained by heating *tropic acid,* $C_6H_5CH(CH_2OH)CO_2H$, which occurs in the alkaloid atropine.

β-Phenylpropionic acid (*hydrocinnamic acid*), m.p. 47°C, may be prepared by reducing cinnamic acid with sodium amalgam:

$$C_6H_5CH = CHCO_2H \xrightarrow{e;\,H^+} C_6H_5CH_2CH_2CO_2H \text{ (85\%)}$$

It may also be synthesised from benzyl chloride and malonic ester:

$$C_6H_5CH_2Cl + [CH(CO_2C_2H_5)_2]^- Na^+ \longrightarrow C_6H_5CH_2CH(CO_2C_2H_5)_2 + NaCl$$

$$\xrightarrow[\text{(ii) boil}]{\text{(i) acid}} C_6H_5CH_2CH_2CO_2H + CO_2 + 2C_2H_5OH$$

β-Phenylpropionic acid is oxidised by chromic acid to benzoic acid.

Cinnamic acid is the *trans*-isomer, and hence is also known as **trans-β-phenylacrylic acid**. The cis-isomer (cis-β-*phenylacrylic acid*) has also been called allocinnamic acid:

$$\begin{array}{ccc} C_6H_5\ \ \ \ H & & C_6H_5\ \ \ \ H \\ \diagdown\ \diagup & & \diagdown\ \diagup \\ C & & C \\ \| & & \| \\ C & & C \\ \diagup\ \diagdown & & \diagup\ \diagdown \\ H\ \ \ \ CO_2H & & HO_2C\ \ \ \ H \end{array}$$

trans-cinnamic acid   cis-cinnamic acid

trans-Cinnamic acid, m.p. 133°C, is the form that occurs naturally (free and as esters) in balsams and resins. It may be prepared:

(i) By Perkin's reaction

$$C_6H_5CHO + (CH_3CO)_2O \xrightarrow{CH_3CO_2Na} C_6H_5CH = CHCO_2H \text{ (85\%)}$$

(ii) By *Knoevenagel's reaction*

$$C_6H_5CHO + CH_2(CO_2H) \xrightarrow[\text{heat}]{NH_3/C_2H_5OH} C_6H_5CH{=}CHCO_2H \text{ (80\%)}$$

(iii) By the *Claisen condensation* between benzaldehyde and ethyl acetate in the presence of sodium ethoxide (since benzaldehyde is one of the reactants, this is also referred to as a *Claisen reaction* :

$$C_6H_5CHO + CH_3CO_2C_2H_5 \xrightarrow{C_2H_5ONa} C_6H_5CH = CHCO_2C_2H_5$$

$$\xrightarrow[\text{(acid)}]{H_2O} C_6H_5CH = CHCO_2H + C_2H_5OH \text{ (68–74\%)}$$

(iv) By the oxidation of benzylideneacetone with sodium hypochlorite (cf. the haloform reaction):

$$C_6H_5CH = CHCOCH_3 \xrightarrow{NaOCl} C_6H_5CH = CHCO_2H$$

Methods (i) and (iv) are used industrially.

Cinnamic acid behaves as an α, β-unsaturated acid, and as a benzene derivative. It is reduced by sodium amalgam to γ-phenylpropionic acid; reduction of its ethyl ester, however, leads to the formation of the bimolecular product, ethyl β, β-diphenyladipate :

$$2C_6H_5CH = CHCO_2C_2H_5 \xrightarrow{e;\,H^+} \begin{array}{c} C_6H_5CHCH_2CO_2C_2H_5 \\ | \\ C_6H_5CHCH_2CO_2C_2H_5 \end{array}$$

When a phenyl group is attached to the β-carbon atom of an α, β-unsaturated carbonyl compound, both the double bond and carbonyl group are reduced by lithium aluminium hydride :

$$C_6H_5CH = CHCO_2H \xrightarrow{LiAlH_4} C_6H_5CH_2CH_2CH_2OH$$

By using the inverse addition at –10°C, the double bond is left intact, *e.g.*,

$$C_6H_5CH = CHCO_2C_2H_5 \xrightarrow[-10\,^\circ C]{LiAlH_4} C_6H_5CH = CHCH_2OH$$

When oxidised with chromic acid, cinnamic acid forms a mixture of benzaldehyde and benzoic acid. Concentrated nitric acid nitrates it to a mixture of o- and p-nitrocinnamic acids, the latter predominating. m-Nitrocinnamic acid may be prepared by the Perkin reaction with m-nitro benzaldehyde (74-77%).

When dry-distilled, cinnamic acid forms styrene:

$$C_6H_5CH = CHCO_2H \longrightarrow C_6H_5CH = CH_2 + CO_2$$

When exposed to sunlight, cinnamic acid dimerises to a mixture of *truxinic acid* (3,4-*diphenylcyclo- butane*-1,2-*dicarboxylic acid*) (I), and *truxillic acid* (2,4-diphenylcyclobutane-l,3-dicarboxylic acid) (II)

$$2C_6H_5CH=CHCO_2H \longrightarrow \begin{array}{l} C_6H_5CH-CHCO_2H \\ \quad | \qquad | \\ C_6H_5CH-CHCO_2H \\ \qquad (I) \end{array} + \begin{array}{l} C_6H_5CH-CHCO_2H \\ \quad | \qquad | \\ HO_2CCH-CHC_6H_5 \\ \qquad (II) \end{array}$$

***cis*-Cinnamic acid.** m.p. 68°C, may be prepared by exposing cinnamic acid in benzene to ultraviolet light, or by the partial reduction of phenylpropiolic acid with hydrogen and colloidal platinum as catalyst:

$$C_6H_5C \equiv CCO_2H + H_2 \xrightarrow{Pt} C_6H_5CH=CHCO_2H$$

cis-Cinnamic acid readily changes into trans-*cinnamic* acid, and under the influence of light dimerises to truxinic acid.

Phenylpropiolic acid (*phenylpropynoic acid*), m.p. 136°C, may be prepared by boiling the ethyl ester of cinnamic acid dibromide with ethanolic potassium hydroxide:

$$C_6H_5CH=CHCO_2C_2H_5 \xrightarrow[(CCl_4)]{Br_2} C_6H_5CHBrCHBrCO_2C_2H_5$$

$$\xrightarrow[(ii)\ acid]{(i)\ KOH/C_2H_5OH} C_6H_5C \equiv CCO_2H$$

Harley-Mason *et al.* (1961) have prepared phenylpropiolic acid by the action of -bromobenzene- sulphonyl chloride (*brosyl chloride*), BsCl, on diethyl sodiobenzoylmalonate, and treating the brosylate with sodium hydroxide in aqueous dioxan at room temperature. This is one of the mildest methods for introducing a triple bond into a compound.

$$PhCOCH(CO_2Et)_2 \xrightarrow[(ii)\ BsCl]{(i)\ EtONa} \underset{\displaystyle OBs}{PhC}=C(CO_2Et)_2$$

$$\xrightarrow{OH^-} \underset{\displaystyle OBs}{PhC}=C\begin{array}{l} C(=O)O^- \\ CO_2^- \end{array}$$

$$\longrightarrow PhC \equiv CCO_2^- + CO_2 + BsO^-$$

On catalytic reduction (using colloidal platinum), phenylpropiolic acid forms *cis*-cinnamic acid. Reduction with zinc and acetic acid gives trans-cinnamic acid, and with sodium amalgam, β-phenyl-propionic acid. This acid and cis-cinnamic acid are produced when di-imide is the reducing agent. When refluxed with barium hydroxide solution, phenylpropiolic acid is converted into phenylacetylene.

$$C_6H_5C \equiv CCO_2H \longrightarrow CCO_2H \longrightarrow C_6H_5C \equiv CH + CO_2$$

o-Coumaric acid (*trans-o-hydroxycimamic acid*), m.p. 208°C, may be prepared by boiling coumarin (see below) with sodium ethoxide, or by diazotising o-aminocinnamic acid and then heating the diazonium sulphate solution. When exposed to ultraviolet light, o-coumaric acid (the stable *trans*-form) is converted into coumarin via coumarinic acid (the unstable cis form of the acid).

Coumarinic acid (*cis-o-hydroxycinnamic acid*) is unstable, spontaneously forming its δ-lactone as soon as it is set free from its salts; the lactone is known as coumarin (1,2-benzopyrone), m.p. 69°C:

H C H CO_2H OH $\xrightarrow{-H_2O}$ O O

Coumarin may be prepared by the Perkin reaction.

CHO OH $\xrightarrow[CH_3CO_2Na]{(CH_3CO)_2O}$ CH=CHCO$_2$Na OCOCH$_3$ $\xrightarrow{H_3O^+}$ O O

One of the most convenient methods for synthesising coumarins is the **Pechmann reaction** (1883) which is the condensation of a phenol with a β-keto-ester. The usual condensing reagents are sulphuric acid, phosphorus pentoxide, etc., but Koo (1955) has shown that polyphosphoric acid is very effective, *e.g.*, resorcinol and ethyl acetoacetate give 7-hydroxy-4-methylcoumarin:

HO OH + CH_2(COCH$_3$)(CO$_2$C$_2$H$_5$) $\longrightarrow$ HO (7) ... CH_3 (4) ... O (1) O (2) (98%)

Coumarin is a natural perfume, and is also used as an artificial flavour. When heated with sodium ethoxide, it is converted into coumaric acid.

BENZENEDICARBOXYLIC ACIDS

Phthalic acid (*benzene*-1,2-*dicarboxylicacid*) may be prepared by the oxidation of any benzene derivative having only two side-chains in the o-positions; permanganate or dilute nitric acid is used as the oxidising agent (chromic acid brings about ring rupture):

$$C_6H_4(CH_3)_2 \xrightarrow{KMnO_4} C_6H_4(CO_2H)_2 \quad (v.g.)$$

Industrially, phthalic acid is prepared by passing naphthalene vapour or o-xylene mixed with air over vanadium pentoxide as catalyst at 400-500°C:

$$C_{10}H_8 \xrightarrow[V_2O_5]{O_2} C_6H_4(CO)_2O + 2CO_2 + 2H_2O \quad (80\%)$$

The anhydride is converted into the acid by heating it with alkali, and then acidifying.

$$C_6H_4(CO)_2O + 2NaOH \longrightarrow C_6H_4(CO_2Na)_2 \xrightarrow{HCl} C_6H_4(CO_2H)_2$$

Phthalic acid is a white crystalline solid, m.p. 231°C (rapid heating), with conversion into its anhydride. It undergoes most of the typical reactions of a dicarboxylic acid. When heated with potassium hydroxide, it is decarboxylated to benzene. It is reduced by sodium amalgam to di- tetra- and hexahydrophthalic acids

Mercuration of phthalic acid is usually carried out by refluxing the sodium salt with aqueous mercuric acetate containing sodium acetate, until no ionic mercury remains in the solution. Phthalic acid may also be mercurated by fusing it with mercuric acetate, or by heating the mercury salt of phthalic acid until the mercury has been transferred to the nucleus:

$$C_6H_4(COO)_2Hg \longrightarrow C_6H_4(CO{-}O{-}Hg) + CO_2$$

Phthalic anhydride. m.p. 128°C, is prepared industrially by the oxidation of naphthalene or o-xylene. It is slowly hydrolysed by water but rapidly by alkalis or acids. When nitrated, phthalic anhydride forms a mixture of 3- and 4-nitrophthalic acids. When these are reduced, their corresponding aminophthalic acids spontaneously decarboxylate to form m-aminobenzoic acid:

$$\text{C}_6\text{H}_3(\text{NH}_2)(\text{CO}_2\text{H})_2 \xrightarrow{-CO_2} \text{C}_6\text{H}_4(\text{NH}_2)\text{CO}_2\text{H} \xleftarrow{-CO_2} \text{NH}_2\text{C}_6\text{H}_3(\text{CO}_2\text{H})_2$$

It is the carboxyl group o- or p- to the amino-group that is lost.

Phthalic anhydride undergoes a large number of condensation reactions (see text). Phthalic anhydride and phthalic acid are used industrially in the preparation of dyes, plastics (glyptals), plasticisers, benzoic acid, etc.

Phthalide may be prepared by reducing phthalic anhydride with zinc dust and aqueous sodium hydroxide:

$$\text{C}_6\text{H}_4(\text{CO})_2\text{O} \xrightarrow{e;\ H^+} \text{C}_6\text{H}_4\begin{matrix}\text{CH}_2\\ \text{CO}\end{matrix}\!\!>\text{O} + \text{H}_2\text{O}$$

It may also be prepared by reducing phthalimide with a zinc-copper alloy and aqueous sodium hydroxide:

$$\text{C}_6\text{H}_4(\text{CO})_2\text{NH} \xrightarrow{e;\ H^+} \text{C}_6\text{H}_4(\text{CO}_2\text{Na})\text{CH}_2\text{OH} \xrightarrow{HCl} \text{C}_6\text{H}_4\begin{matrix}\text{CO}\\ \text{CH}_2\end{matrix}\!\!>\text{O} \quad (66\text{–}71\%)$$

Phthalide, m.p. 75°C, may be converted into *homophthalic acid*, m.p. 175°C, by fusing it with potassium cyanide at 180°C and hydrolysing the product with boiling 50 per cent sulphuric acid:

$$\text{C}_6\text{H}_4\begin{matrix}\text{CH}_2\\ \text{CO}\end{matrix}\!\!>\text{O} \xrightarrow{KCN} \underset{(67\text{–}83\%)}{\text{C}_6\text{H}_4(\text{CH}_2\text{CN})\text{CO}_2\text{K}} \xrightarrow[(H_2SO_4)]{H_2O} \underset{(71\text{–}75\%)}{\text{C}_6\text{H}_4(\text{CH}_2\text{CO}_2\text{H})\text{CO}_2\text{H}}$$

Phthalimide, m.p. 238°C, may be prepared by heating phthalic anhydride with aqueous ammonia:

$$\text{C}_6\text{H}_4(\text{CO})_2\text{O} + \text{NH}_3 \longrightarrow \text{C}_6\text{H}_4(\text{CO})_2\text{NH} + \text{H}_2\text{O} \quad (95\text{–}97\%)$$

Alternatively, it may be prepared by fusing the anhydride with ammonium carbonate. It is weakly acidic; *e.g.*, with ethanolic potassium hydroxide it forms potassium phthalimide:

$$C_6H_4(CO)_2NH + KOH \longrightarrow C_6H_4(CO)_2N^-K^+ + H_2O \quad (68\%)$$

This salt is used in Gabriel's synthesis of primary amines and for preparing many α-aminoacids. The salt is decomposed into phthalimide by carbon dioxide.

Treatment of phthalimide with alkaline sodium hypochlorite results in the formation of anthranilic acid. When hydrolysed with warm aqueous sodium hydroxide, phthalimide forms phthalic acid.

If, however, phthalimide is allowed to stand in cold aqueous potassium hydroxide, or if warmed with barium hydroxide solution, it is converted into phthalamic acid:

$$C_6H_4(CO)_2NH + H_2O \longrightarrow C_6H_4(CO_2H)(CONH_2)$$

Phthaloyl chloride (phthalyl chloride) may be prepared by heating phthalic acid or phthalic anhydride with phosphorus pentachloride at 150°C. It may also be prepared by heating phthalic acid with thionyl chloride in the presence of zinc chloride at 220°C:

$$\underset{(92\%)}{C_6H_4(COCl)_2} \xleftarrow{PCl_5} C_6H_4(CO_2H)_2 \xrightarrow[ZnCl_2]{SOCl_2} \underset{(86\%)}{C_6H_4(COCl)_2}$$

Phthaloyl chloride is a colourless oily liquid, m.p. 15-16°C. When heated with aluminium chloride for some time, it is converted into as-phthaloyl chloride, m.p. 89°C (cf. succinyl chloride):

When reduced with zinc and hydrochloric acid, phthaloyl chloride is converted into phthalide.

$$C_6H_4(COCl)_2 \xrightarrow{AlCl_3} C_6H_4\begin{matrix}CCl_2\\ \diagdown \\ O \\ \diagup \\ CO\end{matrix} \quad (72\%)$$

Phthalaldehydic acid (*o-formylbenzoic acid*), m.p. 100-5°C, may be prepared by oxidising naphthalene with alkaline permanganate and decomposing the product, phthalonic acid, by boiling it in xylene solution:

$$C_{10}H_8 \xrightarrow{[O]} C_6H_4(COCO_2H)(CO_2H) \longrightarrow C_6H_4(CHO)(CO_2H) + CO_2$$

(40–50%) (65%)

It may also be prepared by passing bromine vapour into phthalide and heating the product, 2-bromophthalide with water:

$$C_6H_4\begin{matrix}CO\\CH_2\end{matrix}\!\!>O \xrightarrow{Br_2} C_6H_4\begin{matrix}CO\\CHBr\end{matrix}\!\!>O \xrightarrow{H_2O} C_6H_4(CO_2H)(CHO) + HBr$$

(82–83%) (65–68%)

Phthalaldehyde, m.p. 56°C, may be prepared by ozonolysis of naphthalene, or by the reduction of phthalobisdimethylamide with lithium aluminium hydride (Weygand *et al.*, 1951).

$$C_6H_4(CON(CH_3)_2)_2 \xrightarrow{LiAlH_4} C_6H_4(CHO)_2 \quad (70\%)$$

Monoperphthalic acid may be prepared as follows:

$$C_6H_4\begin{matrix}CO\\CO\end{matrix}\!\!>O + H_2O_2 + Na_2CO_3 \longrightarrow C_6H_4(COO_2Na)(CO_2Na) \xrightarrow{H_2SO_4} C_6H_4(COO_2H)(CO_2H) \quad (78\text{–}86\%)$$

Monoperphthalic acid resembles perbenzoic acid in its properties, but is more stable.

Isophthalic acid (*benzene*-1,1-*dicarboxylic acid*), m.p. 346°C, may be prepared by oxidising m-xylene with permanganate. It does not form an anhydride.

Terephthalic acid (*benzene*-1,4-*dicarboxylic acid*) may be prepared by oxidising p-xylene with permanganate, or by oxidising p-methylacetophenone with concentrated nitric acid at about 300°C.

Commercially, it is prepared by the liquid phase oxidation of p-xylene at 140°C in the presence of a cobalt salt as catalyst.

Terephthalic acid sublimes without melting when heated above 300°C. It does not form an anhydride. It forms polyesters with glycol, and the plastic so obtained is known as *terylene*.

Polycarboxylic acids. Three isomeric tri- and three isomeric tetracarboxylic acids are known; the penta- and the hexacarboxylic acid are also known.

Mellitic acid (*benzenehexacarboxylic acid*) may be prepared by oxidising hexamethylbenzene with permanganate. It is also formed when graphite or wood-charcoal is oxidised with fuming nitric acid.

Mellitic acid is a stable solid, m.p. 288° (with decomposition). It occurs as its aluminium salt in peal and lignite.

AROMATIC SULPHONIC ACIDS

Introduction : One of the most characteristic properties of the aromatic hydrocarbons and their derivatives is the ease with which they can be sulphonated with concentrated or fuming sulphuric acid, or with chlorosulphonic acid. The alkanes are not so readily sulphonated, and so this reaction can be used to separate alkanes from aromatic hydrocarbons.

Aromatic sulphonic acids are usually prepared by direct sulphonation, since this is far more convenient than indirect methods. The usual sulphonating agents are: concentrated sulphuric acid, sulphur trioxide in sulphuric acid, *i.e.*, oleum (fuming sulphuric acid), or in organic solvent such as nitromethane, pyridine, etc.; chlorosulphonic acid in carbon tetrachloride to form the sulphonic acid (using one molecule of reagent) or the sulphonyl chloride (using excess of reagent)

MECHANISM OF SULPHONATION

This is still not settled, one reason being the difficulty in establishing the nature of the sulphonating species (cf. nitration). Thus, various possibilities exist for sulphonation with concentrated sulphuric acid, all arising from different equilibrate, *e.g.*,

(i) $2H_2SO_4 \rightleftharpoons SO_3^+ + H_3O^+ + SO_4^-$

(ii) $SO_3 + H_2SO_4 \rightleftharpoons H_2SO_4(SO_3)$ or $H_2S_2O_7$

(iii) $3H_2SO_4 \rightleftharpoons SO_3H^+ + H_3O^+ + 2HSO_4^-$

Experimental work based on kinetic studies in concentrated sulphuric acid and in oleum strongly favours the theory that the active species is sulphur trioxide. This is also believed to be the case with aqueous sulphuric acid. If we accept this, then it appears that the sequence of the steps and their relative rates depend on the conditions. Thus, in concentrated sulphuric acid, the mechanism proposed is:

$$+ SO_3 \rightleftharpoons \overset{H \;\; SO_3^-}{(+)} \rightleftharpoons \overset{SO_3^-}{} + H^+ \rightleftharpoons \overset{SO_3H}{}$$

On the other hand, in oleum the mechanism proposed is:

$$+ SO_3 \rightleftharpoons \overset{H \;\; SO_3^-}{} \underset{-H^+}{\overset{+H^+}{\rightleftharpoons}} \overset{H \;\; SO_3H}{(+)} \rightleftharpoons \overset{SO_3H}{} + H^+$$

When sulphur trioxide is used in an organic solvent, the evidence is that the mechanism is different from those given above, *e.g.*, according to Cerfontain *et al.* (1968), sulphonation of p-dichlorobenzene with sulphur trioxide in nitromethane proceeds as follows:

$$ArH + SO_3 \rightleftharpoons \overset{+}{Ar}\begin{smallmatrix} SO_3^- \\ H \end{smallmatrix} \overset{SO_3}{\rightleftharpoons} \overset{+}{Ar}\begin{smallmatrix} SO_2SO_6^- \\ H \end{smallmatrix} \longrightarrow ArS_2O_6H$$

In addition to the difficulties mentioned above, there are also the complications that sulphonation is *reversible* and that isomerisation often occurs during the reaction. These factors affect the isomer distribution under different conditions.

Thus, *e.g.*. Wanders el at. (1963) have shown that toluene-p- and m-sulphonic both isomerise when heated in *aqueous* sulphuric acid to yield a mixture of mainly m- and p-sulphonic acid with only a small amount of the o-isomer. Also, the o-isomer rapidly isomerises to the p-isomer, but finally the same equilibrium mixture is obtained as before (*i.e.*, all three isomers). From these results it appears that toluene-m-sulphonic acid is the most and the o-acid the least thermodynamically stable isomer.

Benzenesulphonic acid may be readily prepared by heating benzene with concentrated sulphuric acid at 80°C:

$$C_6H_6 + H_2SO_4 \rightleftharpoons C_6H_5SO_3H + H_2O \; (75\text{–}80\%)$$

During sulphonation a small amount of sulphone is produced as a by-product, *e.g.*, *diphenylsulphone.*

$$2C_6H_6 + H_2SO_4 \longrightarrow (CH_6H_5)_2SO_2 + 2H_2O$$

PROPERTIES OF BENZENESULPHONIC ACID

Benzenesulphonic acid is a colourless crystalline deliquescent solid, m.p. 44°C. It is very soluble in water and the solution is strongly acid (about as strong as sulphuric acid). Benzenesulphonic acid and other aromatic sulphonic acids are useful catalysts in esterification and dehydration, being better than sulphuric acid, since they attack the reaction constituents far less than does sulphuric acid. The sulphonic acids are valuable as synthetic reagents, and since the presence of a sulphonic acid group makes the compound soluble in water, sulphonation is an extremely important process in the preparation of dyes and drugs (converting them into soluble derivatives).

REACTIONS OF BENZENESULPHONIC ACID

The following reactions are typical of all sulphonic acids.

1. As pointed out above, sulphonation is a reversible reaction, but the ease of desulphonation depends on the nature of the aromatic nucleus. Benzenesulphonic acid may be desulphonated by heating with dilute hydrochloric acid under pressure at 150-200°C:

$$C_6H_5SO_3H + H_2O \xrightarrow[150\,°C]{HCl} C_6H_6 + H_2SO_4 \text{ (v.g.)}$$

With some sulphonic acids the sulphonic acid group may be eliminated merely by steam distillation. This desulphonation is very useful for preparing certain isomers, *e.g.*, o-chlorotoluene may be prepared as follows:

$$C_6H_5CH_3 \xrightarrow{H_2SO_4} p\text{-}CH_3C_6H_4SO_3H \xrightarrow{Cl_2/Fe} 2\text{-}Cl\text{-}4\text{-}SO_3H\text{-}C_6H_3CH_3 \xrightarrow[150°C]{HCl} o\text{-}ClC_6H_4CH_3$$

Desulphonation may also be used to separate certain isomers, *e.g.*, the three xylenes. The xylene fraction from coal tar may be treated with cold 80 per cent sulphuric acid. Under these conditions, the m-isomer is readily sulphonated to m-xylene-4-sulphonic acid, the o- and p-isomers remaining unaffected:

$$m\text{-}C_6H_4(CH_3)_2 + H_2SO_4 \longrightarrow C_6H_3(CH_3)_2SO_3H + H_2O$$

Thus, only the m-isomer dissolves, and hence may be separated from the other two. The mixture of o- and -xylenes is then sulphonated with concentrated sulphuric acid (98%), and the resulting o-xylene-4-sulphonic acid and p-xylene-2-sulphonic acid may be separated by fractional crystallisation from the diluted sulphonated mixture; the p-derivative is less soluble than the o-. The xylenes are then regenerated by heating their sulphonic acid derivatives with dilute hydrochloric acid under pressure.

2. When sodium benzenesulphonate is fused with sodamide, aniline is obtained:

$$C_6H_5SO_3^- Na^+ + NaNH_2 \longrightarrow C_6H_5NH_2 + Na_2SO_3 \text{ (f.g.–g.)}$$

3. Fusion with sodium hydroxide converts sodium benzenesulphonate into sodium phenoxide:

$$C_6H_5SO_3^- Na^+ + NaOH \longrightarrow Na_2SO_3 + C_6H_5OH$$

$$\xrightarrow{NaOH} C_6H_5\bar{O}Na+ \text{ (g.–v.g.)}$$

4. When sodium benzenesulphonate is fused with sodium cyanide, phenyl cyanide is formed:

$$C_6H_5SO_3^- Na^+ + NaCN \longrightarrow C_6H_5CN + Na_2SO_3 \text{ (f.g.)}$$

5. When the potassium salt of benzenesulphonic acid is fused with potassium hydrogen sulphide thiophenol is formed:

$$C_6H_5SO_3^- K^+ + KSH \longrightarrow C_6H_5SH + K_2SO_3 \text{ (f.g)}$$

6. The sulphonic acid group is often readily replaced by a nitro-group. This offers a means of preparing nitro-derivatives of compounds that are easily oxidised by nitric acid, since the, sulphonic acid derivatives are not easily oxidised; *e.g.*, picric acid from phenol:

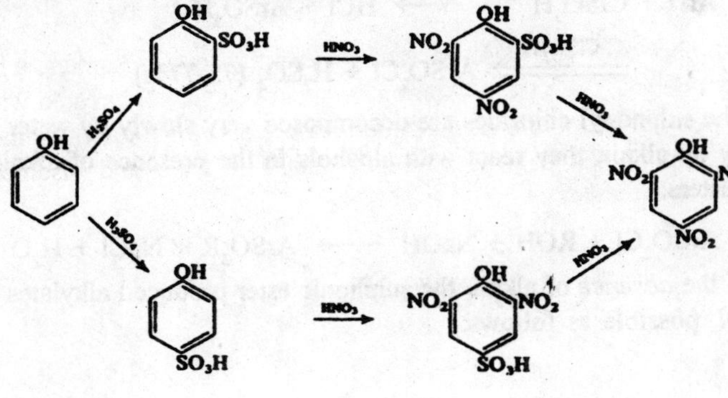

OH, SO₃H, HNO₃, NO₂, H₂SO₄

Halogen may also replace a sulphonic acid group which is either o- or p- to a hydroxyl or to an amino-group; *e.g.*, when treated with bromine water, sulphanilic acid forms s-tribromo-aniline:

$$\text{p-}NH_2C_6H_4SO_3H \xrightarrow{Br_2/H_2O} 2,4,6\text{-}Br_3C_6H_2NH_2$$

Sulphonic acids form many derivatives that are analogous to those of the carboxylic acid; *e.g.*, salts, esters, acid chlorides, amides, etc.

The acid chloride may be prepared by treating a sulphonic acid or its sodium salt with phosphorus pentachloride or phosphoryl chloride, *e.g.*, benzenesulphonyl chloride (b.p. 246°C):

$$C_6H_5SO_3H + PCl_5 \longrightarrow C_6H_5SO_2Cl + HCl + POCl_3 \quad (g. -ex)$$

$$C_6H_5SO_3^- Na^+ + PCl_5 \longrightarrow C_6H_5SO_2Cl + NaCl + POCl_3 \ (75–80\%)$$

$$2C_6H_5SOO_3^- Na^+ + POCl_3 \longrightarrow 2C_6H_5SO_2Cl + NaCl + NaPO_3 \ (74–87\%)$$

Thionyl chloride has no action on sulphonic acids (and some carboxylic acids), but in the presence dimethylformamide, the sulphonyl chloride is obtained in excellent yield (Bosschard *et al.*, 1959). The reaction proceeds through the amide chloride Me_2NCHCl_2.

Sulphonyl chlorides may also be prepared by reaction between an arene and sulphuryl) chloride in the presence of aluminium chloride (cf. the Friedel-Crafts reaction).

The aromatic sulphonyl chlorides, however, are usually best prepared by treating an aromatic compound with excess of chlorosulphonic acid:

$$ArH + ClSO_3H \longrightarrow HCl + ArSO_3H$$

$$\xrightleftharpoons{ClSO_3H} ArSO_2Cl + H_2SO_4 \ (75\text{-}77\%)$$

The sulphonyl chlorides are decomposed very slowly by water but rapidly by alkali; they react with alcohols in the presence of alkali to form esters:

$$ArSO_2Cl + ROH + NaOH \longrightarrow ArSO_3R + NaCl + H_2O$$

In the *absence* of alkali, the sulphonic ester produced alkylates the alcohol, possible as follows:

$$ArSO_2\text{—}OR + H^+ \rightleftharpoons ArSO_2\overset{\overset{\displaystyle H}{|}}{O^+}\text{—}R$$

$$ArSO_2\text{—}\overset{\overset{\displaystyle H}{|}}{O^+}\text{—}R \;\; :\overset{\overset{\displaystyle H}{|}}{O}\text{—}R \rightleftharpoons ArSO_2OH + R_2\overset{+}{O}\text{—}H \longrightarrow H^+$$

Thus, sulphonic acids esters are very good alkylating agents for alcohols and amines. It is because of this alkylating action on alcohols that sulphonic esters cannot be prepared by direct esterification.

When shaken with concentrated ammonia, the sulphonyl chlorides form sulphonamides *e.g.*, benzenesulphonamide from benzenesulphonyl chloride:

$$C_6H_5SO_2Cl + 2NH_3 \longrightarrow C_6H_5SO_2NH_2 + NH_4Cl$$

These sulphonamides are well-defined crystalline solids, and so are used to characterise the sulphonic acids. Better derivatives for characterising the sulphonic acids are their S-benzyliso-thiouronium salts:

$$C_6H_5CH_2SC\begin{matrix}\nearrow \overset{+}{N}H_2 \\ \searrow NH_2\end{matrix} \; Cl^- + ArSO_3Na$$

$$\longrightarrow C_6H_5CH_2SC\begin{matrix}\nearrow \overset{+}{N}H_2 \\ \searrow NH_2\end{matrix} \quad ArSO_3^- + NaCl$$

Sulphonyl chlorides also react with primary and secondary amines to form N-substituted sulphonamides. Sulphonamides are more difficult to hydrolyse than carbonamides, and form salts with alkalis (see the Hinsberg separation). Sulphonyl chlorides may be used in the Friedel-Crafts reaction to give sulphones.

Sulphonic acid anhydrides have been prepared by heating the acid with excess of phosphorus pentoxide (Field *et al.*, 1952), *e.g.*,

$$2C_6H_5SO_3H \xrightarrow[(-H_2O)]{P_2O_5} (C_6H_5SO_2)_2O \; (70\%)$$

Benzenedisulphonic acids. When heated with excess of fuming sulphuric acid at 200°C, benzene forms benzene-m-disulphonic acid as the main product and the p-isomer in a small amount:

$$C_6H_6 \xrightarrow{\text{oleum}} C_6H_4(SO_3H)_2 \text{ (m-)} + C_6H_4(SO_3H)_2 \text{ (p-)}$$

In the presence of mercuric sulphate as catalyst, the yield of p-disulphonic acid is increase (30-34%). Benzene-o-disulphonic acid may be prepared by sulphonating m-aminobenzene sulphonic acid, and then replacing the amino-group by hydrogen (diazotising, etc.).

When fused with potassium hydroxide, o-, m- and p-benzene-disulphonic acids form resorcinol (m-dihydroxybenzene); benzene-o-disulphonic acid forms catechol (o-dihydrcxybenzene).

1,3,5-Benzenetrisulphonic acid may be prepared by heating benzene-m-disulphonic acid with fuming sulphuric acid.

Toluenesulphonic acids. When toluene is treated with concentrated sulphuric acid, the o- and p-toluenesulphonic acids are formed, low temperatures (below 100°C) favouring the formation of the o-isomer, and high temperatures (above 100°C) the p-:

$$C_6H_5CH_3 \xrightarrow[\text{(below 100°C)}]{H_2SO_4} \underset{(50\%)}{o\text{-}CH_3C_6H_4SO_3H} + \underset{(45\%)}{p\text{-}CH_3C_6H_4SO_3H}$$

Both isomers are crystalline solids, the o- melting at 67.5°, and the p- at 106°C.

A much better yield of o-toluenesulphonic acid may be obtained by treating toluene with chlorosulphonic acid at low temperatures:

$$C_6H_5CH_3 \xrightarrow{ClSO_3H} o\text{-}CH_3C_6H_4SO_2Cl + p\text{-}CH_3C_6H_4SO_2Cl \xrightarrow[\text{(ii) NaOH (iii) }H^+]{\text{(i) separate}} p\text{-}CH_3C_6H_4SO_3H$$

o-Toluenesulphonyl chloride is used in the preparation of saccharin; p-toluene- sulphonic acid is used in the preparation of antiseptics, chloramine T and dichloramine T. **Chloramine** T is the sodium salt of N-chloro-p-toluenesulphonamide and may be prepared as follows:

$$p\text{-}CH_3C_6H_4SO_3H \xrightarrow{PCl_5} p\text{-}CH_3C_6H_4SO_2Cl \xrightarrow{NH_3} p\text{-}CH_3C_6H_4SO_2NH_2 \xrightarrow[NaOH]{NaOCl} p\text{-}CH_3C_6H_4SO_2NCl^-Na^+$$

In addition to being used as an antiseptic, chloramine T is also used as a laboratory reagent instead of hypochlorite salts, since it is stable and liberates hypochlorous acid when acidified:

$$p\text{-}CH_3C_6H_4SO_2NCl^-Na^+ \xrightarrow{HCl} p\text{-}CH_3C_6H_4SO_2NHCl \xrightarrow{H_2O} p\text{-}CH_3C_6H_4SO_2NH_2 + HOCl$$

When treated with a large excess of sodium hypochlorite solution, p-toluenesulphonamide forms **dichloramine T** (N,N-dichloro-p-*toluenesulphonamide*)

$$p\text{-}CH_3C_6H_4SO_2NH_2 \xrightarrow{NaOCl} p\text{-}CH_3C_6H_4SO_2NCl_2$$

The p-toluenesulphonyl group is often referred to as the tosyl group, and is denoted by Ts. Tosylates are useful for preparing, *e.g.*, ethers (with ROH), cyanides (with KCN), thiols (with KSH) thioethers (with K_2S), thiocyanates (with KSCN), alkyl iodides (with NaI), alkanes (with Grignard reagents), etc.

m-*Toluenesulphonic* acid (an oil) may be prepared by replacing the amino-group in p-toluidine-m-sulphonic acid by hydrogen:

$$CH_3C_6H_3(SO_3H)NH_2 \xrightarrow{\text{diazotise, etc.}} m\text{-}CH_3C_6H_4SO_3H$$

Jacobsen rearrangement (1886). During sulphonation polyalkyl-benzenes, halogenated polyalkyl-benzenes or polyhalogenated benzenes fairly readily undergo isomerisation due to the migration of an alkyl group or a halogen atom. This is known as the Jacobsen rearrangement, and in practice works best for tetra-alkylbenzenes. Furthermore, the rearrangement is mainly confined to the migration of methyl and ethyl groups; larger groups, particularly isopropyl and t-butyl, are eliminated as alkenes. In addition to rearrangement, disproportionation also occurs, *e.g.*, durene gives prehnitenesulphonic acid (2,3,4,5-tetramethylbenzene-sulphonic acid) as the major product, together with small amounts of 2,4,5-trimethyl- and pentamethylbenzenesulphonic acid, and hexamethyl-benzene.

An example of halogen migration is:

The mechanism of the Jacobsen rearrangement is uncertain. It is widely held that it is the sulphonic acid that rearranges, but a free-radical mechanism not involving the sulphonic acid has also been proposed.

Sulphonic acids. The general methods of preparing the aromatic sulphonic acids may be illustrated by the preparation benzenesulphinic acid:

(i) By the reduction of benzenesulphonyl chloride with zinc dust and water, aqueous sodium sulphide, or alkaline sodium sulphite, *e.g.*,

$$C_6H_5SO_2Cl \xrightarrow{Zn/H_2O} (C_6H_5SO_2)_2Zn \xrightarrow{HCl} 2C_6H_5SO_2H \quad (g)$$

(ii) By the action of copper powder on a solution of a diazonium salt saturated with sulphur dioxide:

$$C_6H_5N_2^+HSO_4^- + SO_2 + Cu \longrightarrow C_6H_5SO_2H + N_2 + CuSO_4 \quad (g)$$

(iii) By the action of sulphur dioxide on benzene in the presence of anhydrous aluminium chloride:

$$C_6H_6 + SO_2 \xrightarrow{AlCl_3} C_6H_5SO_2H \quad (g)$$

(iv) By the action of sulphur dioxide on an arylmagnesium bromide.

$$C_6H_5MgBr + SO_2 \xrightarrow{-40\,°C} C_6H_5SO_2MgBr \xrightarrow{H_2O} C_6H_5SO_2H(g)$$

The usual by-product in this reaction is the sulphoxide.

$$C_6H_5SO_2MgBr + C_6H_5MgBr \longrightarrow (C_6H_5)_2SO + (MgBr)_2O$$

(v) Sulphonic acids may be prepared by fusion of sulphones with potassium hydroxide at 200°C.

The sulphonic acids are unstable solids which readily oxidise in the air to sulphonic acids. They are reduced by zinc and hydrochloric acid to thiols; *e.g.*, benzenesulphinic acid gives thiophenol:

$$C_6H_5SO_2H \xrightarrow{Zn/HCl} C_6H_5SH + 2H_2O$$

The most useful reaction of the sulphinic acids is the ease with which the sulphinic acid group is replaced by a chloromercuri-group (this is an example of indirect mercuration); *e.g.*, p-tolylmercuric chloride may be prepared by boiling an aqueous solution of sodium p-toluenesulphinate with mercuric chloride:

$$CH_3C_6H_4SO_2Na + HgCl_2 \longrightarrow CH_3C_6H_4HgCl + SO_2 + NaCl \quad (51\text{–}57\%)$$

Thiophenol (b.p. 170°C, with a nauseating odour) may be prepared by reducing benzenesulphonyl chloride with zinc and sulphuric acid, or with lithium aluminium hydride, *e.g.*,

$$C_6H_5SO_2Cl \xrightarrow{e;H^+} C_6H_5SH + HCl + 2H_2O \ (96\%)$$

A general method for preparing thiophenols is by reaction between diazonium salts and ethyl potassium xanthate or sodium hydrogen sulphide.

Thiophenol undergoes the usual reactions of thiols (p. 399); it is also a stronger acid than its oxygen analogue, phenol.

Sulphanilic acid (p-*aminobenzenesulphonic acid*) is formed as the main product when aniline is sulphonated with oleum (containing 10 per cent sulphur trioxide) at 180°C; some metanilic and a little orthanilic acid are also produced. Sulphanilic acid is prepared commercially by the 'baking process'; the acid sulphate of aniline (prepared by mixing about equal weights of aniline and concentrated sulphuric acid) is heated for some time at 200°C.

The mechanism of the baking process is uncertain. According to Bamberger (1897), the acid sulphate of aniline is converted into *phenylsulphamic acid* (I), which then rearranges to *orthanilic acid* (II) which, in turn, rearranges to sulphanilic acid (cf. the Hofmann-Martius rearrangement):

$C_6H_5NH_2 \cdot H_2SO_4 \xrightarrow{-H_2O} C_6H_5NHSO_3H$ (I) $\longrightarrow$ $o\text{-}NH_2C_6H_4SO_3H$ (II) $\longrightarrow$ $p\text{-}NH_2C_6H_4SO_3H$ (III)

This rearrangement has been generally accepted as being intramolecular, but Scott *et al.* (1968) have shown that at least in dioxan-sulphuric acid solution, the rearrangement is intermolecular. These workers heated potassium phenylsulphamate in a mixture of dioxan and [^{35}S] sulphuric acid and obtained labelled sulphanilic acid. They also showed that orthanilic acid did not rearrange under these conditions. Thus, the rearrangement is intermolecular and does not proceed via orthanilic acid as an intermediate. The mechanism proposed (for the rearrangement under these conditions) is rapid hydrolysis to aniline and sulphuric acid followed by slow sulphonation. It might also be noted that phenylsulphamic acid has been postulated as, but never proved to be, an intermediate in the baking process.

Sulphanilic acid, m.p. 288°C (with decomp.), forms salts with bases but does not combine with acids. The latter is due to sulphanilic acid existing as a dipolar ion, p-$H_3\overset{+}{N}C_6H_4SO_3^-$ (sulphonic acids are as strong as the inorganic acids). When sulphanilic acid is treated with nitric acid, the sulphonic acid group is replaced by a nitro-group to form p-nitroaniline. Similarly bromine water attacks sulphanilic acid to form 2,4,6-tribromoaniline. Sulphanilic acid is a very important intermediate in dye chemistry, and its substituted amides form the sulphanilamide drugs.

Orthanilic acid (*o-aminobenzenesulphonic acid*) may be prepared as follows:

$o\text{-}NO_2C_6H_4Cl \xrightarrow[EtOH]{Na_2S/S} (o\text{-}NO_2C_6H_4)S\text{—}S(C_6H_4NO_2\text{-}o)$ (58–66%) $\xrightarrow{HNO_3/Cl_2} o\text{-}NO_2C_6H_4SO_2Cl$ (84%)

$\xrightarrow[H_2O]{Na_2CO_3} o\text{-}NO_2C_6H_4SO_2ONa \xrightarrow{Fe/CH_3CO_2H} o\text{-}NH_2C_6H_4SO_3H$ (57%)

Orthanilic acid is a crystalline solid which exists as a dipolar ion.

Metanilic acid (m-*aminobenzenesulphonic acid*) may be prepared by reducing m-nitrobenzenesuphonic acid. It is a crystalline solid (dipolar ion structure) and is used in the manufacture of dyes.

EXERCISES

1. Arensulphonic esters are hydrolysed by aqueous alkali. There are two possible mechanisms. What are they? Devise an experiment to differentiate between them.
2. Complete the equations:

   (i) $p\text{-}MeC_6H_5SO_2Cl \longrightarrow p\text{-}MeC_6H_4SO_2H$

   (ii) $PhSO_2Cl \xrightarrow{?} PhSH$

   (iii) $PhSO_3^- Na^+ + PCl_5 \longrightarrow ?$

   (iv) $p\text{-}MeC_6H_5SO_2Cl + PhH \xrightarrow{AlCl_3}$

   (v) $PhSO_3Et + 2MeNH_2 \longrightarrow$

   (vi) $PhSO_3Et + EtMgI \xrightarrow{Et_2O}$
3. The sodium salt of sulphanilic acid can be readily acetylated with Ac_2O, but not the free acid. Explain.

   $ArH + SO_2Cl_2 \xrightarrow{AlCl_3} ArSO_2Cl + HCl$
4. Starting from p-chloronitrobenzene, prepare bis(4-aminopheyl) sulphone.
5. Acetyl chloride is rapidly decomposed by water, but benzoyl chloride only very slowly. Suggest an explanation.
6. Complete the equations:

   (i) $PhCO(CH_2)_2CO_2H \xrightarrow{?} Ph(CH_2)_3CO_2H$

   (ii) $Ph_2C(OH)CO_2H \xrightarrow[heat]{I_2/P}$

   (iii) (2,5-dihydroxybenzoic acid: ring with CO_2H, OH, OH) $+ Br_2 \xrightarrow{AcOH} ?$

(iv) o-$HOC_6H_4CO_2H$ + 2ICl ⟶ ?

(v) o-$H_2NC_6H_4CO_2H$ + KNCO $\xrightarrow[H_2O]{AcOH}$?

(vi) Phthalic anhydride ($C_6H_4(CO)_2O$) + PCl_5 $\xrightarrow{heat}$?

(vii) PhCOCl + HF (gas) ⟶ ?

(viii) PhCHO + ? ⟶ $PhCHOHCH_2CO_2Et$ $\xrightarrow{heat}$?

(ix) o-$HO_2CC_6H_4CH_2CO_2H$ $\xrightarrow{Ac_2O}$?

(x) p-ClC_6H_4Br $\xrightarrow[Et_2O]{Mg}$? $\xrightarrow[(ii)\ H^+]{(i)\ CO_2}$?

(xi) $CH_2CH_2CH_2CO$ (cyclic, –O– linking the ends) $\xrightarrow[(ii)\ SOCl_2]{(i)\ HCl}$? $\xrightarrow[AlCl_3]{PhH}$?

$\xrightarrow[MeOH]{KOH}$ cyclopropane derivative.

(xii) PhH + ? $\xrightarrow{AlCl_3}$ $PhCOCH = CHCO_2H$

(xiii) PhCOMe $\xrightarrow{HNO_2}$? $\xrightarrow[AcOH]{Br_2}$? $\xrightarrow[H_2O]{boil\ in}$?

7. Deduce the structures of A, B and C:

A, C_9H_{12} $\xrightarrow{KMnO_4}$ B, $C_8H_6O_4$ $\xrightarrow[Fe]{Br_2}$

C, $C_8H_5BrO_4$ (1 product only)

8. Convert:

(i) p-$MeCOC_6H_4CO_2H$ ⟶ p-$MeCHOHC_6H_4CO_2H$;

(ii) PhMe ⟶ o-BrC_6H_4COCl;

(iii) PhH ⟶ p-$ClC_6H_4COC_6H_4Br$-p;

(iv) PhH $\longrightarrow$ 4-NH_2-2-OH-benzoic acid.

9. Convert toluene into m-toluic acid by two methods.

10. The sodium salt of 2-OH-cis- cinnamic acid (coumarinic acid), whereas the sodium salt of 2-OH-3-nitro-cis-cinnamic acid gives the free acid. Suggest an explanation.

11. Starting from benzene and any other compounds you like, synthesise:

    (i) $Ph(CH_2)_3CO_2H$;

    (ii) $PhCH_2COPh$;

    (iii) 4-OH-3-MeO-cinnamic acid.

5

AROMATIC ALCOHOLS, ALDEHYDES AND KETONES

AROMATIC ALCOHOLS

Aromatic alcohols are compounds containing a hydroxyl group in a *side-chain*, and may be regarded as aryl derivatives of the aliphatic alcohols. Aromatic alcohols may be classified as primary, secondary or tertiary alcohols, and their methods of preparation are similar to those used for aliphatic alcohols. Only primary alcohols will be discussed in this section; secondary and tertiary aromatic alcohols are dealt with in the chapter on polynuclear hydrocarbons.

Benzyl alcohol (*phenylcarbinol*), b.p.205°C, may be prepared:

(i) By hydrolysing benzyl chloride with aqueous sodium hydroxide:

$$C_6H_5CH_2Cl + NaOH \longrightarrow C_6H_5CH_2OH + NaCl \ (70\%)$$

This method is used commercially.

(ii) By reducing benzaldehyde with zinc and hydrochloric acid:

$$C_6H_5CHO \xrightarrow{e;\,H^+} C_6H_5CH_2OH$$

(iii) By means of the Cannizzaro reaction:

$$2C_6H_5CHO + NaOH \longrightarrow C_6H_5CH_2OH + C_6H_5CO_2Na \ (90\%)$$

It may also be prepared by means of a crossed Cannizzaro reaction

$$C_6H_5CHO + NaOH \longrightarrow C_6H_5CH_2OH + HCO_2Na$$

The reactions of benzyl alcohol are similar to those of the primary aliphatic alcohols; *e.g.*, on oxidation, it forms benzaldehyde and finally benzoic acid:

$$C_6H_5CH_2OH \xrightarrow{[O]} C_6H_5CHO \xrightarrow{[O]} C_6H_5CO_2H$$

Benzyl alcohol forms esters, many of which are used in perfumery, *e.g.*, benzyl acetate (prepared from benzyl alcohol and acetic anhydride) occurs in oil of amine. Benzyl alcohol reacts with sodium to form sodium benzyl oxide.

In addition to the above reactions, benzyl alcohol exhibits aromatic properties due to the presence of the ring, *e.g.*, it can be nitrated, sulphonated, etc. Care, however, must be taken to avoid reaction with the hydroxyl group; it is therefore generally better, when preparing nuclear-substituted derivatives of benzyl alcohol, to use benzyl chloride (the CH_2Cl and CH_2OH groups are both o-p-orienting), and then hydrolyse to the alcohol. Benzyl alcohol may be catalytically (palladium) reduced to toluene:

$$C_6H_5CH_2OH + H_2 \xrightarrow{Pd/C} C_6H_5CH_3 + H_2O$$

This reaction occurs when the benzyl group is attached to O, N or S, and may also be effected by sodium amalgam or lithium aluminium hydride. *Debenzylation* is a very useful reaction since a benzyl group may be introduced into a compound to protect a sensitive group and then removed at the end of the reaction.

Hydroxybenzyl alcohols may be prepared by the Lederer-Manasse reaction. **o-Hydroxybenzyl alcohol, salicyl alcohol** (*saligenin*), o-$HOC_6H_4CH_2OH$, occurs in the glucoside *salicin*. It is a crystalline solid, m.p. 87°C, and is used in medicine as an antipyretic.

The NMR spectrum of benzyl alcohol.

Mass spectra. Benzyl alcohol shows an intense peak for the molecular ion, which undergoes the following fragmentations:

M^+ 108 $\xrightarrow{-H_2}$ m/e 106 $\xrightarrow{-H\cdot}$ m/e 105

M^+ 108 $\xrightarrow{-H\cdot}$ m/e 107 (CHOH) ≡ (hydroxytropylium) $\xrightarrow{-CO}$ m/e 79 $\xrightarrow{-H_2}$ m/e 77

m/e 105 $\xrightarrow{-CO}$ m/e 77

Fig. 5.1

o-Substituted benzyl alcohols may behave differently from the m- and p-isomers in that they can also undergo elimination of a molecule of water by rearrangement (A = CH_2, O):

$$\left[\text{(o-}C_6H_4\text{)}(CH_2OH)(AH)\right]^{+} \xrightarrow{-H_2O} \left[C_6H_4(=CH_2)(=A)\right]^{+}$$

M^+ $\qquad$ $M-18$

2-Phenylethanol (β-*phenylethyl alcohol*), b.p. 220°C, may be prepared by reducing phenylacetic ester (sodium and ethanol, LAH):

$$C_6H_5CH_2CO_2H_5 \longrightarrow C_6H_5CH_2CH_2OH + C_2H_5OH$$

It is prepared industrially by reaction between ethylene oxide and benzene in the presence of aluminium chloride

$$C_6H_6 + \underset{\diagdown O \diagup}{CH_2—CH_2} \xrightarrow{AlCl_3} C_6H_6 + CH_2—CH_2OH$$

When heated with alkali, it forms styrene.

AROMATIC ALDEHYDES

Aromatic aldehydes fall into two groups: those in which the aldehyde group is directly attached to the nucleus, and those in which it is attached to the side-chain. The former group comprises the aromatic aldehydes: the latter, which behave as aliphatic aldehydes, are best regarded as aryl-substituted aliphatic aldehydes.

The infrared absorption region of the carbonyl group (C = O stretch) in aromatic aldehydes is 1715-1695 cm $^{-1}$(s); this is lower than for aliphatic aldehydes.

Also, the benzene nucleus and the pattern of orientation may be identified. There may be complications, however, due to intramolecular hydrogen bonding in certain o-derivatives.

Benzaldehyde (*benzenecarbonal*) is also known *as oil of bitter almonds*, since it is found in the glucoside *amygdalin* which occurs in bitter almonds. Amygdalin may be hydrolysed by dilute acids or the enzyme emulsin to benzaldehyde, glucose and hydrogen cyanide:

$$C_{20}H_{27}O_{11}N + 2H_2O \longrightarrow C_6H_5CHO + 2C_6H_{12}O_6 + HCN$$

Benzaldehyde may be prepared by any of the following methods, which are general for its homologues as well.

1. By the hydrolysis of benzylidene chloride with aqueous acid (this is also a commercial method).

$$C_6H_5CHCl_2 + H_2O \longrightarrow C_6H_5CHO + 2HCl \ (76\%)$$

2. By boiling benzyl chloride with aqueous copper or lead nitrate in a current of carbon dioxide

The mechanism of this reaction is uncertain; one suggestion is:

$$C_6H_5CH_2Cl \xrightarrow{Pb(NO_3)_2} C_6H_5CH(H){-}O{-}\overset{+}{N}(=O){-}O^-$$

$$\longrightarrow C_6H_5CHO + HNO_2$$

3. Benzaldehyde may be conveniently prepared in the laboratory by oxidising toluene with chromium trioxide in acetic anhydride. As the benzaldehyde is formed, it is converted into benzylidene acetate, thereby preventing further oxidation of the benzaldehyde. Hydrolysis of the acetate with dilute sulphuric or hydrochloric acid gives benzaldehyde:

$$C_6H_5CH_3 \xrightarrow[(CH_3CO)_2O]{CrO_3} C_6H_5CH(OCOCH_3)_2$$

$$\xrightarrow{H^+} C_6H_5CHO + 2CH_3CO_2H \ (40\text{-}50\%)$$

A better yield of benzaldehyde may be obtained by oxidising benzyl alcohol with chromium trioxide in acetic anhydride (yield 90%) or with acid dichromate (yield 80-95%).

An interesting oxidising agent is chromyl chloride (**Etard's reaction.** 1877). In this method, toluene is treated with chromyl chloride in carbon tetrachloride solution and the complex, which is precipitated is decomposed with water:

$$C_6H_5CH_3 + 2CrO_2Cl_2 \longrightarrow C_6H_5CH(OCrCl_2OH)_2$$

$$\xrightarrow{H_2O} C_6H_5CHO \quad (80\%)$$

Toluene is the main substrate that has been studied. The oxidation of n-propylbenzene gives propiophenone and benzyl methyl ketone- (Miller *et al.*, 1890), and Wiberg *et al.* (1962) showed that the latter compound is produced by a rearrangement.

$$PhCH_2CH_2CH_3 \xrightarrow{CrO_2Cl_2} PhCOCH_2CH_3 + PhCH_2COCH_3$$

When more than one methyl group is present, only one is oxidised, *e.g.*, m-xylene gives m-tolualdehyde (80%).

Commercially, the oxidation of toluene to benzaldehyde is carried out in the vapour phase or in the liquid phase. In *vapour phase oxidation*, toluene is catalytically oxidised with air diluted with nitrogen to prevent complete oxidation of the hydrocarbon. The temperature may be as high as 500°C, and the catalyst is the oxide of metals such as manganese, molybdenum, zirconium, etc.:

$$C_6H_5CH_3 + O_2 \longrightarrow C_6H_5CHO + H_2O$$

In *liquid phase oxidation*, toluene is oxidised with manganese dioxide and 65 per cent sulphuric acid at 40°C.

4. Gattermann-Koch aldehyde synthesis (1897). Benzaldehyde may be synthesised by bubbling a mixture of carbon monoxide and hydrogen chloride through a solution of nitrobenzene or ether containing benzene and a catalyst consisting of aluminium chloride and a small amount of cuprous chloride:

$$C_6H_6 + CO + HCl \xrightarrow{AlCl_3} C_6H_5CHO + HCl$$

The mechanism of this reaction is uncertain, but it appears likely that the formyl cation is the active species.

$$CO + HCl + AlCl_3 \longrightarrow [H\overset{+}{C}{=}O \longleftrightarrow HC \equiv \overset{+}{O}] + AlCl_4^-$$

It also appears likely that the cuprous chloride forms a complex with the carbon monoxide thereby increasing its local concentration.

When there are substituents in the ring, *e.g.*, a methyl group, the aldehyde group is introduced into the p-position only. The Gattermann-Koch aldehyde synthesis is not applicable to phenols and their ethers, or when the substituent is strongly deactivating, *e.g.* nitrobenzene (cf the Friedel-Crafts reaction).

5. Gattermann aldehyde synthesis (1906). When benzene is treated with a mixture of hydrogen cyanide and hydrogen chloride in the presence of aluminium chloride, and the complex so produced decomposed with water, benzaldehyde is produced (in low yield).

Several mechanisms have been proposed for this reaction. A widely accepted one is that imidoformyl chloride is formed as an intermediate:

$$HCl + HCN \xrightarrow{AlCl_3} HN = CHCl$$

$$C6H6 + HN = CHCl \xrightarrow{AlCl_3} HCl + \underset{\text{aryl imine}}{C_6H_5CH = NH}$$

The Gattermann reaction is applicable to phenols and phenolic ethers, but not to nitrobenzene.

6. Sommelet's reaction (1913). Benzaldehyde is produced when benzyl chloride is refluxed with hexamethylenetetramine in aqueous ethanolic solution, followed by acidification and steam distillation:

$$C_6H_5CH_2Cl + (CH_2)_6N_4 \longrightarrow C_6H_5CHO\ (60\text{–}70\%)$$

The mechanism of this reaction is uncertain. According to Angyal *et al.* (1953), the mechanism is

$$\text{(i) } C_6H_5CH_2Cl + C_7H_{12}N_4 \longrightarrow [C_6H_5CH_2C_6H_{12}N_4]^+ Cl^-$$

$$\longrightarrow C_6H_5CH_2N = CH_2 \overset{H^+}{\rightleftharpoons} C_6H_5CH_2\overset{+}{N}H = CH_2$$

$$\text{(ii) } C_6H_5CH_2\overset{+}{N}H = CH_2 + C_6H_5\underset{|}{\underset{H}{C}}H—\ddot{N}H_2 \rightleftharpoons$$

$$C_6H_5CH_2NHCH_3 + C_6H_5CH = \overset{+}{N}H_2 \overset{H_2O}{\rightleftharpoons} C_6H_5CHO + NH_3$$

Methylenebenzylamine, formed by hydrolysis of the quaternary compound, adds on a proton, and the ion thus formed reacts with benzylamine (formed as an intermediate) with transfer of a *hydride* ion from the latter to the former.

7. Rosenmund reduction (1918). Benzaldehyde is produced by the catalytic reduction of benzoyl chloride in the presence of a quinoline-sulphur poison.

$$C_6H_5COCl + H_2 \xrightarrow{Pd} C_6H_5CHO + HCl\ \text{(v.g.)}$$

This method may be used to prepare hydroxybenzaldehydes provided the hydroxyl group is protected, *e.g.*, by acetylation.

8. Stephen's method (1925). When phenyl cyanide is reduced with stannous chloride and hydrochloric acid in ethereal solution, and then the product hydrolysed with water, benzaldehyde is formed:

$$C_6H_5CN \xrightarrow[\text{(ii) } H_2O]{\text{(i) } SnCl_2/HCl} C_6H_5CHO\ \text{(v.g.)}$$

This method is a general one except for o-substituted cyanides, *e.g.*, o-tolyl cyanide, in which the yields are negligible due, probably, to steric hindrance.

Aryl cyanides may also be reduced to aldehydes by sodium triethoxyaluminium' hydride, and sterically hindered cyanides by moist Raney nickel (Staskun *et al.*, 1964).

9. Benzaldehyde may be prepared by the reaction between phenylmagnesium bromide an ethyl formate, or better, ethyl orthoformate:

$$C_6H_5MgBr + HCO_2C_2H_5 \longrightarrow C_6H_5CHO + MgBr\,(OC_2H_5)$$

10. Benzaldehyde may be obtained from aniline via the diazonium salt and formaldoxime (Beech 1954):

$$C_6H_5N_2^+Cl^- \xrightarrow{CH_2=NOH} C_6H_5CH = NOH \xrightarrow{H_2O} C_6H_5CHO\ (40\%)$$

Benzaldehyde is a colourless liquid, b.p. 179°C, with a smell of almonds. It is used for flavouring purposes, in perfumery and in the dye industry.

Benzaldehyde (and aromatic aldehydes in general) resembles aliphatic aldehydes in the following reactions:

(i) It gives the Schiff's reaction.

(ii) It is readily oxidised, *i.e.*, it is a strong reducing agent; *e.g.*, it reduces ammoniacal silver nitrate to silver, itself being oxidised to benzoic acid. Benzaldehyde oxidises to benzoic acid when exposed to air.

The mechanism of this autoxidation is not certain, but it is accepted that peroxy-acids are involved.

$$PhCHO \xrightarrow{-H\cdot} Ph\dot{C}{=}O \xrightarrow{O_2} PhC(=O)O{-}O\cdot$$

$$\xrightarrow{PhCHO} Ph\dot{C}{=}O\ \ PhC(=O)O{-}OH \xrightarrow{PhCHO} 2PhC(=O)OH$$

(iii) Benzaldehyde forms a bisulphite compound, and may be prepared pure via this compound:

$$C_6H_5CHO + NaHSO_3 \longrightarrow C_6H_5CH(OH)SO_3Na$$

(iv) Benzaldehyde forms a cyanohydrin (*mandelonitrile*), $C_6H_5CH(OH)CN$, with hydrogen cyanide, and an oxime (and phenylhydrazorref with hydroxylamine (and phenylhydrazine). The latter compounds exist in two geometrical isomeric forms, *e.g.*, benzaldoxime:

$$\underset{\text{syn-form}}{\begin{matrix} C_6H_5 & & H \\ & \diagdown\;\diagup & \\ & C & \\ & \| & \\ & N & \\ & \diagdown & \\ & & OH \end{matrix}} \qquad \underset{\text{anti-form}}{\begin{matrix} C_6H_5 & & H \\ & \diagdown\;\diagup & \\ & C & \\ & \| & \\ & N & \\ & \diagup & \\ HO & & \end{matrix}}$$

With hydrazine, benzaldehyde forms benzylideneazine,

$$C_6H_5CH= NN = CHC_6H_5$$

(v) Benzaldehyde reacts with phosphorus pentachloride to form benzylidene chloride:

$$C_6H_5CHO + PCl_5 \longrightarrow C_6H_5CHCl_2 + POCl_3$$

(vi) Benzaldehyde readily undergoes condensation with many aromatic and aliphatic compounds (see below).

Benzaldehyde (and other aromatic aldehydes) differs from aliphatic aldehydes in the following ways:

(i) It does not reduce Fehling's solution.

(ii) It does not readily polymerise; *e.g.*, it does not resinify with sodium hydroxide, but undergoes the Cannizzaro reaction due to its not having an a-hydrogen atom:

$$2C_6H_5CHO + NaOH \longrightarrow C_6H_5CH_2OH + C_6H_5CO_2Na$$

The mechanism involves hydride ion transfer, one possibility being as follows:

$$\text{Ph}-\overset{\overset{\displaystyle O}{\|}}{C}-\text{H} \underset{\text{fast}}{\overset{OH^-}{\rightleftharpoons}} \text{Ph}-\underset{\underset{\displaystyle OH}{|}}{\overset{\overset{\displaystyle O^-}{|}}{C}}-\text{H}\;\underset{\underset{\displaystyle OH}{|}}{\overset{\overset{\displaystyle O}{\|}}{C}}-\text{Ph} \xrightarrow{\text{slow}}$$

$$\text{Ph}-\underset{\underset{\displaystyle OH}{|}}{\overset{\overset{\displaystyle O}{\|}}{C}} + \text{H}-\underset{\underset{\displaystyle H}{|}}{\overset{\overset{\displaystyle O^-}{|}}{C}}-\text{Ph} \xrightarrow{\text{fast}} \text{PhCO}_2^- + \text{PhCH}_2\text{OH}$$

(iii) Benzaldehyde does not yield a simple addition product with ammonia, but forms a complex product, *hydrobenzamide* (90%).

The mechanism is uncertain, but the reaction is believed to proceed via benzaldimine:

$$C_6H_5CHO \xrightarrow{NH_3} [C_6H_5CHOHNH_2] \xrightarrow{-H_2O} C_6H_5CH=NH$$

$$\xrightarrow{3\ mol} C_6H_5CH=NCH\ (C_6H_5)N=CHC_6H_5 + NH_3$$

Benzaldehyde also reacts with primary aromatic amines to form *anils* or *Schiff bases*.

(iv) Reduction of benzaldehyde with zinc and hydrochloric acid or with sodium amalgam gives *hydrobenzoin* as well as benzyl alcohol (cf. pinacol,):

$$2C_6H_5CHO \xrightarrow{e;H^+} C_6H_5CHOHCHOHC_6H_5$$

(v) When chlorinated in the absence of a halogen carrier, benzaldehyde forms benzoyl chloride (no α-hydrogen present):

$$C_6H_5CHO + Cl_2 \longrightarrow C_6H_5COCl + HCl$$

CONDENSATION REACTIONS OF BENZALDEHYDE

1. Claisen reaction. Benzaldehyde, in the presence of dilute alkali, condenses with aliphatic aldehydes or ketones containing α-hydrogen; *e.g.*, with acetaldehyde it form; *cinnamaldehyde*:

$$C_6H_5CHO + CH_3CHO \xrightarrow{NaOH} C_6H_5CH=CHCHO + H_2O$$

With acetone, benzaldehyde forms benzylideneacetone (m.p. 42°C; used in perfumery):

$$C_6H_5CHO + CH_3COCH_3$$

$$\xrightarrow{NaOH} C_6H_5CH=CHCOCH_3 + H_2O \quad (65\text{-}78\%)$$

If the reaction is carried out in aqueous ethanolic sodium hydroxide, *dibenzylideneacetone* (m.p. 112°C) is produced by interaction of benzylideneacetone and another molecule of benzaldehyde:

$$C_6H_5CH=CHCOCH_3 + C_6H_5CHO \longrightarrow$$

$$C_6H_5CH=CHCOCH=CHC_6H_5 + H_2O \quad (90\text{–}94\%)$$

Benzaldehyde also condenses with acetophenone to form *phenyl styryl ketone* (*benzylidene acetophenone, chalcone*).

$$C_6H_5CHO + CH_3COC_6H_5 \xrightarrow{NaOH} C_6H_5CH{=}CHCOC_6H_5 + H_2O\ (85\%)$$

Phenyl styryl ketone may be reduced catalytically (platinum) in ethyl acetate solution or by zinc and acetic acid to *benzylacetophenone*:

$$C_6H_5CH = CHCOC_6H_5 + H_2 \xrightarrow{Pt} C_6H_5CH_2CH_2COC_6H_5\ (81\text{–}95\%)$$

The above condensation reactions, apart from their synthetic value, illustrate the use of benzaldehyde to detect the presence of a —CH_2CO— group in carbonyl compounds.

Benzylideneacetophenone (and other α, β-unsaturated ketones) may be introduced into an aryl nucleus by means of the Friedel-Crafts reaction.

Benzaldehyde condenses with crotonaldehyde, in the presence of pyridine acetate, to fom-5-phenylpentadienal :

$$C_6H_5CHO + CH_3CH = CHCHO \longrightarrow C_6H_5CH{=}CHCH{=}CHCHO + H_2O$$

This reaction is an example of vinylogy, and was introduced by Kuhn (1929).

Benzaldehyde also condenses with nitromethane to form m-nitrostyrene; the methylene group is made 'active' by the adjacent nitro-group:

$$C_6H_5CHO + CH_3NO_2 \xrightarrow{NaOH} C_6H_5CH = CHNO_2 + H_2O\ (80\text{–}83\%)$$

2. Perkin reaction (1877). When benzaldehyde (or any other aromatic aldehyde) is heated with the anhydride of an aliphatic acid (containing two α-hydrogen atoms) in the presence of its sodium salt, condensation takes place to form a β-arylacrylic acid; *e.g.*, with acetic anhydride and sodium acetate, cinnamic acid is formed:

$$C_6H_5CHO + (CH_3CO)_2O \xrightarrow{CH_3CO_2Na} C_6H_5CH = CHCO_2H$$

With propionic anhydride and sodium propionate, a-methylcinnamic acid is formed:

$$C_6H_5CHO + (CH_3CH_2CO)_2O \xrightarrow{CH_3CO_2CO_2Na} C_6H_5CH = C(CH_3)CO_2H$$

The mechanism may be as follows (cf. the aldol condensation):

$$CH_3COOCOCH_3 + CH_3CO_2^- \rightleftharpoons \overset{-}{C}H_2COOCOCH_3 + CH_3CO_2H$$

$$C_6H_5\overset{\overset{O}{\|}}{\underset{H}{C}} + \bar{C}H_2COOCOCH_3 \rightleftharpoons C_6H_5\overset{\overset{\bar{O}}{|}}{\underset{H}{C}}CH_2COOCOCH_3$$

$$\xrightleftharpoons{H^+} C_6H_5\overset{\overset{OH}{|}}{\underset{H}{C}}CH_2COOCOCH_3 \xrightleftharpoons{-H_2O}$$

$$C_6H_5CH{=}CHCOOCOCH_3 \xrightarrow{H_2O} C_6H_5CH{=}CHCO_2H + CH_3CO_2H$$

It should be noted that only the α-hydrogen atoms of the anhydride are involved in the condensation. The acyloxyl ion acts as the base, and in some cases triethylamine as base gives better yields.

The Perkin reaction does not usually take place with aliphatic aldehydes.

Benzaldehyde undergoes the Knoevenagel reaction with malonic acid in ethanolic ammonia to form cinnamic acid:

$$C_6H_5CHO + CH_2(CO_2H)_2 \longrightarrow C_6H_5CH = CHCO_2H + CO_2 + H_2O$$

A special case of the Perkin reaction is the condensation of benzaldehyde with cyclic anhydrides, *e.g.*, succinic anhydride.

3. Benzoin condensation. When refluxed with aqueous ethanolic potassium cyanide, benzaldehyde forms benzoin:

$$2C_6H_5CHO \xrightarrow{KCN} C_6H_5CHOHCOC_6H_5 \ (83\%)$$

4. Benzaldehyde condenses with phenols and tertiary aromatic amines in the presence of sulphuric acid or zinc chloride to form triphenylmethane derivatives; *e.g.*, with dimethylaniline it forms *malachite green.*

DERIVATIVES OF BENZALDEHYDE

A large variety of substituted benzyl alcohols can be oxidised to the corresponding aldehydes by means of dinitrogen tetroxide solutions (yields: 91-98%). The benzyl alcohols (except the nitro-compounds) may be obtained by reduction of acid chloride or methyl esters with lithium aluminium hydride (Field *et al.*, 1955), *e.g.*,

$$o\text{-}C_6H_4(OCH_3)COCl \xrightarrow[93\%]{LiAlH_4} o\text{-}C_6H_4(OCH_3)CH_2OH \xrightarrow[96\%]{N_2O_4} o\text{-}C_6H_4(OCH_3)CHO$$

Various substituted benzaldehydes may also be obtained via diazonium salts (Beech, 1954), *e.g.*

$$o\text{-}C_6H_4(NH_2)NO_2 \longrightarrow o\text{-}C_6H_4(N_2^+Cl^-)NO_2 \xrightarrow[\text{(ii) } H_2O]{\text{(i) } CH_2{=}NOH} o\text{-}C_6H_4(CHO)NO_2 \quad (33\%)$$

Nitrobenzaldehydes. Mixed acid nitrates benzaldehyde to give mainly the m-isomer, m.p. 58°C (about 50%), together with the o-isomer (about 20%).

o-*Nitrobenzaldehyde* may be prepared by oxidising o-nitrocinnamic acid with cold aqueous potassium permanganate:

$$o\text{-}C_6H_4(CH{=}CHCO_2H)NO_2 \xrightarrow{KMnO_4} o\text{-}C_6H_4(CHO)NO_2$$

Alternatively, it may be prepared by dissolving o-nitrotoluene in glacial acetic acid containing acetic anhydride, and adding chromium trioxide and sulphuric acid. The product is the diacetate and this, on hydrolysis, gives the aldehyde:

$$o\text{-}C_6H_4(CH_3)NO_2 \longrightarrow \underset{(23\text{–}24\%)}{o\text{-}C_6H_4(CH(OCOCH_3)_2)NO_2} \longrightarrow \underset{(74\%)}{o\text{-}C_6H_4(CHO)NO_2}$$

A better yield may be obtained by oxidising the o-nitrotoluene with manganese dioxide and sulphuric acid.

o-Nitrobenzaidehyde is a yellow solid, m.p. 44°C. Its most important reaction is its conversion into indicating when heated with acetone and sodium hydroxide:

$$2\,o\text{-}C_6H_4(NO_2)CHO + 2CH_3COCH_3 \xrightarrow{NaOH} C_6H_4\langle^{NH}_{CO}\rangle C{=}C\langle^{CO}_{NH}\rangle C_6H_4$$

p-*Nitrobenzaldehyde* (m.p. 106°C) may be prepared by oxidising p-nitrocinnamic acid or p-nitrotoluene; it may also be prepared by oxidising p-nitrobenzyl chloride with aqueous lead nitrate.

All three nitrobenzaldehydes may be reduced to the corresponding aminobenzaldehydes with ferrous sulphate and ammonium hydroxide solution or by stannous chloride and hydrochloric acid. On the other hand, sodium borohydride, lithium aluminium hydride-aluminium chloride, and aluminium isopropoxide (MPV reduction) reduce nitrobenzaldehydes to the corresponding nitrobenzyl alcohols.

o-*Aminobenzaldehyde* readily condenses with compounds containing a —CH_2CO— group to form quinoline compounds; *e.g.*, with acetaldehyde it forms *quinoline* m- and p-*Amino- benzaldehydes* are used in the preparation of dyes. o- and p-Aminobenzaldehydes are prepared commercially by the oxidation of the corresponding aminobenzyl alcohols.

2,4-*Dinitrobenzaldehyde* (m.p. 72°C) may be prepared as follows:

$C_6H_5N(CH_3)_2 \xrightarrow{HNO_2} p\text{-}(CH_3)_2NC_6H_4NO$

$(CH_3)_2NC_6H_4NO + CH_3C_6H_3(NO_2)_2 \xrightarrow[heat]{Na_2CO_3}$

$(CH_3)_2NC_6H_4N{=}CHC_6H_3(NO_2)_2 \xrightarrow[heat]{HCl} (CH_3)_2NC_6H_4NH_2 + OHCC_6H_3(NO_2)_2$ (24–32%)

Fig. 5.2

Halogen derivatives of benzaldehyde. m-Halogenobenzaldehydes may be prepared from m-nitrobenzaldehyde by reduction and subsequent replacement of the diazo-group by halogen (Sandmeyer reaction), *e.g.*, m-chlorobenzaldehyde:

$m\text{-}O_2NC_6H_4CHO \xrightarrow{SnCl_2/HCl} m\text{-}H_2NC_6H_4CHO \xrightarrow[HCl]{NaNO_2} m\text{-}N_2^+Cl^-C_6H_4CHO \xrightarrow[HCl]{CuCl} m\text{-}ClC_6H_4CHO$ (75–79%)

o- and p-Halogenobenzaldehydes may be prepared by brominating the corresponding halogenotoluenes at their boiling points in the presence of light to give the benzylidene bromide derivative, and hydrolysing this with calcium hydroxide; *e.g.*, p-bromobenzaldehyde from p-bromotoluene

$$CH_3C_6H_4Br \xrightarrow{Br_2} CHBr_2C_6H_4Br \xrightarrow{H_2O} CHOC_6H_4Br \quad (60\text{–}69\%)$$

Alternatively, the halogenotoluene may be oxidised with manganese dioxide and sulphuric acid

2,6-*Dichlorobenzaldehyde* (m.p. 72°C) is important in the preparation of triphenylmethane dyes; it may be prepared as follows:

$$o\text{-}CH_3C_6H_4NO_2 \xrightarrow[Fe]{Cl_2} \xrightarrow{Fe/HCl} (NH_2) \xrightarrow[HCl]{NaNO_2} (N_2^+Cl^-) \xrightarrow[HCl]{CuCl} (2,6\text{-}Cl_2C_6H_3CH_3) \xrightarrow[(at\ b.p.)]{Cl_2} (CHCl_2) \xrightarrow[(con.\ H_2SO_4)]{H_2O} (CHO)$$

Fig. 5.3

Concentrated sulphuric acid must be used to hydrolyse the o, o′-dichlorobenzylidene chloride: alkali has very little effect (this is an example of steric hindrance). The overall yield of 2,6-dichloro-benzaldehyde is very small; this is to be expected from the large number of steps involved. This method illustrates the important point that although a particular method gives a small yield of the desired product, it may be the only worthwhile method to use.

Benzaldehydesulphonic acids. Sulphonation of benzaldehyde gives mainly the m-derivative. The o- and p-isomers are prepared indirectly.

Cinnamaldehyde *(3-phenylpropenal)* is the chief constituent of cinnamon oil from which it may be isolated by means of its bisulphite compound. It may be prepared synthetically by the Claisen reaction between benzaldehyde and acetaldehyde:

$$C_6H_5CHO + CH_3CHO \xrightarrow{NaOH} C_6H_5CH = CHCHO + H_2O$$

Cinnamaldehyde is an oil, b.p. 252 C. which slowly oxidises in air to cinnamic acid. This acid may also be obtained by oxidising cinnamaldehyde with ammoniacal silver nitrate:

$$C_6H_5CH = CHCHO + [O] \longrightarrow C_6H_5CH = CHCO_2H$$

Vigorous oxidising agents, *e.g.*, acid permanganate, convert cinnamaldehyde into benzoic acid.

Cinnamaldehyde forms the normal bisulphite compound with sodium hydrogen sulphite but on prolonged treatment with this reagent, the sodium salt of a disulphonic acid is formed, possibly as follows :

$$C_6H_5CH = CHCHO + NaHSO_3 \rightleftharpoons C_6H_5CH = CH(OH)SO_3Na$$

$$\downarrow$$

$$\underset{\substack{|\\ SO_3Na \\ (1,4\text{-addition})}}{C_6H_5CHCH_2CHO} \xrightleftharpoons{NaHSO_3} \underset{\substack{|\\ SO_3}}{C_6H_5CHCH_2CH(OH)SO_3Na}$$

The normal addition is reversible: the 1,4-addition is irreversible, and hence all the cinnamaldehyde is gradually converted into the disulphonic acid. Cinnamaldehyde adds on bromine to the double bond to form the dibromide. $C_6H_5CHBrCHBrCHO$, and is reduced by aluminium isopropoxide to cinnamyl alcohol, $C_6H_5CH = CHCH_2OH$. This reduction may also be carried out with aluminium hydride (Jorgenson, 1962): this reagent does *not* attack the double bond (cf. lithium aluminium hydride).

PHENOLIC ALDEHYDES

Phenolic aldehydes (*hydroxyaldehydes*) contain an aldehyde group and one or more hydroxyl groups directly attached to the nucleus.

General Methods of Preparation

1. Gattermann's aldehyde synthesis (cf. benzaldehyde). When phenol or a phenolic ether is treated with a mixture of hydrogen cyanide and hydrogen chloride in the presence of aluminium chloride, and the complex produced then decomposed with water, p-hydroxy-(or alkoxy) benzaldehyde is the main product:

$$HOC_6H_5 + ClCH{=}NH \xrightarrow{AlCl_3} HOC_6H_4CH{=}NH + HCl \xrightarrow{H_2O} HOC_6H_4CHO + NH_3$$

For more reactive compounds, *e.g.*, m-di- or trihydric phenols (o- and p-positions strongly activated), zinc chloride is a better catalyst. In either case, to avoid the use of hydrogen cyanide, zinc cyanide and hydrogen chloride are very convenient for generating the hydrogen cyanide *in situ.*

2. Reimer-Tiemann reaction (1876). This reaction is carried out by refluxing an alkaline solution of phenol with chloroform at 60°C, distilling off the excess of chloroform, acidifying the residual liquid with sulphuric acid and then steam-distilling it. Unchanged phenol and o-hydroxy-benzaldehyde distil over, leaving behind p-hydroxy-benzaldehyde.

The mechanism of the Reimer-Tiemann reaction is believed to involve the formation of dichloromethylene (produced from the chloroform;

CCl_2; OH^-; H_2O; $\bar{C}Cl_2$; H; H_2O; H; $CHCl_2$; $-H^+$; O^-; $CHCl_2$; OH^-; H_2O; O^-; CHO

Fig. 5.3

Phenols with blocked p-positions give cyclohexadienones containing the dichloromethyl group, *e.g.*,

OH; CH_3; NaOH; $CHCl_3$; OH; CHO; CH_3; +; O; CH_3 $CHCl_2$

In the Reimer-Tiemann reaction, the o-isomer predominates, but if one of the o-positions is occupied, the aldehyde group tends to go to the p-position *e.g.*, guaiacol forms vanillin:

OH; OCH_3; $CHCl_3$; NaOH; OH; OCH_3; CHO

Phenols may be selectively formylated in the o-position by heating the phenoxymagnesium halide with ethyl orthoformate (Casnati *et al.*, 1965), *e.g.*,

$$\text{Me-C}_6\text{H}_4\text{-OMgBr} \xrightarrow[\text{(ii) H}^+]{\text{(i) HC(OEt)}_3} \text{Me-C}_6\text{H}_3\text{(OH)CHO}$$

3. o- and -Hydroxybenzyl alcohols may be prepared by the Lederer-Manasse reaction. These alcohols may be oxidised to the corresponding hydroxyaldehydes (aminobenzaldehydes may be prepared similarly from aniline). This method is used industrially, the oxidation being carried out with nitrosobenzene or phenylhydroxylamine, in the presence of copper as catalyst:

$$\text{HO-C}_6\text{H}_4\text{-CH}_2\text{OH} + C_6H_5NO \xrightarrow{\text{Cu}} \text{HO-C}_6\text{H}_4\text{-CH}=NC_6H_5 \xrightarrow{\text{H}^+} \text{HO-C}_6\text{H}_4\text{-CHO} \quad (40\text{–}50\%)$$

An extension of the above method is the chloral condensation. This is a modified Lederer-Manasse reaction, chloral being used instead of formaldehyde, *e.g.*,

$$C_6H_5OH + CCl_3CHO \xrightarrow{\text{aq. } K_2CO_3} \text{HO-C}_6\text{H}_4\text{-CHOHCCl}_3 \xrightarrow[\text{NaOH}]{Na_2Cr_2O_7} \text{HO-C}_6\text{H}_4\text{-CHO} \quad (40\text{–}60\%)$$

In this reaction, the chloral always enters the p-position unless it is occupied, in which case o-substitution takes place, but to a lesser extent.

4. Duff's reaction (1932). When phenol is heated with a mixture of hexamethylenetetramine, glycerol and boric acid, the mixture then acidified with sulphuric acid and steam distilled, o-hydroxybenzaldehyde is formed:

$$C_6H_5OH \xrightarrow{(CH_2)_6N_4} \text{o-HO-C}_6\text{H}_4\text{-CH}=NCH_3 \xrightarrow{\text{H}^+} \text{o-HO-C}_6\text{H}_4\text{-CHO} \quad (15\text{–}20\%)$$

This method gives only the o-compound, and is hindered by the presence of a –I group in the ring.

5. Formylation with N-methylformanilide (Vilsmeier and Haack, 1927). Provided the aromatic compound has an activated nucleus, it can be formylated by means of N-methylformanilide. The method works well for the o- and p-positions of phenolic ethers (and dialkyl anilines);

it does not work with hydrocarbons, except with anthracene, in which case the aldehyde group enters the 9-position. The formylation is carried out by treating the compound with N-methylformanilide and phosphoryl chloride, and when the reaction is complete, adding aqueous sodium acetate and steam distilling:

$$C_6H_5OCH_3 + C_6H_5N(CH_3)CHO \longrightarrow p\text{-}OCH_3C_6H_4CHO + C_6H_5NHCH_3$$

Dimethylformamide is often used instead of N-methylformanilide.

A possible mechanism for formylation with phosphoryl chloride and dimethylformamide is

$$Me_2NCHO + POCl_3 \longrightarrow [Me_2N{-}\overset{+}{C}HCl \longleftrightarrow Me_2\overset{+}{N}{=}CHCl]OPOCl_2^-$$

$$MeO{-}C_6H_5 + \overset{+}{C}H(Cl)NMe_2 \rightleftharpoons Me\overset{+}{O}{=}C_6H_5(H)CH(Cl)NMe_2 \xrightarrow{-HCl}$$

$$MeO{-}C_6H_4{-}CH{=}\overset{+}{N}Me_2 \xrightarrow{OH^-} MeO{-}C_6H_4{-}CHO$$

N-Methylformanilide (and, in general, disubstituted formamides) reacts with Grignard reagents to form aldehydes (Bouveault, 1904; Smith et at., 1941), *e.g.*,

$$o\text{-}CH_3C_6H_4Br \xrightarrow{Mg} o\text{-}CH_3C_6H_4MgBr \xrightarrow{C_6H_5NMeCHO} o\text{-}CH_3C_6H_4CHO \quad (50\%)$$

6. By means of organolithium compounds. When a substituted phenyl-lithium compound is heated with ethyl orthoformate or A-methylformanilide, an intermediate compound is obtained which, on hydrolysis with acid, gives a substituted benzaldehyde in 70 per cent yield, *e.g.*,

$$p\text{-}CH_3OC_6H_4Li + HC(OC_2H_5)_3 \longrightarrow p\text{-}CH_3OC_6H_4CH(OC_2H_5)_2 \xrightarrow{H^+} p\text{-}CH_3OC_6H_4CHO$$

Salicylaldehyde (o-*hydroxybenzaldehyde*) occurs in certain essential oils. It may be prepared by any of the general methods applicable to o-hydroxyaldehydes. It is manufactured by the oxidation of o-hydroxybenzyl alcohol (obtained by method 3, above), and by the oxidation of o-methylphenyl benzenesulphonate with manganese dioxide and sulphuric acid:

$$o\text{-}CH_3C_6H_4OSO_2C_6H_5 \xrightarrow{[O]} o\text{-}OHCC_6H_4OSO_2C_6H_5 \xrightarrow{H^+} o\text{-}HOC_6H_4CHO$$

Salicylaldehyde is an oil, b.p. 197°C, and its aqueous solution gives a violet coloration with ferric chloride. Salicylaldehyde may be oxidised to salicylic acid and reduced to o-hydroxybenzyl alcohol. Oxidation with alkaline hydrogen peroxide converts it into catechol.

The hydroxyl group of salicylaldehyde is not so reactive as that in the m- and p-isomers. This is probably due to hydrogen bonding, which disc accounts for the high volatility of salicylaldehyde (compared with the m- and p-compounds).

m-Hydroxybenzaldehyde may be prepared as follows:

$$m\text{-}O_2NC_6H_4CHO \xrightarrow[\text{(ii) } NaNO_2-HCl]{\text{(i) } Sn-HCl} m\text{-}Cl^-\,N_2^+C_6H_4CHO \xrightarrow[\text{boil}]{H_2O} m\text{-}HOC_6H_4CHO \quad (51\text{–}56\%)$$

Anisaldehyde (p-*methoxybenzaldehyde*) occurs in various essential oils. It is prepared industrially by oxidising anethole (by ozonolysis or with acid dichromate):

$$p\text{-}CH_3O C_6H_4CH{=}CHCH_3 \xrightarrow[H_2SO_4]{K_2Cr_2O_7} p\text{-}CH_3OC_6H_4CHO$$

Anisaldehyde may be prepared synthetically by methylating with hydroxybenzaldehyde with methyl sulphate and aqueous sodium hydroxide or with methyl iodide and ethanolic potassium hydroxide, or by introducing an aldehyde group into the p-position of anisole.

Anisaldehyde, b.p.. 248°C, may be oxidised to anisic acid and reduced to anisyl alcohol.

Protocatechualdehyde (3,4-*dihydroxybenzaldehyde*) may be prepared by heating vanillin with hydrochloric acid:

$$\text{OH, OCH}_3\text{, CHO} \xrightarrow{\text{HCl}} \text{OH, OH, CHO}$$

It may also be prepared by adding phosphorus pentachloride to piperonal, then treating the product with cold water, and finally boiling the solution:

$$\text{O—CH}_2\text{—O, CHO} \xrightarrow{\text{PCl}_5} \text{O—CCl}_2\text{—O, CHO} \xrightarrow{\text{H}_2\text{O}} \text{O—CO—O, CHO} \xrightarrow{\text{H}_2\text{O}} \text{OH, OH, CHO}$$

Another preparation is by means of the Reimer-Tiemann reaction using catechol

$$\text{HO, HO} + CHCl_3 \xrightarrow{\text{NaOH}} \text{HO, HO, CHO}$$

Protocatechualdehyde, m.p. 153°C, gives a green colour with ferric chloride, and reduces ammoniacal silver nitrate. When methylated with approximately one equivalent of methyl sulphate, protocatechu-aldehyde gives a mixture of vanillin (1) and isovanillin (II); excess of methyl sulphate give; veratraldehyde (III):

$$\text{OH, OH, CHO} \xrightarrow[\text{NaOH}]{(CH_3)_2SO_4} \underset{\text{(I)}}{\text{OH, OCH}_3\text{, CHO}} + \underset{\text{(II)}}{\text{OCH}_3\text{, OH, CHO}} \longrightarrow \underset{\text{(III)}}{\text{OCH}_3\text{, OCH}_3\text{, CHO}}$$

Vanillin (p-hydroxy-m-methoxybenzaldehyde) occurs in many substances of plant origin, *e.g.*, the vanilla bean. It may be prepared synthetically from guaiacol by means of the Reimer-Tiemann or the Gattermann reaction. Industrially, it is prepared:

(i) By oxidising isoeugenol (from eugenol) with nitrobenzene:

$$\text{OH, OCH}_3\text{, } CH_2CH{=}CH_2 \xrightarrow{\text{KOH}} \text{OH, OCH}_3\text{, } CH{=}CHCH_3 \xrightarrow{[O]} \text{OH, OCH}_3\text{, CHO}$$

If the oxidation is carried out with acid dichromate, it is necessary to protect the hydroxyl group (by acetylation, etc.). On the other hand, ozonolysis may be used without protecting the hydroxyl group.

(ii) By the Lederer-Manasse reaction as follows:

OH, OCH_3 + HCHO $\xrightarrow{NaOH}$ OH, OCH_3, CH_2OH $\xrightarrow{C_6H_5NHOH}$ OH, OCH_3, $CH{=}NC_6H_5$ $\xrightarrow{H^+}$ OH, OCH_3, CHO

(iii) The liquors from the extraction of lignin from pulp-wood contain vanillin, and are now used as a source of vanillin.

Vanillin is a crystalline solid, m.p. 81°C. When heated with hydrochloric acid, it forms protocatechualdehyde; with methyl sulphate it forms veratraldehyde. Oxidation by silver oxide in aqueous sodium hydroxide converts vanillin into vanillic acid (96-97%).

Piperonal (*heliotropin*, 3, 4-*methylenedioxybenzaldehyde*) is the methylene ether of protocatechu-aldehyde, and is obtained when piperic acid (from the alkaloid piperine) is oxidised. It may be prepared synthetically by treating protocatechualdehyde with methylene di-iodide and sodium hydroxide:

OH, OH, CHO $+ CH_2I_2 + 2NaOH \longrightarrow$ O—CH_2—O, CHO $+ 2NaI + 2H_2O$

It is manufactured by the oxidation of isosafrole (from safrole) by ozonolysis or with acid dichromate:

O—CH_2—O, $CH_2CH{=}CH_2$ $\xrightarrow{KOH}$ O—CH_2—O, $CH{=}CHCH_3$ $\xrightarrow{[O]}$ O—CH_2—O, CHO

Piperonal, m.p. 37°C, may be oxidised to piperonylic acid and reduced to piperonyl alcohol. When treated with phosphorus pentachloride and then with water, piperonal forms protocatechualdehyde. This aldehyde is also obtained, together with formaldehyde or methanol, when piperonal is heated with dilute hydrochloric acid at 200°C under pressure.

Protocatechualdehyde (3,4-*dihydroxybenzaldehyde*) may be prepared by heating vanillin with hydrochloric acid:

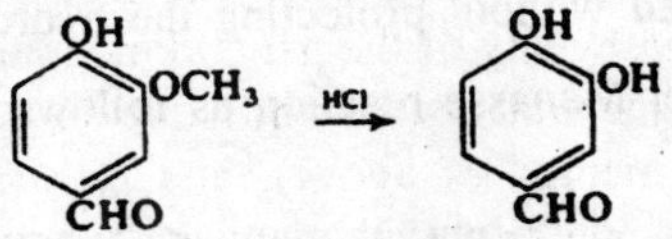

It may also be prepared by adding phosphorus pentachloride to piperonal, then treating the product with cold water, and finally boiling the solution:

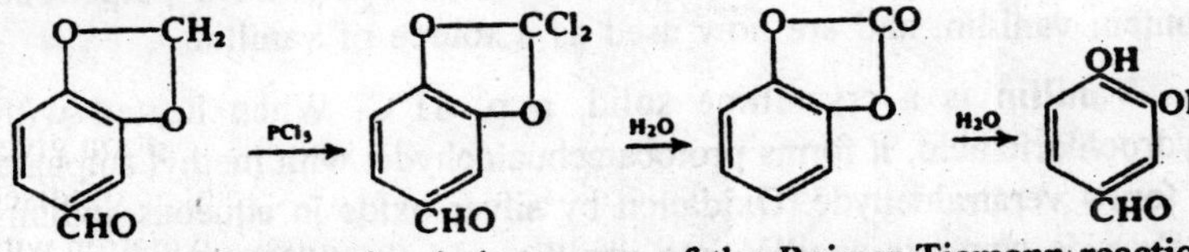

Another preparation is by means of the Reimer-Tiemann reaction using catechol

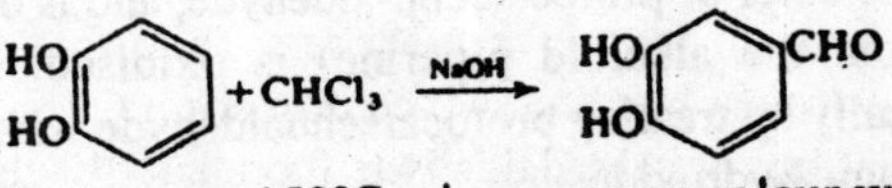

Protocatechualdehyde, m.p. 153°C, gives a green colour with ferric chloride, and reduces ammoniacal silver nitrate. When methylated with approximately one equivalent of methyl sulphate, protocatechu-aldehyde gives a mixture of vanillin (1) and isovanillin (II); excess of methyl sulphate give; veratraldehyde (III):

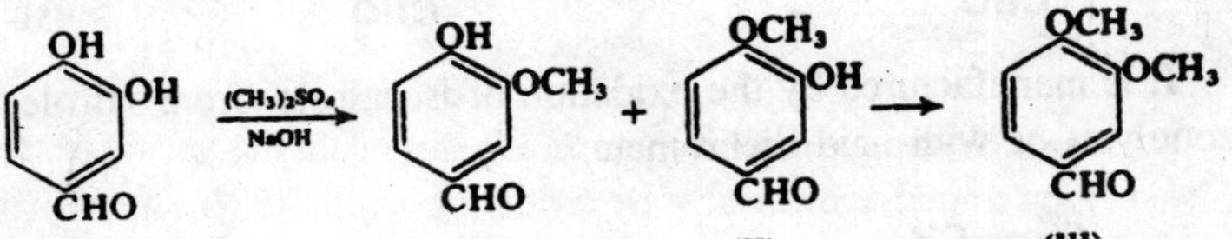

Vanillin (p-hydroxy-m-methoxybenzaldehyde) occurs in many substances of plant origin, *e.g.*, the vanilla bean. It may be prepared synthetically from guaiacol by means of the Reimer-Tiemann or the Gattermann reaction. Industrially, it is prepared:

(i) By oxidising isoeugenol (from eugenol) with nitrobenzene:

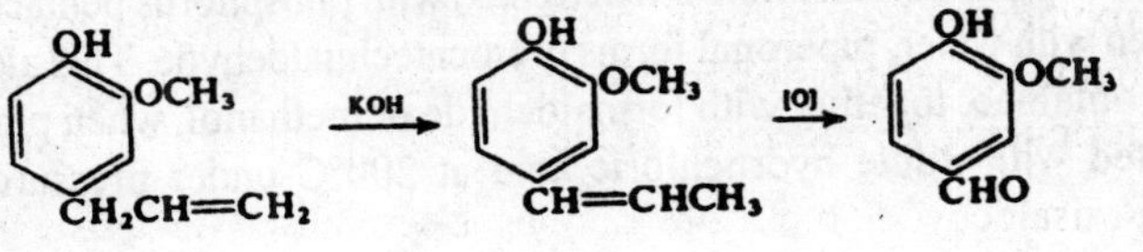

If the oxidation is carried out with acid dichromate, it is necessary to protect the hydroxyl group (by acetylation, etc.). On the other hand, ozonolysis may be used without protecting the hydroxyl group.

(ii) By the Lederer-Manasse reaction as follows:

$$\text{(OH, OCH}_3\text{-benzene)} + HCHO \xrightarrow{NaOH} \text{(OH, OCH}_3\text{, CH}_2\text{OH)} \xrightarrow{C_6H_5NHOH} \text{(OH, OCH}_3\text{, CH=NC}_6\text{H}_5\text{)} \xrightarrow{H^+} \text{(OH, OCH}_3\text{, CHO)}$$

(iii) The liquors from the extraction of lignin from pulp-wood contain vanillin, and are now used as a source of vanillin.

Vanillin is a crystalline solid, m.p. 81°C. When heated with hydrochloric acid, it forms protocatechualdehyde; with methyl sulphate it forms veratraldehyde. Oxidation by silver oxide in aqueous sodium hydroxide converts vanillin into vanillic acid (96-97%).

Piperonal (*heliotropin*, 3, 4-*methylenedioxybenzaldehyde*) is the methylene ether of protocatechu-aldehyde, and is obtained when piperic acid (from the alkaloid piperine) is oxidised. It may be prepared synthetically by treating protocatechualdehyde with methylene di-iodide and sodium hydroxide:

$$\text{(OH, OH, CHO)} + CH_2I_2 + 2NaOH \longrightarrow \text{(O–CH}_2\text{–O, CHO)} + 2NaI + 2H_2O$$

It is manufactured by the oxidation of isosafrole (from safrole) by ozonolysis or with acid dichromate:

$$\text{(O–CH}_2\text{–O, CH}_2\text{CH=CH}_2\text{)} \xrightarrow{KOH} \text{(O–CH}_2\text{–O, CH=CHCH}_3\text{)} \xrightarrow{[O]} \text{(O–CH}_2\text{–O, CHO)}$$

Piperonal, m.p. 37°C, may be oxidised to piperonylic acid and reduced to piperonyl alcohol. When treated with phosphorus pentachloride and then with water, piperonal forms protocatechualdehyde. This aldehyde is also obtained, together with formaldehyde or methanol, when piperonal is heated with dilute hydrochloric acid at 200°C under pressure.

AROMATIC KETONES

Aromatic ketones may be either alkaryl ketones or diaryl ketones.

The infrared absorption region of the carbonyl group (C=O stretch) in alkaryl ketones is 1700-1680 cm^{-1} (s), and in diaryl ketones is 1 670-1660 cm^{-1} (s). In both cases, the frequencies are lower than for acyclic and cyclic ketones. At the same time, the benzene nucleus and the nature of the orientation may be identified. There may, however, be complications due to intramolecular hydrogen bonding in certain o-derivatives.

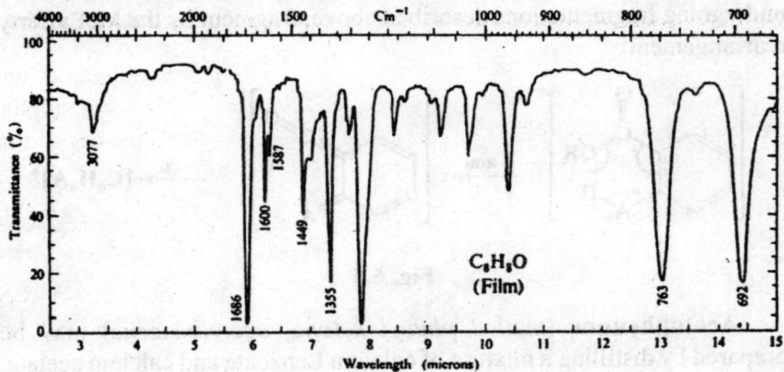

Fig. 5.4

1 686 cm^{-1} : C=0 (str.) in alkaryl ketones. 1600, 1 587. and 1 449 cm^{-1}: C=C (str.) in aromatic nuclei. 3077 cm^{-1}: =C—H (str.) in aromatics. 1355 cm^{-1}: C—H (def.) in Me in MeCO. 763 and 692 cm^{-1}: Monosubstituted benzenes.

The infra-red spectrum of acetophenone (as a film) is shown in Fig. 5.4.

Mass spectra. The fragmentation patterns of aromatic aldehydes (R = H), ketones (R = R), acids (R = OH), methyl esters (R = OMe), and amides (R = NH_2) are similar in that all under; β-cleavage with loss of R, and formation of the benzoyl cation. All give strong peaks for the molecule ion.

$$[C_6H_5COR]^{+\cdot} \xrightarrow{-R\cdot} C_6H_5C\equiv\overset{+}{O} \xrightarrow{-CO} C_6H_5^{+} \xrightarrow{-C_2H_2} C_4H_3^{+}$$

$M^{+\cdot}$ m/e 105 m/e 77 m/e 51

Esters in which the alkyl group is ethyl or higher also eliminate alkenes (cf. ethers):

$$\left[C_6H_5C(=O)-O-CHR-CH_2-H\right]^{+} \longrightarrow \left[C_6H_5C(=O)OH\right]^{+} + RCH{=}CH_2$$

o-Substituted acids and esters (A = CH_2, O, NH), in addition to undergoing fragmentations described above, fragment by the McLafferty rearrangement:

$$\left[o\text{-}C_6H_4(COOR)(AH)\right]^{+} \xrightarrow{-ROH} \left[C_6H_4(=C{=}O)(=A)\right]^{+} \xrightarrow{-CO} [C_6H_4A]^{+}$$

Fig. 5.5

Acetophenone (*methyl phenyl ketone, acetylbenzene*) may be prepared by distilling a mixture of calcium Lenzoate and calcium acetate, or better, by heating a mixture of the corresponding iron salts (Granito *et al.*, 1963).

$$(C_6H_5CO_2)_2Ca + (CH_3CO_2)_2Ca \longrightarrow 2C_6H_5COCH_3 + 2CaCO_3$$

A convenient laboratory method is by means of the Friedel-Crafts reaction:

$$C_6H_6 + CH_3COCl \xrightarrow{AlCl_3} C_6H_5COCH_3 + HCl\ (50\%)$$

Acetophenone is manufactured by the catalytic oxidation of ethylbenzene with air at 126°C under pressure in the presence of manganous acetate as catalyst.

Alkaryl ketones may be prepared from diazonium salts and aldoximes other than formaldoxime.

Acetophenone, m.p. 20°C, is used as an hypnotic (under the name of *hypnone*) and in perfumery. When reduced with sodium and ethanol, acetophenone gives *phenylmethylmethanol*, $C_6H_5CHOHCH_3$; reduction by Clemmensen's method gives ethylbenzene. Ethylbenzene is also produced when the reducing agent is lithium aluminium hydride containing aluminium chloride. Oxidation with cold potassium permanganate gives phenylglyoxylic acid (benzoylformic acid which, on further oxidation, is converted into benzoic acid:

$$C_6H_5COCH_3 \xrightarrow{[O]} C_6H_5COCO_2H \xrightarrow{[O]} C_6H_5CO_2H$$

Phenylglyoxylic acid may be prepared as follows:

$$PhCOCl \xrightarrow{CuCN} \underset{(60\text{–}65\%)}{PhCOCN} \xrightarrow{HCl} \underset{(73\text{–}77\%)}{PhCOCO_2H}$$

Acetophenone is oxidised by selenium dioxide to *phenylglyoxal:*

$$C_6H_5COCH_3 + SeSO_2 \longrightarrow C_6H_5COCHO + Se + H_2O \ (69\text{–}72\%)$$

Acetophenone undergoes the Baeyer-Villiger oxidation to form predominantly, with perbenzoic acid, phenyl acetate (63%).

This is in keeping with the general observation that it is the more strongly electron-releasing group that migrates.

Acetophenone can be chloromethylated, and it is readily halogenated in the ω-position *e.g.*, *phenacyl bromide* (ω-bromoacetophenone) is formed when acetophenone is treated with bromine in ether at 0°C in the presence of a small amount of aluminium chloride:

$$C_6H_5COCH_3 + Br_2 \longrightarrow C_6H_5COCH_2Br + HBr \ (64\text{–}66\%)$$

Phenacyl bromide (m.p. 51°C) is used to identify acids, with which it forms well-defined crystalline esters. It is oxidised to phenylglyoxal by dissolving it in a large excess of dimethyl sulphoxide (Kornblum *et al.*, 1957).

When treated with two equivalents of bromine, acetophenone forms ω, ω-dibromoacetophenone (*phenacylidene bromide*). This compound undergoes rearrangement on treatment with alkali to form *mandelic acid*. A probable mechanism is one involving an *intramolecular* Cannizzaro reaction (cf. glyoxal):

$$C_6H_5\text{—CO—}CHBr_2 \xrightarrow{OH^-} C_6H_5\text{—}\overset{\overset{\Large O}{\|}}{C}\text{—}C(=O)\text{—}H$$

$$\rightleftharpoons C_6H_5\text{—}\overset{\overset{\Large O}{\|}}{C}\text{—}\underset{\underset{\Large H}{|}}{\overset{\overset{\Large OH}{|}}{C}}\text{—}O^- \xrightarrow[\text{shift}]{H^-} C_6H_5\text{—}\underset{\underset{\Large H}{|}}{\overset{\overset{\Large O^-}{|}}{C}}\text{—}\overset{\overset{\Large OH}{|}}{C}\text{=C} \longrightarrow$$

$$C_6H_5-\underset{H}{\overset{OH}{\overset{|}{C}}}-\overset{O^-}{\overset{|}{C}}=O \xrightarrow{HCl} C_6H_5-CHOH-CO_2H$$

This rearrangement does not take place if both o-positions are occupied, *e.g.*, in the ω, ω-dibromo-derivative of acetomesitylene. This is probably due to a steric effect.

Phenacyl chloride, m.p. 59°C, may be conveniently prepared by the Friedel-Crafts condensation between chloroacetyl chloride and benzene:

$$C_6H_6 + ClCH_2COCl \xrightarrow{AlCl_3} C_6H_5COCH_2Cl + HCl$$

Acetophenone forms an oxime, phenylhydrazone and cyanohydrin but does not form bisulphite compound. It reacts with ammonia to form acetophenoneammonia PhCMe=NCMePhN=CMePh (cf. benzaldehyde), and with Grignard reagents in the usual way. When however, both o-positions in acetophenone are occupied, the compound exhibits steric hindrance: *e.g.* acetomesitylene does not readily form an oxime, and does not react with Grignard reagents in the usual way.

The acetyl group is w-orienting and so when acetophenone undergoes nuclear substitution, the main product is the m-derivative.

[Structure: benzene ring with $COCH_3$ at top, CH_3 at both ortho positions and CH_3 para]

Nuclear halogenation is difficult, since side-chain substitution occurs so readily.

However, Pearson *et al.* (1958) have prepared m-bromoacetophenone (75%) by bromination in the presence excess of aluminium chloride (which forms a complex with the carbonyl group: cf. Friedel-Crafts reaction). These conditions are referred to as 'swamping' catalytic conditions.

o- and p-Substituted acetophenones are prepared indirectly, *e.g.*

$$o\text{-}MeC_6H_4NO_2 \xrightarrow[(ii)\ EtOH-HCl]{(i)\ Na_2Cr_2O_7-H_2SO_4} o\text{-}NO_2C_6H_4CO_2Et \xrightarrow[EtONa]{MeCO_2Et} o\text{-}NO_2C_6H_4COCH_2CO_2Et \xrightarrow[(ii)\ H_3O^+]{(i)\ KOH} o\text{-}NO_2C_6H_4COMe$$

In the presence of aluminium t-butoxide, acetophenone undergoes condensation to form **dypnone** (b.p. 340-345°C):

$$C_6H_5C(CH_3)=O + CH_3COC_6H_5 \longrightarrow C_6H_5C(CH_3)=CHCOC_6H_5 + H_2O \quad (77\text{–}82\%)$$

In the presence of hydrochloric acid, 1,3,5-triphenylbenzene (m.p. 172°C) is formed

$$3C_6H_5COCH_3 \longrightarrow 1,3,5\text{-}(C_6H_5)_3C_6H_3$$

Acetophenone condenses with acetic anhydride in the presence of boron trifluoride to form *benzoylacetone*, m.p. 61°C (Meerwein *et al.*, 1934):

$$C_6H_5COCH_3 + (CH_3CO)_2O \xrightarrow{BF_3} C_6H_5COCH_2COCH_3 + CH_3CO_2H \quad (83\%)$$

This is also formed by the condensation of acetophenone with ethyl acetate in the presence of sodium ethoxide:

$$C_6H_5COCH_3 + CH_3CO_2C_2H_5 \xrightarrow{C_2H_5ONa} C_6H_5COCH_2COCH_3 \quad (66\%)$$

Acetophenone condenses with ethyl benzoate in the presence of sodium ethoxide to form *dibenzoylmethane*, m.p. 77° C.

$$C_6H_5CO_2C_2H_5 + CH_3COC_6H_5 \xrightarrow{C_2H_5ONa} C_6H_5COCH_2COC_6H_5 \quad (62\text{–}71\%)$$

It also undergoes the Mannich reaction:

$$C_6H_5COCH_3 + HCHO + (CH_3)_2NH{\cdot}HCl \longrightarrow C_6H_5COCH_2CH_2N(CH_3)_2.HCl \quad (60\%)$$

Aromatic aldehydes and ketones generally behave as bases in concentrated sulphuric acid; cryoscopic experiments show that *i* (van't Hoff factor) is 2, and this value agrees with the following equation (R = H or R):

$$ArCOR + H_2SO_4 \rightleftharpoons [ArC(=\overset{+}{O}H)R]\,[HSO_4^-]$$

The carbonyl compound may be recovered by dilution with water.

Wiligerodt reaction (1887). This is the name given to those reactions in which a carbonyl compound is converted into an amide with the same number of carbon atoms. The reaction was originally carried out by heating an aryl alkyl ketone with an aqueous solution of yellow ammonium polysulphide: *e.g.*, acetophenone forms the amide of phenylacetic acid together with a small amount of the ammonium salt:

$$C_6H_5COCH_3 \xrightarrow{(NH_4)_2S_x} C_6H_5CH_2CONH_2 + C_6H_5CH_2CO_2NH_4$$

The amido-group is always formed at the end of the chain whatever the size of the R group in C_6H_5COR *e.g.*, butyrophenone forms γ-phenylbutyramide:

$$C_6H_5COCH_2CH_2CH_3 \longrightarrow C_6H_5CH_2CH_2CH_2CONH_2$$

The Wiligerodt reaction is very useful for preparing aryl-substituted aliphatic acids.

A modified technique is to use equimolecular amounts of sulphur and a secondary amine (Kindler, 1923, 1927), particularly morpholine.

NH O The final product is a thioamide and this, on acid or alkaline hydrolysis, gives an acid; or alternatively, on electrolytic reduction the thioamide gives an amine.

According to Mayer *et al.* (1964), the reaction proceeds via examine intermediates (I and II), *e.g.*,

$$ArCOCH_2CH_3 + R_2NH \longrightarrow \underset{\text{(I)}}{ArC(NR_2){=}CHCH_3}$$

$$\xrightleftharpoons{\text{isomn. by S}} \underset{\text{(II)}}{ArCH_2C(NR_2){=}CH_2} \xrightleftharpoons{S}$$

$$ArCH_2C(NR_2){=}CHSH \xrightarrow{\text{isomn.}} ArCH_2CH_2C({=}S)NR_2$$

$$ArCH_2CH_2C({=}S)NR_2 \xrightarrow{H_3O^+} ArCH_2CH_2CO_2H$$

$$ArCH_2CH_2C({=}S)NR_2 \xrightarrow{e;\ H^+} ArCH_2CH_2Nr_2$$

It should be noted that isomerisation occurs along the side-chain until the amino-group reaches the *end* carbon atom.

Phenolic ketones may be prepared by the **Houben-Hoesch synthesis** (1927). This is the condensation of cyanides with polyhydric phenols, particularly m-compounds, in the presence of zinc chloride and hydrogen chloride; *e.g.*, phloroglucinol condenses with methyl cyanide form *phloroacetophenone* (m.p. 219°C)

$$C_6H_3(OH)_3 + CH_3CN \xrightarrow[HCl]{ZnCl_2} \left[C_6H_2(OH)_3C(CH_3){=}NH \right] \xrightarrow{H^+} C_6H_2(OH)_3COCH_3 \quad (74\text{–}87\%)$$

Fig. 5.6

This method is really an extension of Gattermann's phenolic aldehyde synthesis. It is applicable to phenol itself.

It is probable that the imido chloride is an intermediate:

$$MeC{\equiv}N + HCl \rightleftharpoons MeCCl = NH$$

Phenolic ketones may also be prepared by the Fries rearrangement , and by heating polyhydric phenols with aliphatic acids in the presence off used zinc chloride; *e.g.*, pyrogallol and acetic acid form *gallacetophenone* (m.p. 173°C):

$$C_6H_3(OH)_3 + CH_3CO_2H \xrightarrow{ZnCl_2} C_6H_2(OH)_3COCH_3 + H_2O$$

Benzophenone (*diphenyl ketone*) may be prepared in a similar manner to acetophenone, *e.g.*,

(i) By heating calcium benzoate:

$$(C_6H_5CO_2)_2Ca \longrightarrow C_6H_5COC_6H_5 + CaCO_3 \ (30\%)$$

(ii) By the Friedel-Crafts condensation between benzoyl chloride and benzene:

$$C_6H_6 + C_6H_5COCl \xrightarrow{AlCl_3} C_6H_5COC_6H_5 + HCl \ (80\%)$$

Carbonyl chloride may be used instead of benzoyl chloride:

$$2C_6H_6 + COCl_2 \xrightarrow{AlCl_3} C_6H_5COC_6H_5 + 2HCl$$

The Friedel-Crafts reaction may also be carried out with carbon tetrachloride, followed by steam distillation of the dichloro-compound produced (this method is used commercially):

$$2C_6H_6 + CCl_4 \xrightarrow{AlCl_3} (C_6H_5)_2CCl_2 \xrightarrow{H_2O} (C_6H_5)_2CO\ (80\text{–}89\%)$$

Benzophenone exists in two solid forms, a stable, m.p. 49°C, and an unstable form, m.p. 26°C. It cannot be chloromethylated (cf. acetophenone). It is reduced by zinc and ethanolic potassium hydroxide **to benzhydrol** (*diphenylcarbmol*, m.p. 68°C) and by zinc and acetic acid to **benzopinacol** (m.p. 188°C):

$$\underset{(90\%)}{(C_6H_5)_2C(OH)C(OH)(C_6H_5)_2} \xleftarrow[CH_3CO_2H]{Zn} (C6H5)2CO$$

$$\xrightarrow{Zn/KOH} \underset{(90\%)}{(C_6H_5)_2CHOH}$$

When benzophenone, dissolved in isopropanol to which a drop of acetic acid has been added, is exposed to bright sunlight, benzopinacol is formed.

$$2(C_6H_5)_2CO + (CH_3)_2CHOH$$
$$\xrightarrow{h\nu} (C_6H_5)_2C(OH)(C_6H_5)_2 + (CH_3)_2CO \quad (93\text{–}94\%)$$

Benzopinacol is also obtained by the reduction of benzophenone with a mixture of magnesium and magnesium iodide.

$$2(C_6H_5)_2CO \xrightarrow{MgI_2} (C_6H_5)_2C(O^-)\text{—}C(O^-)(C_6H_5)_2\ \ Mg^{2+}$$

$$\xrightarrow{H^+} (C_6H_5)_2C(OH)C(OH)(C_6H_5)_2$$

When heated in acetic acid solution in the presence of iodine as catalyst, benzopinacol undergoes the pinacol-pinacolone rearrangement to form benzopinacolone (m.p. 179°C):

When distilled with zinc dust, or reduced with a mixture of lithium aluminium hydride and aluminium chloride, benzophenone forms diphenylmethane, and when fused with potassium hydroxide, it forms benzene and potassium benzoate:

$$(C_6H_5)_2CO + KOH \longrightarrow C_6H_6 + C_6H_5CO_2K$$

Benzophenone forms an oxime, but does not form a cyanohydrin or a bisulphite compound.

Benzophenone in ether dissolves sodium without the evolution of hydrogen, forming a blue solution containing radical anions known as **ketyls**.

$$(C_6H_5)_3CO + Na \longrightarrow (C_6H_5)_2\dot{C}—\bar{O}\,Na^+$$

Aromatic ketyls are stabilised by resonance:

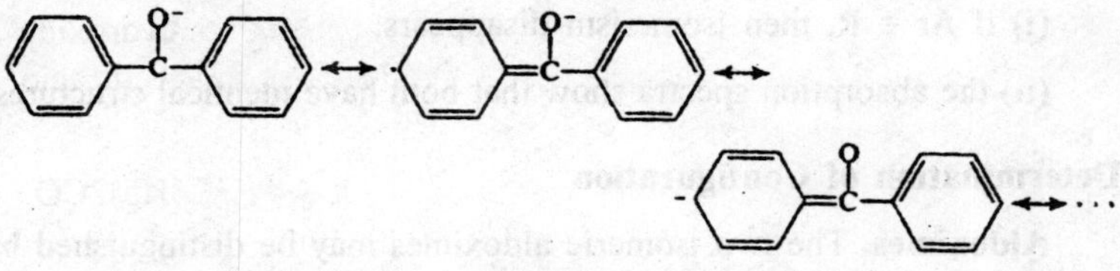

Fig. 5.7

Dilute acid converts the ketyl into benzopinacol; water produces benzophenone and benzhydrol; and iodine regenerates' benzophenone.

If an excess of sodiums added to benzophenone in ether, the violet dianion, $(C_6H_5)_2C^-—O^-$, produced by addition of a second electron.

A very important derivative of benzophenone is Michler's ketone, m.p. 172°C, which may be prepared by treating dimethylaniline with carbonyl chloride in the presence of zinc chloride.

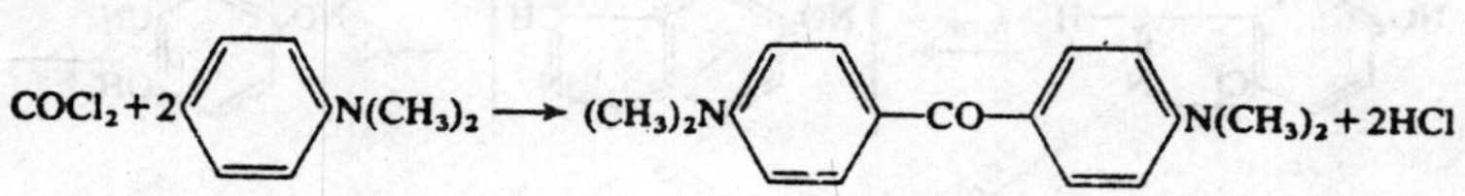

$$COCl_2 + 2C_6H_5N(CH_3)_2 \longrightarrow (CH_3)_2NC_6H_4—CO—C_6H_4N(CH_3)_2 + 2HCl$$

Michler's ketone is used in the preparation of certain triphenyl-methane dyes.

STEREOCHEMISTRY OF ALDOXIMES AND KETOXIMES

Many aromatic aldoximes and ketoximes exist in two isomeric forms, and this was explained by Hantzsch and Wemer (1890) as being due to geometrical isomerism. According to these authors, nitrogen is tervalent (in oximes) and is situated at one corner of a tetrahedron with its three valencies directed towards the other three corners; consequently the three valencies are not coplanar. These authors also assumed that there is no free rotation about the C=N bond and therefore proposed configurations (I) and (II) for the two isomers:

```
Ar   R          Ar   R
  \ /             \ /
   C               C
   ||              ||
   N               N
    \             /
     OH         HO
   (I)            (II)
```

Many facts are in favour of geometrical isomerism, *e.g.*,

(i) if Ar = R, then isomerism disappears;

(ii) the absorption spectra show that both have identical structures.

Determination of Configuration

Aldoximes. The two isomeric aldoximes may be distinguished by the behaviour of their acetyl derivatives towards aqueous carbonate; one gives the oxime back again, and the other forms the cyanide ArCN.

Brady *et al.* (1925) showed that the cyanide is formed by *anti*-elimination. These-authors found that only one of the two isomers of 2-chloro-5-nitro-benzaldoxime readily gave ring closure on treatment with sodium hydroxide; this isomer is therefore the *anti*-(*trans*-) isomer. Furthermore, it was this isomer that gave the cyanide, thus showing that anti-elimination must have occurred.

Fig. 5.8

Actually, the ring compound produced, the 5-nitrobenzisoxazole, is unstable, and rearranges to nitrosalicylonitrile.

Thus the general reaction may be formulated as follows, the *anti*-oxime giving the cyanide and the *syn*-(*cis*-)oxime regenerating the oxime:

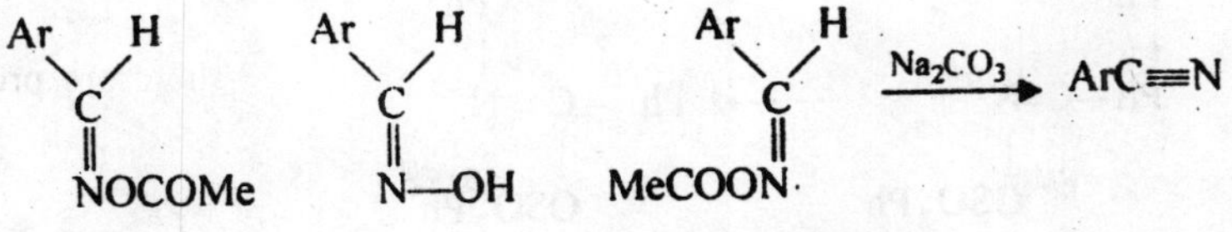

Ar H / C = NOCOMe (syn-form); Ar H / C = N—OH; Ar H / C = MeCOON (anti-form) $\xrightarrow{Na_2CO_3}$ ArC≡N

syn-form anti-form

Ketoximes: Beckmann rearrangement (1886). When treated with reagents such as phosphorus pentachloride, sulphuric acid, polyphosphoric acid, etc., aromatic ketoximes undergo the Beckmann rearrangement to form an acid amide:

$$Ar_2C{=}NOH \longrightarrow ArCONHAr$$

Meisenheimer (1921, 1925) showed that the rearrangement occurs by *anti*-rearrangement, *e.g.*, he (1925) showed that the α-oxime of 2-bromo-5-nitroacetophenone is unaffected by sodium hydroxide, whereas the β-isomer undergoes ring closure to form 3-methyl-5-nitrobenzisoxazole; thus the α-oxime is the *syn*-methyl isomer (I), and the β-oxime the *anti*-methyl isomer (II). When treated with sulphuric acid or phosphorus pentachloride, the α-oxime underwent the Beckmann rearrangement to give the N-substituted acetamide; thus the exchange occurs in the *anti*-positions.

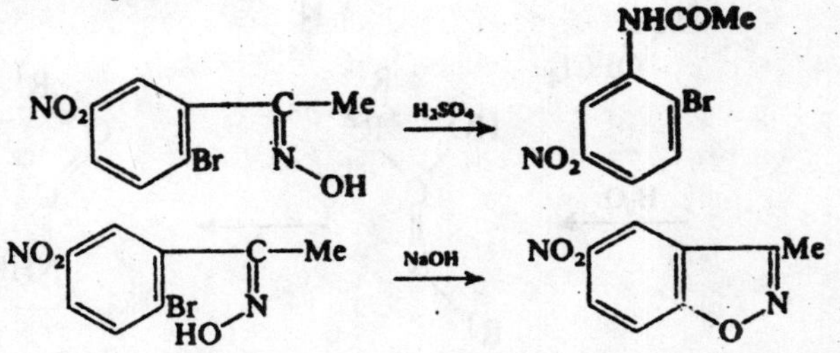

NO₂-C₆H₃(Br)—C(Me)=N—OH $\xrightarrow{H_2SO_4}$ NHCOMe, Br, NO₂ (N-substituted acetamide)

(II) NO₂-C₆H₃(Br)—C(Me)=N(HO) $\xrightarrow{NaOH}$ NO₂, Me, N, O (3-methyl-5-nitrobenzisoxazole)

Hence, by identifying the amide and using the fact that the rearrangement occurs in the *anti*-position, the configuration of the ketoxime can be determined.

The mechanism of the Beckmann rearrangement is still the subject of much discussion. Since the oxime itself does not rearrange, it is reasonable to suppose that some intermediate is former between oxime and reagent, and it is this intermediate which rearranges. Kuhara *et al.* (1914, 1916) prepared the benzenesulphonate of benzophenone oxime and showed this readily underwent rearrangement in neutral solvents to give an isomeric compound (III) which gave benzanilide on hydrolysis; thus:

$$\underset{\substack{|\\ OSO_2Ph}}{\overset{\substack{Ph\\ |}}{Ph{-}C{=}N}} \longrightarrow \underset{\substack{|\\ OSO_2Ph}}{Ph{-}\overset{}{C}{=}\overset{\substack{Ph\\ |}}{N}} \xrightarrow{H^+} PhCONHPh + PhSO_3H$$

The structure of (III) was assigned to it on the basis of the similarity of its absorption spectrum with that of synthetic compounds of similar structure. Chapman (1934) showed the rate of rearrangement of benzophenone oxime picryl ester is more rapid in polar solvents than in non- polar. This is strong evidence that the rate-determining step is the ionisation of the intermediate Furthermore, Kenyon el al. (1946) found that when (+)-α-phenylethyl methyl ketoxime is rearranged with sulphuric acid, the product is almost 100 per cent optically pure (retention). Thus the rearrangement is intramoiecular. A mechanism which fits these facts is the 1,2-shift a follows (for PCl_5):

$$R^1R^2C{=}N{-}OPCl_4 \longrightarrow R^2\overset{+}{C}{=}N{-}R^1 + OPCl_4^- \longrightarrow Cl_4PO{-}C(R^2){=}N{-}R^1 \xrightarrow{H_2O} HO{-}C(R^2){=}N{-}R^1 \longrightarrow O{=}C(R^2){-}NHR^1$$

(When sulphuric acid is used, replace $OPCl_4$ in the above equation by OH_2^+) When the migrating group is aryl and contains an electron-releasing group in the p-position, the rearrangement is accelerated. This may cited as evidence for the formation of a bridged ion.

[Structure: bridged ion with + in ring, C–R and N]

Other mechanisms have also been suggested (Stephen *et al.*).

EXERCISES

1. Distinguish between PhCOEt and p-MeC_6H_4COMe by:

   (i) a chemical method;

   (ii) a physical method.

2. Complete the following equations:

   (i) $Ph_2CO \xrightarrow[NaOH]{Zn} ?$

   (ii) p-MeC6H4CHO $\xrightarrow[KOH]{HCHO}$?

   (iii) PhCHO + NaCN + ? $\xrightarrow[H_2O]{MeOH}$ PhCH(CN)CH(CN)Ph

   (iv) PhCH=CHCOPh + KCN $\xrightarrow{AcOH}$?

   (v) $PhCH_2OH + ? \longrightarrow PhCH_2OCOCl$

   (vi) PhCHBrCHBrCHO $\xrightarrow[heat]{K_2CO_3}$?

   (vii) PhCH(CN)COMe $\xrightarrow[heat]{aq.H_2SO_4}$?

   (viii) PhCOMe $\xrightarrow{?}$ PhCOCHO

   (ix) $PhCO_2H \xrightarrow{SOCl_2} ? \xrightarrow{Me_2NH} ? \xrightarrow{?} PhCHO$

3. Which of the carbonyl groups in p-$MeOC_6H_4COMe$ and p-$NO_2C_6H_4COMe$ protonates more readily in acid solution and why?

4. Starting with PhH. $EtCO_2H$ and $PhNH_2$, synthesise $PhCH_2CH_2$-$CONH_3PHh$.

5. Convert PhCHO into PhCH = CHCOPh by two methods.

6. Convert $Br(CH_2)_3CHO$ into $PhCHOH(CH_2)_3CHO$.

7. What are the products formed when the syn-phenyl and anti-phenyl monoximes of bezil undergo the acid-catalysed Beckmann rearrangement?

8. Suggest a mechanism for the rearrangement: $Ph_2CHCHO \xrightarrow{H^+} PhCH_2COPh$.

9. Discuss the two reactions:

$$PhCOMe \xrightarrow[NaOH]{NaOCl} PhCO_2^- + CHCl_3$$

$$2,6\text{-}Me_2C_6H_3COMe \xrightarrow[NaOH]{NaOCl} 2,6\text{-}Me_2C_6H_3COCCl_3$$

10. Suggest what A and B could be in the reactions:

o-$H_2NC_6H_4CHO$ $\xleftarrow{A}$ o-$O_2NC_6H_4CHO$ $\xrightarrow{B}$ o-$O_2NC_6H_4CH_2OH$

11. The deamination of $Ph_2C(OH)CH_2NH_2$ ($NaNO_2$—HCl) gives a product which, on oxidation, gives benzoic acid only. Formulate the steps involved.

12. Convert:

    (i) p-xylene ⟶ 2 ,5-$Me_2C_6H_3CHO$;

    (ii) Ph ⟶ m-nitroacetophenone;

    (iii) $PhNO_2$ ⟶ m-bromo-benzaldehyde;

    (iv) PhH ⟶ p-$OCHC_6H_4CHO$;

    (v) PhMe ⟶ p-MeC_6H_4CH = $CHNO_2$.

    (vi) pHMe ⟶ p-MeC_6H_4CH=$CHNO_2$.

13. Compound A, $C_8H_6O_2$, on treatment with aq. NaOH followed by acidification, gives B, $C_8H_8O_3$, which on oxidation, gives benzoic acid only. Suggest structure for A and B, and formulate the reactions.

14. The mechanism of the Cannizzaro reaction involves addition of hydrogen to a benzaldehyde molecule. Devise an experiment to show whether the hydrogen comes from the aqueous solvent or not.

15. Suggest paths whereby o-methylbenzyl alcohol can produce the following ions in its mass spectrum: 122 (30%), 120 (1%), 119(4%), 104(100%), 91(36%).

16. Account for the following ions in the mass spectrum of methyl salicylate (methyl o-hydroxybenzoate): 152 (41%), 121 (32%), 120 (100%), 93 (13%), 92 (58%), 65(26%), 64 (17%).

6

POLYNUCLEAR HYDROCARBONS AND THEIR DERIVATIVES

INTRODUCTION

Polynuclear hydrocarbons may be divided into two groups, those in which the rings are isolated, *e.g.*, biphenyl, diphenylmethane, etc.; and those in which two or more rings are fused together in the o-positions, *e.g.*, naphthalene, anthracene, etc.

ISOLATED SYSTEMS

Biphenyl (diphenyl), m.p. 71°C, occurs in small quantities in coal-tar. It may be prepared:

(i) **By Fittig's reaction** (1863). This is carried out by treating bromobenzene with sodium in ethereal solution:

$$2C_6H_5Br + 2Na \longrightarrow C_6H_5C_6H_5 + 2NaBr \ (20–30\%)$$

(ii) When bromobenzene in ethanolic solution is made alkaline and refluxed with hydrazine in the presence of a palladium catalyst on a calcium carbonate support, biphenyl is obtained (Busch *et al.*, 1936):

$$2C_6H_5Br \longrightarrow C_6H_5C_6H_5 + 2HBr \ (80\%)$$

(iii) (a) By warming benzenediazonium sulphate in ethanol with copper powder (see also the Gomberg reaction):

$$2C_6H_5N_2^+HSO_4^- \xrightarrow[C_2H_5OH]{Cu} C_6H_5C_6H_5 + 2N_2 + 2H_2SO_4 \ (25\%)$$

The yield has been improved (to 50%) by warming a mixture of aniline, pentyl nitrite and benzene and finally refluxing (Cadogan, 1962). This method may also be used for preparing biphenyl derivatives.

(b) By diazotising benzidine and allowing the diazonium salt solution to stand in contact with hypophosphorous acid:

$$H_2N-C_6H_4-C_6H_4-NH_2 \xrightarrow{NaNO_2/HCl} \bar{Cl}\overset{+}{N_2}-C_6H_4-C_6H_4-\overset{+}{N_2}\bar{Cl} \xrightarrow{H_3PO_2} C_6H_5-C_6H_5 \quad (60\%)$$

(iv) By the **Ullmann biaryl synthesis** (1903). Iodobenzene is heated with copper powder in a sealed tube:

$$2C_6H_5I + 2Cu \longrightarrow C_6H_5C_6H_5 + 2CuI \qquad (80\%)$$

Aryl chlorides and bromides do not usually react unless there is – I group o- and/or p- to the halogen atom, *e.g.*, o-chloronitrobenzene forms 2,2'-dinitrobiphenyl:

$$2\,(o\text{-}NO_2C_6H_4Cl) \xrightarrow{Cu} (o\text{-}NO_2C_6H_4)-(C_6H_4NO_2\text{-}o) \quad (52\text{–}61\%)$$

Kornblum *et al.* (1952) have shown that dimethylformamide is a good solvent for this synthesis, and the yields are higher.

There is still uncertainty about the mechanism of the Ullmann biaryl synthesis; both a free-radical and an ionic mechanism have been proposed.

(v) Biphenyl is formed by reaction between phenylmagnesium bromide and bromobenzene in the presence of a small amount of cobaltous chloride.

$$C_6H_5MgBr + C_6H_5Br \xrightarrow{CoCl_2} C_6H_5C_6H_5 + MgBr_2 \text{ (g.)}$$

Industrially, biphenyl is prepared by passing benzene vapour through a red-hot tube, preferably packed with pumice (600-800°C):

A more recent method is to mix benzene vapour (preheated to 650°C) with superheated steam (1000–1100°C), and passing the mixture into steel vessels coated internally with a film of Fe_3O_4.

Biphenyl is increasing in use as a heat transfer medium; chlorinated biphenyls are used as plasticisers.

Biphenyl undergoes the usual nuclear substitution reactions, the phenyl group being o- and p-orienting (one phenyl group behaving as an electron-releasing group and the other as an electron-acceptor). The first substituent enters mainly the 4-position, and to a lesser extent the 2-position. Whether this group is electron-releasing or withdrawing, introduction of a second substituent usually takes place in the *unsubstituted* ring in the p-and o-positions, *e.g.*, on nitration, the main product is 4-nitrobiphenyl, together with a small amount of 2-nitrobiphenyl.

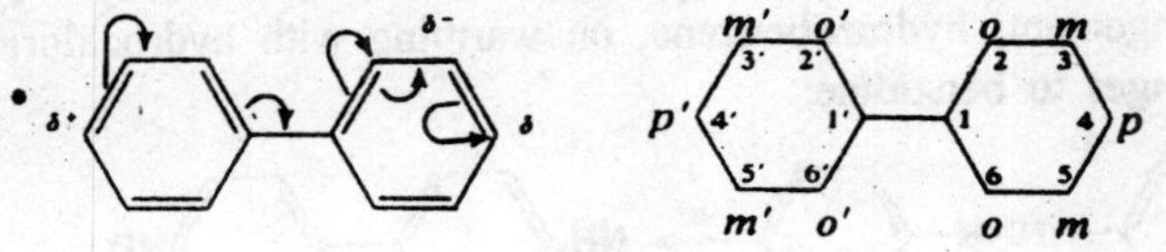

Fig. 6.1

Further nitration gives 4,4'-dinitrobiphenyl (and some 2,4'- and 2,2'-dinitrobiphenyls). Biphenyl behaves similarly on halogenation and sulphonation.

According to the charge distribution approach, a p-nitro-group would have been expected to direct to the m-position in the other ring:

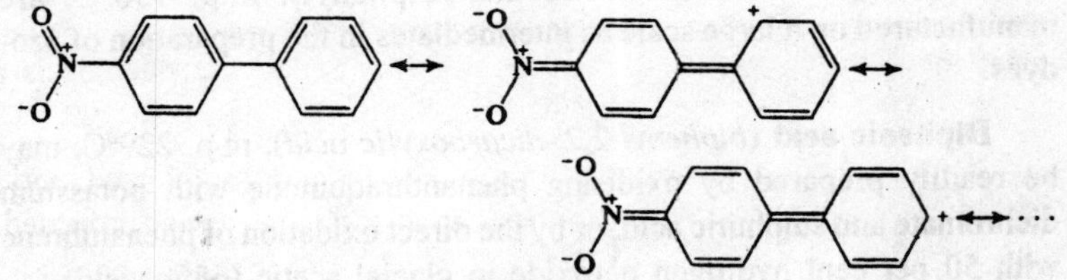

Fig. 6.2

On the other hand, if the stabilities of the carbonium ion intermediates are examined, it will be seen that there is greater dispersal of charge in the -(or o-) ion than in the m-. Consequently, the former is more stable and so this orientation is favoured.

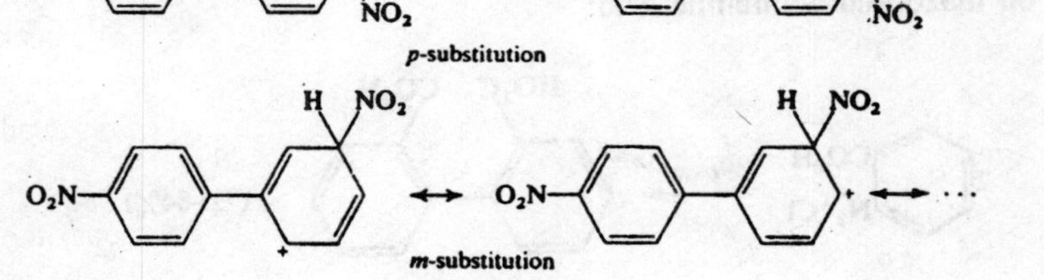

p-substitution

m-substitution

Fig. 6.3

When oxidised with chromic acid, biphenyl forms a small amount of benzoic acid, most of the hydrocarbon being oxidised completely to carbon dioxide and water.

On the other hand ozonolysis of biphenyl at – 20°C gives an 86 per cent yield of benzoic acid (Copeland *et al.* 1961).

Benzidine (4,4-*diaminobiphenyl*) may be prepared by the benzidine rearrangement; hydrazobenzene, on warming with hydrochloric acid, rearranges to benzidine:

$$C_6H_5\text{-NHNH-}C_6H_5 \xrightarrow{HCl} NH_2C_6H_4\text{-}C_6H_4NH_2 \quad (90\%)$$

Benzidine is a colourless solid, m.p. 127°C. Its hydrochloride is soluble, but its sulphate is sparingly soluble in water. It is very important commercially since it is used in the preparation of azo-dyes, *e.g.*, Congo red.

o-Tolidine (4,4'-diammo-3,3'-*dmethylbiphenyl*), m.p. 129°C, and dianisidine (4,4'-diamino-3'-dimethoxybiphenyl), m.p. 138°C, are manufactured on a large scale as intermediates in the preparation of azo-dyes.

Diphenic acid (*biphenyl-2,2'-dicarboxylic acid*), m.p. 229°C, may be readily prepared by oxidising phenanthraquinone with potassium dichromate and sulphuric acid, or by the direct oxidation of phenanthrene with 50 per cent hydrogen peroxide in glacial acetic (68% yield).

$$\text{Phenanthraquinone} \xrightarrow{[O]} HO_2C\text{-}C_6H_4\text{-}C_6H_4\text{-}CO_2H$$

It may also be prepared by the action of ammoniacal cuprous oxide on diazotised anthranilic acid:

$$2\,C_6H_4(CO_2H)N_2^+Cl^- \xrightarrow{Cu_2O} HO_2C\text{-}C_6H_4\text{-}C_6H_4\text{-}CO_2H \quad (72\text{–}84\%)$$

The methyl ester may be obtained by heating methyl o-iodobenzoate with copper powder (Ullmann synthesis):

$$2\,C_6H_4(I)CO_2CH_3 \xrightarrow{Cu} CH_3O_2C\text{-}C_6H_4\text{-}C_6H_4\text{-}CO_2CH_3$$

When heated with acetic anhydride, diphenic acid forms diphenic anhydride, and when heated with concentrated sulphuric acid, fluorenone-4-carboxylic acid:

$$\text{fluorenone-4-carboxylic acid} \xleftarrow{H_2SO_4} HO_2C\text{-}C_6H_4\text{-}C_6H_4\text{-}CO_2H \xrightarrow{(CH_3CO)_2O} \text{diphenic anhydride}$$

When the calcium salt of diphenic acid is distilled, fluorenone is obtained:

$$(C_6H_4CO_2)_2Ca \longrightarrow \text{fluorenone} + CaCO_3$$

When oxidised with potassium permanganate, diphenic acid forms phthalic acid. Distillation with soda-lime gives biphenyl.

If at least three of the positions 2, 2', 6 and 6' are occupied by sufficiently large groups, free rotation about the single bond joining the two phenyl groups is no longer possible. Provided each ring has no vertical plane of symmetry, this restricted rotation gives rise to optical activity due to the molecule being *asymmetric as a whole*; *e.g.*, 6-*nitrobiphenyl*-2,2'-*dicarboxylic acid* (1), and 6,6'-*diamino*-2,2'-*dimethylbiphenyl* (II), have been resolved.

If the substituent group is large enough, then only two groups in the o- and o'-positions will cause restricted rotation, *e.g.*, *biphenyl-2,2-disulphonic* acid (III), has -been shown to be optically active.

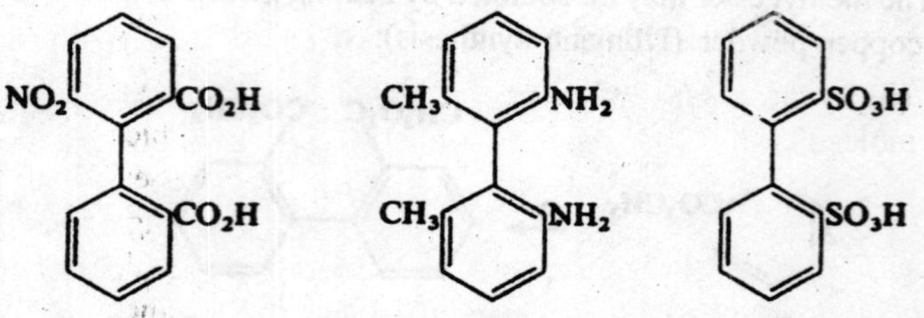

NO_2 CO_2H CO_2H (I) CH_3 NH_2 CH_3 NH_2 (II) SO_3H SO_3H (III)

Fig. 6.4

The cause of the restricted rotation is due mainly to the stearic effects of the groups in the o- and o'-positions.

This type of stereoisomerism, arising from restricted rotation about a single bond and where the stereoisomers can be isolated, is called **atropisomerism**; the isomers are known as *atropisomers*.

Since the optical activity of biphenyl compounds arises from restricted rotation, it might be expected that racemisation would not be possible. In practice, however, many optically active biphenyls can be racemised, *e.g.*, by boiling in solution. The general theory of these racemisations is that heating increases the amplitude of vibrations of the substituents and also the amplitude of vibration of the interannular bond. In addition to these bond stretchings, the various valency angles between ring and substituents are also deformed. Thus the nuclei will pass through a common plane, and hence, when the molecule returns to its restricted state, it will do so equally in the (+)- and (–)-forms, *i.e.*, the molecules are racemised.

A number of polyphenyls and some of their derivatives have been prepared, *e.g.*, *terphenyl*, *quater-phenyl*, *quinquephenyl* and *sexiphenyl*.

Diphenylmethane, m.p. 26°C. Some related compounds of diphenylmethane have already been discussed, *e.g.*, benzophenone. Diphenylmethane may be prepared:

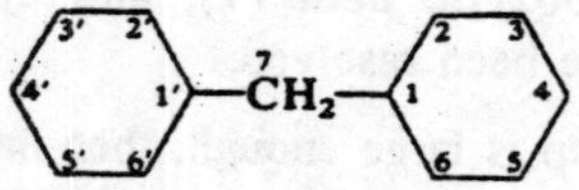

3' 2' 4' 1' 5' 6' $\overset{7}{C}H_2$ 2 3 1 4 6 5

(i) By the Friedel-Crafts condensation between benzyl chloride and benzene:

$$C_6H_5CH_2Cl + C_6H_6 \xrightarrow{AlCl_3} C_6H_5CH_2C_6H_5 + HCl \ (50\%)$$

(ii) By the condensation between one molecule of formaldehyde and two molecules of benzene in the presence of concentrated sulphuric acid:

$$2C_6H_6 + CH_2O \xrightarrow{H_2SO_4} C_6H_5CH_2C_5H_5 + H_2O$$

(iii) By heating benzophenone with hydriodic acid and red phosphorus at 160°C under pressure, by the Wolff-Kishner reduction, or by LAH–$AlCl_3$.

$$C_6H_5COC_6H_5 \xrightarrow{HI/P} C_6H_5CH_2C_6H_5 \ (95\%)$$

(iv) By the following Grignard reaction:

$$C_6H_5MgBr + C_6H_5CH_2Cl \xrightarrow[60-70\,°C]{PhH} C_6H_5CH_2C_6H_5 \ (ex)$$

The reactions of diphenylmethane are similar to those of biphenyl, *e.g.*, on nitration, the nitro-group enters mainly the 4-positions; the second nitro-group then enters the 4'-position (the benzyl group is o-p-orienting).

Since the hydrogen of the methylene group is very active (due to each ring being electron-attracting), bromination of diphenyl-methane gives rise to substitution at the methylene carbon atom, and not in the ring, to form, *e.g.*, diphenylmethyl bromide:

$$C_6H_5CH_2C_6H_5 + Br_2 \longrightarrow C_6H_5CHBrC_6H_5 + HBr$$

Diphenylmethyl bromide is readily hydrolysed to alkali to *benzhydrol (diphenlcarbinol;)*. When oxidised with chromic acid, diphenylmethane forms benzophenone. When its vapour is passed through a red-hot tube, diphenylmethane forms fluorene:

CH_2 → + H

Triphenylmethane (tritane) may be prepared by the following methods:

$$2C_6H_5 + C_6H_5\ CHCl_2 \xrightarrow{AlCl_3} (C_6H_5\ )_3CH + 2HCl$$

$$3C_6H_5\ + CHCl_3 \xrightarrow{AlCl_3} (C_6H_5\ )_3CH + 3HCl \ (33\%)$$

$$C_6H_5\ CHO + 2C_6H_6 \xrightarrow{AnCl_2} (C_6H_5\ )_3CH + H_2O$$

Triphenylmethane, m.p. 93°C, is the parent substance of the triphenylmethane dyes. When brominated, it forms tripheylmethyl bromide, $(C_6H_5)_3CBr$ (cf. diphenylmethane), and may be oxidised directly to triphenylcarbinol. Triphenylmethane is acidic (more so than diphenylmethane), the conjugate base being stabilised by resonance. Thus, when treated with sodium in ethereal solution (or in liquid ammonia), triphenylmethane forms triphenyl-methylsodium:

$$(C_6H_5)_3CH + Na \longrightarrow (C_6H_5)_3\ \bar{C}Na^+ + \frac{1}{2}Na$$

Triphenylmethylsodium is also formed when triphenylmethyl chloride is treated with sodium amalgam.

Derivatives of triphenylmethane are best prepared synthetically.

Triphenylmethyl chloride (trityl chloride), m.p. 112°C, may be prepared from triphenyl-carbinol and acetyl chloride (see below), or by the Friedel-Crafts condensation between benzene and carbon tetrachloride; tetraphenylmethane is not formed:

$$3C_6H_6 + CCl_4 \xrightarrow{AlCl_3} (C_6H_5)_3CCl + 3HCl\ (68\text{-}84\%)$$

The halogen atom is extremely reactive, *e.g.*, when boiled with water, triphenylmethyl chloride forms the corresponding alcohol, and with ethanol it forms the ethyl ether:

$$(C_6H_5)_3COC_2H_5 \xleftarrow{C_2H_5OH} (C_6H_5)_3CCl$$

$$\xrightarrow{H_2O} (C_6H_5)_3COH$$

Triphenylmethyl chloride may be converted into triphenylmethane as follows:

$$(C_6H_5)_3CCl \xrightarrow{Mg} (C_6H_5)_3CMgCl$$

$$\xrightarrow{H_2O} (C_6H_5)_3COH$$

This reaction is particularly interesting since the formation of the Grignard reagent from triphenylmethyl bromide was shown to occur via a free-radical mechanism (Gomberg and Bachmann, 1930). These authors showed that when half the total amount of magnesium had reacted, the solution contained no Grignard reagent:

$$2(C_6H_5)_3CBr \longrightarrow 2(C_6H_5)_3C\cdot + MgBr$$

$$Mg + MgBr_2 \rightleftharpoons 2\cdot MgBr$$

$$(C_6H_5)C\cdot + \cdot MgBr \longrightarrow (C_6H_5)_3CMgBr$$

When triphenylmethyl chloride is dissolved in liquid sulphur dioxide, it ionises:

$$Ph_3CCl \rightleftharpoons Ph_3C^+ + Cl^-$$

Triphenylcarbinol, m.p. 165°C; may be prepared by the oxidation of triphenylmethane with chromium trioxide:

$$(C_6H_5)_3CH \xrightarrow{CrO_3} (C_6H_5)_3COH\ (85\%)$$

It may also be prepared from benzophenone or ethyl benzoate as follows:

$$(C_6H_5)_2CO \xrightarrow[(ii)\,H_2O^+]{(i)\,C_6H_5MgBr} (C_6H_5)_3COH$$

$$\xleftarrow[(ii)\,H_3O^+]{(i)\,2C_6H_5MgBr} C_6H_5CO_2C_2H_5$$

Triphenylcarbinol reacts extremely rapidly with hydrochloric acid or acetyl chloride to form triphenylmethyl chloride (cf. tertiary alcohols):

$$(C_6H_5)_3COH + CH_3COCl \longrightarrow (C_6H_5)_3CCl + CH_3CO_2H\ (93\text{–}95\%)$$

It also reacts rapidly with alcohols in the presence of a trace of acid or iodine to form ethers, *e.g.*,

$$(C_6H_5)_3COH \overset{H^+}{\rightleftharpoons} (C_6H_5)_3C\text{—}OH_2^+ \rightleftharpoons H_2O + (C_6H_5)_3C^+$$

$$\xrightarrow{C_2H_5OH} (C_6H_5)_3COC_2H_5 + H^+$$

Triphenylcarbinol is reduced to triphenylmethane by formic acid or LAH—$AlCl_3$, and condenses with aniline hydrochloride to form p-aminotetraphenylmethane.

Tetraphenylmethane, m.p. 282°C, may be prepared as follows:

$$(C_6H_5)_3CCl + C_6H_5MgBr \longrightarrow (C_6H_5)_4C + MgClBr\ \ (v.p.)$$

It may also be prepared by diazotising p-aminotetraphenylmethane in ethanolic solution and then boiling (see also triphenylcarbinol, above):

$$(C_6H_5)_3COH + C_6H_5NH_2\cdot HCl \longrightarrow (C_6H_5)_3CC_6H_4NH_2\cdot HCl$$

$$\xrightarrow[C_2H_5OH]{NaNO_2/H_2SO_4} (CH_5)_4C$$

Tetraphenylmethane *cannot* be prepared by reaction between benzene and carbon tetrachloride in the presence of aluminium chloride (see triphenylmethyl chloride, above).

The reason for this is not certain, but it is likely that the steric effect plays a large part in hindering the reaction.

When triphenylmethyl chloride, in benzene solution, is treated with silver *in the absence of air*, and the yellow benzene solution then evaporated, the residue is a white solid, m.p. 145-147°C. This was believed to be *hexaphenylethane*, $(C_6H_5)_6C_2$, the yellow colour being due to its dissociation into triphenylmethyl free radicals (Gomberg, 1900). Lankamp *et al.* (1968) and Guthrie *et al.* (1969), however, have now shown that the dimer of triphenylmethyl radical is not hexaphenylethane but is **1-diphenylmethylene-4-tritylcyclohexa-2,5-diene:**

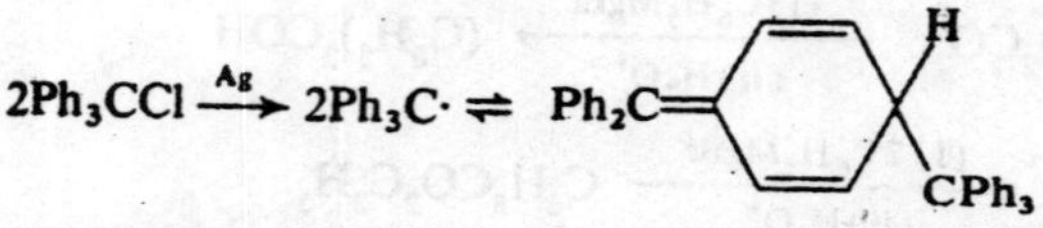

$$2Ph_3CCl \xrightarrow{Ag} 2Ph_3C\cdot \rightleftharpoons Ph_2C{=}C_6H_4(H)(CPh_3)$$

The stability of this free radical is attributed to resonance, a large number of resonating structures contributing to the resonance hybrid:

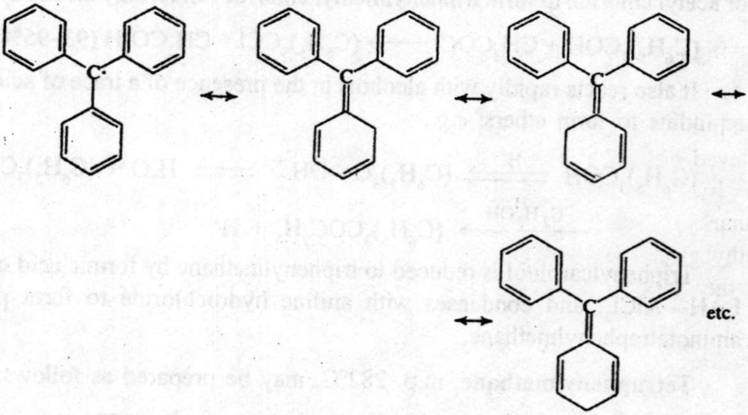

$\leftrightarrow$ $\leftrightarrow$ $\leftrightarrow$ $\leftrightarrow$ etc.

Fig. 6.5

Triphenylmethyl free radical reacts immediately with a number of reagents to form triphenyl-methyl derivatives, *e.g.*, with oxygen, it forms a colourless peroxide:

$$(C_6H_5)_3C\cdot \xrightarrow{O_2} (C_6H_5)_3C{-}O{-}O\cdot$$

$$\xrightarrow{(C_6H_5)_6C_2} (C_6H_5)_3C{-}O{-}O{-}C(C_6H_5)_3 + (C_6H_5)_3C\cdot$$

With iodine it forms triphenylmethyl iodide, and with nitric oxide, nitrosotriphenylmethane (the formation of the latter compound appears to be uncertain):

$$2(C_6H_5)_3C\cdot + I_2 \rightleftharpoons 2(C_6H_5)_3CI$$

$$(C_6H_5)_3C\cdot + NO \longrightarrow (C_6H_5)_3CNO$$

It also combines with sodium to form triphenylmethylsodium, a brick-red solid and an electrical conductor:

$$(C_6H_5)3C\cdot + Na \longrightarrow (C_6H_5)_3\ \bar{C}Na^+$$

In addition to the above reactions, triphenylmethyl free radical can act as a powerful reducing agent, and will reduce the salts of silver, gold and mercury to the metals, *e.g.*,

$$(C_6H_5)_3C\cdot + AgCl \longrightarrow (C_6H_5)_3CCl + Ag$$

The triphenylmethyl free radical is an example of a long-life free radical. The problem here is why does 'hexaphenylethane' dissociate so readily (as compared to ethane)? The answer is believed to be as follows. The free radical is stabilised by resonance, which therefore acts as a 'driving force' in the dissociation.

At the same time, steric factors also operate. In 'hexaphenylethane' (sp^3 hybridisation) there is steric strain due to crowding, and this is relieved by dissociation to form a planar or almost planar free radical (sp^2 hybridisation; resonance requires complete or almost complete planarity). That the steric effect is operating is supported by the fact that methyl groups in the o-positions increase the dissociation. Further support for the steric effect is the case of schitschibabin's hydrocarbon:

$$Ph_2\dot{C}-C_6H_4-C_6H_4-\dot{C}Ph_2 \rightleftharpoons Ph_2C{=}C_6H_4{=}C_6H_4{=}CPh_2$$

This compound contains about 4.5 per cent free radical, but when all four o-positions are occupied by chlorine the free-radical content is almost 100 per cent. In the chlorinated compound coplanarity of the two benzene rings is prevented, *i.e.*, this is an example of steric inhibition of resonance.

When dissolved in liquid sulphur dioxide, the solution shows a high conductivity (Walden, 1903), and this was explained by assuming that the 'hexaphenylethane' dissociates into positive and negative ions:

$$(C_6H_5)_6C_2 \rightleftharpoons (C_6H_5)_3C^+ + (C_6H_5)_3C^-$$

The existence of the triphenylmethyl carbonium ion has definitely been established by Dauben *et al.*, (1960), who prepared the crystalline perchlorate, $Ph_3C^+ClO_4^-$, and the fluoroborate, $Ph_3C^+BF_4^-$.

These carbonium ions have been shown to have the property of abstracting hydride ions from hydrocarbons of low nucleophilic power:

$$Ph_3C^+ClO_4^- + RH \rightleftharpoons Ph_3CH + R^+\ ClO_4^-$$

This property thus affords a means of preparing new carbonium salts.

Bibenzyl (1,2-*diphenylethane*), m.p. 52°C, may be prepared by the Friedel-Crafts condensation between benzene and ethylene dichloride:

$$2C_6H_6 + ClCH_2CH_2Cl \xrightarrow{AlCl_3} C_6H_5CH_2CH_2C_6H_5 + 2HCl$$

Another method of preparation is by the action of sodium or copper on benzyl bromide:

$$2C_6H_5CH_2Br \xrightarrow{Na} C_6H_5CH_2CH_2C_6H_5$$

Bibenzyl is oxidised by permanganate or chromic acid to benzoic acid.

trans-stilbene (trans-1, 2-diphenylethylene) may be prepared:

(i) By treating benzylmagnesium bromide with benzaldehyde and dehydrating the product, benzylphenylmethanol:

$$C_6H_5CHO + C_6H_5CH_2MgBr \longrightarrow C_6H_5CH\begin{matrix} \diagup OMgBr \\ \diagdown CH_2C_6H_5 \end{matrix} \xrightarrow{H^+}$$

$$C_6H_5CHOHCH_2C_6H_5 \xrightarrow[(heat)]{-H_2O} C_6H_5CH{=}CHC_6H_5$$

(ii) By reducing benzoin with amalgamated zinc and an ethanolic solution of hydrogen chloride:

$$C_6H_5CHOHCOC_6H_5 \longrightarrow C_6H_5CH = CHC_6H_5\ (53\text{-}57\%)$$

(iii) By heating α-phenylcinnamic acid in quinoline in the presence of copper chromite

$$C_6H_5CHO + C_6H_5CH_2CO_2K \xrightarrow{(CH_3CO)_2O}$$

$$C_6H_5CH = C(C_6H_5)CO2H \longrightarrow C_6H_5CH = CHC_6H_5$$

(iv) Stilbene may also be prepared by the Meerwein reaction.

trans-Stilbene, m.p. 124°C, is the stable isomer; cis-stilbene (isostilbene), b.p. 145° (13mm), is the unstable isomer;

trans-stilbene: $C_6H_5(H)C{=}C(H)C_6H_5$ — cis-stilbene: $C_6H_5(H)C{=}C(H)C_6H_5$

trans-stilbene cis-stilbene

trans-Stilbene is reduced by sodium and ethanol to bibenzyl. It adds on bromine to form stilbene dibromide which, on heating with ethanolic potassium hydroxide, forms **tolan** (*diphenylacetylene*), m.p. 62°C:

$$C_6H_5CH{=}\ CHC_6H_5 \xrightarrow{Br_2} C_6H_5CHBrCHBrC_6H_5$$
$$\xrightarrow{KOH} C_6H_5C{\equiv}CC_6H_5$$

Stilbene dibromide reacts with silver acetate to form two isomeric diacetates. These, on hydrolysis, form hydrobenzoin and isohydrobenzoin :

$$C_6H_5CH(OCOCH_3)CH(OCOCH_3)C_6H_5$$
$$\xrightarrow{H_2O} C_6H_5CHOHCHOHC_6H_5$$

cis-Stilbene may be prepared by reducing diphenylacetylene with zinc dust and ethanol, or by irradiating trans-stilbene with ultra-violet light.

$$C_6H_5C \equiv CC_6H_5 \xrightarrow{e;\,H^+} C_6H_5CH = CHC_6H_5,$$

cis-Stilbene is readily converted into trans-stilbene under the catalytic influence of traces of hydrogen bromide and peroxides.

Benzoin, m.p. 137°C, is usually prepared by the benzoin condensation, this reaction being carried out by refluxing benzaldehyde with aqueous ethanolic potassium cyanide:

$$2C_6H_5CHO \xrightarrow{KCN} C_6H_5CHOHCOC_6H_5\ (83\%)$$

The generally accepted mechanism is

$$C_6H_5\overset{O}{\overset{\|}{C}}H + :CN^- \rightleftharpoons \left[C_6H_5\overset{O^-}{\overset{|}{C}}H\text{—}CN\right] \rightleftharpoons \left[C_6H_5\text{—}\overset{OH}{\overset{|}{\ddot{C}}^-}\text{—}CN\right] \xrightleftharpoons{C_6H_5CHO}$$

$$\left[C_6H_5\underset{CN}{\overset{OH}{C}}\text{—}\underset{CN}{\overset{O^-}{C}}\text{—}C_6H_5\right] \rightleftharpoons \left[C_6H_5\underset{CN}{\overset{O^-}{C}}\text{—}\underset{H}{\overset{OH}{C}}\text{—}C_6H_5\right]$$

$$\rightleftharpoons C_6H_5\overset{O}{\overset{\|}{C}}\text{—}\underset{H}{\overset{OH}{C}}\text{—}C_6H_5 + :CN^-$$

Fig. 6.6

When a mixture of aldehydes is treated with aqueous ethanolic potassium cyanide, 'mixed' benzoins (as well as the 'single' benzoins) are obtained:

$$Ar^1CHO + Ar^2CHO \xrightarrow{KCN} Ar^1CHOHCOAr^2 + Ar^2CHOHCOAr^1$$

Furthermore, the reversibility of the benzoin condensation is indicated by the fact that when the benzoin $Ar^1CHOHCOAr^1$ is heated with the aldehyde Ar2CHO in the presence of potassium cyanide, a mixed benzoin is obtained (Buck and Ide, 1933):

Aliphatic aldehydes do not undergo the benzoin condensation.

Benzoins (and acyloins) tautomerise to *enediols*:

$$\text{—CHOHCO—} \rightleftharpoons HO\text{—}\overset{|}{C}\text{=}\overset{|}{C}\text{—}OH$$

There is much evidence to support this, *e.g.*, acetylation produces the enediol diacetate. The enediol is usually less stable than its corresponding a-hydroxyketone; in certain cases the former has been isolated.

Benzoin is reduced to **deoxybenzoin** (*desoxybenzoin*), m.p. 60°C, which may also be prepared by the following Friedel-Crafts reaction.

$$C_6H_5CH_2OCl + C_6H_6 \xrightarrow{AlCl_3} C_6H_5CH_2COC_6H_5 + HCl\ (82\text{-}83\%)$$

The prefix *deoxy-* (formerly *desoxy-*) indicates the replacement of a hydroxyl group by hydrogen. Deoxybenzoin contains an active methylene group; *e.g.*, it forms a sodio-derivative, condenses with aldehydes in the presence of piperidine, and adds on to a, unsaturated carbonyl compounds (cf. ethyl malonate).

Benzoin is an α-hydroxyketone; hence it reduces Fehling's solution and forms an osazone.

When reduced by sodium amalgam, benzoin forms mainly **hydrobenzoin**, and a small amount of **isohydrobenzoin**:

$$C_6H_5CHOHCOC_6H_5 \xrightarrow{e;\,H^+} C_6H_5CHOHCHOHC_6H_5$$

This formula (tolylene glycol) contains two identical asymmetric carbon atoms, and hence the compound can exist in the dextro, laevo and meso-forms (cf. tartaric acid). The racemic modification is known as hydrobenzoin (m.p. 139°C), and the meso-form as isohydrobenzoin (m.p. 121°C). Both forms give benzoic acid on oxidation.

Benzoin is oxidised by chromic acid to a mixture of benzaldehyde and benzoic acid, and by nitric acid to **benzil**;

$$C_6H_5CHOHCOC_6H_5 \xrightarrow[HNO_3]{[O]} C_6H_5COCOC_6H_5\ (96\%)$$

Purer benzil is obtained by oxidation with copper sulphate in aqueous pyridine (86% yield).

Benzil is a yellow crystalline solid, m.p. 95°C. It behaves as a typical α-diketone, *e.g.*, it is oxidised by hydrogen peroxide in acetic acid to benzoic acid, forms a monoxime and dioxime etc.

Two monoximes of benzil are known: α-(cis-), m.p. 134°C, and β-(trans), m.p. 113°C. Three dioximes are possible and all are known: α-(cis-cis-), m.p. 237°C, β-(trans-trans-), m.p. 207°C, and γ-or *amphi-* (*cis-trans-*), m.p. 166°C.

Benzil condenses with o-phenylenediamine to form 2,3-*diphenylquinoxaline*:

$$\begin{array}{l} C_6H_5C{=}O \\ \quad | \\ C_6H_5C{=}O \end{array} + C_6H_4(NH_2)_2 \longrightarrow \text{2,3-diphenylquinoxaline } (C_6H_5)_2C_8H_4N_2 \quad (ex.)$$

The nature of the reduction products of benzil depends on the reducing agent used. Thus reduction with sodium amalgam gives hydrobenzoin (and some isohydrobenzoin); amalgamated zinc and hydrogen chloride in ethanol, stilbene; sodium hyposulphite in ethanol benzoin; amalgamated tin and hydrogen chloride in ethanol deoxybenzoin: and catalytic reduction using nickel at 230°C, bibenzyl.

When heated with ethanolic potassium hydroxide, benzil undergoes the **benzilic acid rearrangement:**

$$C_6H_5COCOC_6H_5 + KOH \longrightarrow (C_6H_5)_2C(OH)CO_2K \ (77\text{-}79\%)$$

The rearrangement has been shown to be first order with respect to both benzil and hydroxide ion (Westheimer, 1936), and it has also been shown that when benzil is heated for a very short time with methanolic sodium hydroxide in water containing ^{18}O, the benzil recovered contained ^{18}O (Roberts and Urey, 1938). This is in keeping with the assumption that the first step is the rapid *reversible* addition of hydroxide ion to benzil.

$$Ph{-}\overset{O}{\overset{\|}{C}}{-}\overset{O}{\overset{\|}{C}}{-}Ph + \bar{O}H^- \rightleftharpoons Ph{-}\overset{O}{\overset{\|}{C}}{-}\underset{OH}{\overset{O^-}{\overset{|}{\underset{|}{C}}}}{-}Ph \rightleftharpoons$$

$$Ph{-}\overset{O}{\overset{\|}{C}}{-}\underset{O^-}{\overset{OH}{\overset{|}{\underset{|}{C}}}}{-}Ph \rightleftharpoons Ph{-}\overset{O}{\overset{\|}{C}}{-}\underset{O}{\underset{\|}{C}}{-}Ph + OH^-$$

All of these results are in keeping with the mechanism proposed by Ingold in 1928:

$$Ph{-}\overset{O}{\overset{\|}{C}}{-}\underset{Ph}{\underset{|}{\overset{O}{\overset{\|}{C}}}}\ \bar{O}H \overset{fast}{\rightleftharpoons} Ph{-}\overset{O}{\overset{\|}{C}}{-}\underset{Ph}{\underset{|}{\overset{\bar{O}}{\overset{|}{C}}}}{-}OH \xrightarrow{slow}$$

$$Ph{-}\overset{O^-}{\overset{|}{C}}{-}\overset{O}{\overset{\|}{C}}{-}OH \xrightarrow{fast} Ph{-}\underset{Ph}{\underset{|}{\overset{OH}{\overset{|}{C}}}}{-}\overset{O}{\overset{\|}{C}}{-}O^-$$

Fig. 6.7

This is an example of the 1,2-shift, and since a phenyl group is involved, it is possible that a phenonium ion is formed as an intermediate. A prediction that can be made from the above mechanism is that when the two aryl groups are different, the one that is more electron-releasing (due to the presence of a suitable substituent, *e.g.*, p-MeO) will tend to neutralise the positive charge on the carbon atom to which it is attached when the C=O bond is polarised. Thus, it will be the *other* CO group that will link with the hydroxide ion, and consequently it will be the aryl group attached to this 'other' CO group that migrates preferentially. This prediction is borne out in practice, *e.g.*, in p-methoxybenzil, the methoxyphenyl group migration was about 32 per cent. (Roberts *et al.*, 1951), and Eastham, *et al.* (1958), using labelled p-methylbenzil, showed that the phenyl group migrated almost exclusively.

Furthermore, it has also been shown that the hydroxide ion is ***not*** a specific catalyst; sodium methoxide and potassium t-butoxide produce the corresponding esters of benzilic acid (Doering *et al.*, 1956). Formation of these esters is strong evidence that the first step in the rearrangement is the addition of the nucleophilic reagent.

$$\mathrm{Ph{-}\overset{O}{\overset{\|}{C}}{-}\underset{Ph}{\underset{|}{\overset{O}{\overset{\|}{C}}}}\;\;OMe^-} \longrightarrow \mathrm{Ph{-}\underset{Ph}{\underset{|}{\overset{O^-}{\overset{\|}{C}}}}{-}\overset{O}{\overset{\|}{C}}{-}OMe}$$

$$\xrightarrow{H_2O} Ph_2C(OH)CO_2Me$$

Benzilic acid (m.p. 150°C) forms diphenylacetic acid when refluxed in acetic acid solution with hydriodic acid and red phosphorus:

$$(C_6H_5)_2C(OH)CO_2H \xrightarrow{HI/P} (C_6H_5)_2CHCO_2H + H_2O \quad (94\text{–}97\%)$$

When oxidised with chromic acid, benzilic acid forms benzophenone.

CONDENSED SYSTEMS

Naphthalene

Naphthalene, $C_{10}H_8$, is the largest single constituent of coal-tar (9%). It is obtained by cooling the middle and heavy oils, whereupon naphthalene crystallises out. The oil is pressed free from the naphthalene, the crude naphthalene cake melted, treated with concentrated sulphuric acid (to remove basic impurities), washed with water, and then treated with aqueous sodium hydroxide (to remove acidic impurities). Finally the naphthalene is distilled to give the pure product.

The modern tendency for isolating naphthalene is to replace the 'hot-pressing process' by the continuous washing or distillation processes. Naphthalene is also now being made synthetically from petroleum. Petroleum fractions are passed over a heated catalyst, *e.g.*, copper, at 680°C at atmospheric pressure: naphthalene and methylnaphthalenes are obtained, and the latter are hydrodealkylated to naphthalene (cf. benzene from toluene).

STRUCTURE OF NAPHTHALENE

Erlenmeyer (1866) proposed the symmetrical formula (1), for naphthalene, and Graebe (1869) proved that it did consist of two benzene rings fused together in the o-positions. He used several methods, but the line of approach was the same, *e.g.*, he found that on oxidation, naphthalene gave phthalic acid. Thus naphthalene contains the group (II), *i.e.*, a benzene ring with two side-chains in the o-positions. When nitrated, naphthalene gave nitronaphthalene which, on oxidation, gave o-nitrophthalic acid.

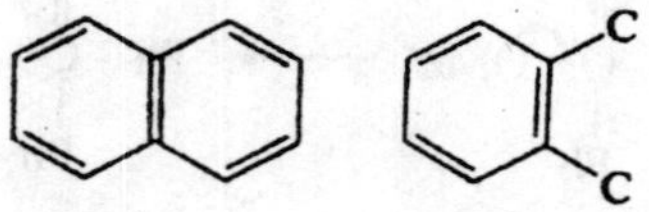

(I) (II)

This indicates that the nitro-group is in the benzene ring, and that it is the side-chains which are oxidised. When nitronaphthalene was reduced and the corresponding aminonaphthalene oxidised, phthalic acid was obtained. As we have seen, an amino-group attached to the nucleus renders the latter extremely sensitive to oxidation.

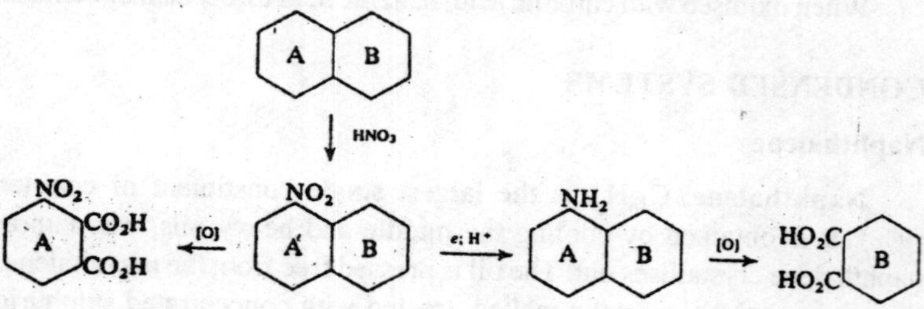

Fig. 6.8

Hence, the inference is that the benzene ring in phthalic acid obtained by oxidation of aminonaphthalene is not the same ring as that originally containing the nitro-group in nitronaphthalene, *i.e.*, naphthalene contains two benzene rings. The above facts fit the following scheme:

This structure for naphthalene has been confirmed by many syntheses, *e.g.*:

(i) When 4-phenylbut-l-ene is passed over red-hot calcium oxide, naphthalene is formed:

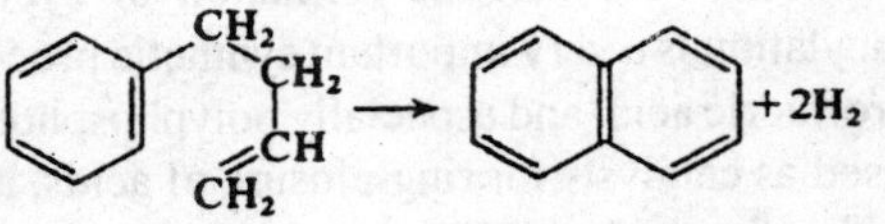

CH_2 CH_2 CH CH_2 → + $2H_2$

(ii) When 4-phenylbut-3-enoic acid is warmed with sulphuric acid, 1-naphthol is formed and this, on distillation with zinc dust, gives naphthalene:

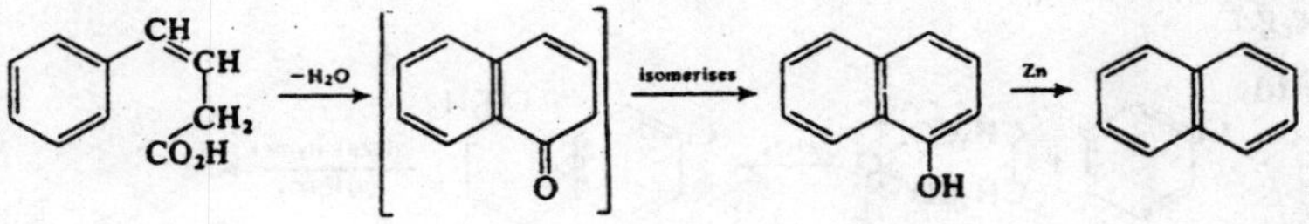

CH CH CH_2 CO_2H $\xrightarrow{-H_2O}$ [O] $\xrightarrow{\text{isomerises}}$ OH $\xrightarrow{Zn}$

Fig. 6.9

(iii) **Haworth synthesis** (1932). Benzene is treated with succinic anhydride in the presence of aluminium chloride, and the ketonic acid produced (Burcker, 1882) is reduced by the Clemmensen method. The ring is closed by heating with concentrated sulphuric acid and the product, α-tetralone, reduced to tetrahydronaphthalene by the Clemmensen method. This compound is then dehydrogenated to naphthalene by heating it with selenium or palladised charcoal:

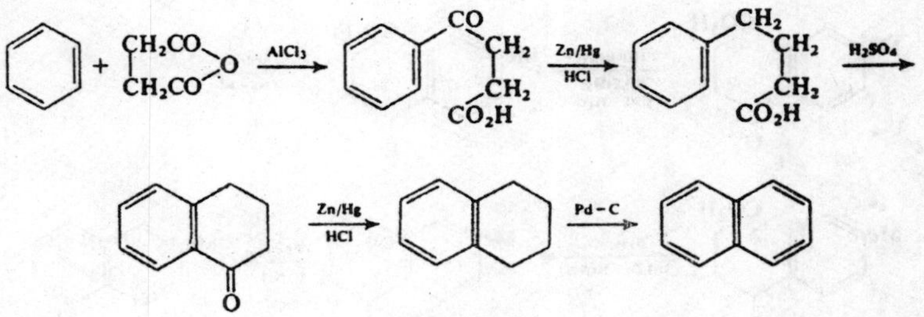

+ CH_2CO CH_2CO O $\xrightarrow{AlCl_3}$ CO CH_2 CH_2 CO_2H $\xrightarrow[HCl]{Zn/Hg}$ CH_2 CH_2 CH_2 CO_2H $\xrightarrow{H_2SO_4}$

O $\xrightarrow[HCl]{Zn/Hg}$ $\xrightarrow{Pd-C}$

Fig. 6.10

Ring closure may also be effected by means of a Friedel-Crafts reaction on the acid chloride as follows:

CH_2 CH_2 CH_2 CO_2H $\xrightarrow{SOCl_2}$ CH_2 CH_2 CH_2 COCl $\xrightarrow{AlCl_3}$ O

It should be noted here that the formation of cyclic ketones by *intramolecular* acylation is a very important synthetic process. Sulphuric, phosphoric, hydrofluoric acid, and especially polyphosphoric acid (PPA), are commonly used as catalysts for ring-closure of acids, and aluminium or stannic chlorides for acid chlorides.

Various substituted naphthalenes may also be prepared by the Haworth synthesis. With o/p-orienting groups present in the benzene ring. the main product from the Friedel-Crafts reaction is the *p*-compound *e.g.*:

(i)

Cl + CH_2CO O CH_2CO $\xrightarrow{AlCl_3}$ Cl CO_2H O $\xrightarrow[\text{(ii) } H_2SO_4]{\text{(i) Zn - Hg/HCl}}$

Cl O $\xrightarrow{MeMgI}$ Cl HO Me $\xrightarrow[(H_2SO_4)]{-H_2O}$ Cl Me $\xrightarrow[(-H_2)]{Pd-C}$ Cl Me

(ii)

Me CO_2H O $\xrightarrow[\text{(iii) } -H_2O]{\text{(i) EtOH/HCl (ii) EtMgI}}$ Me CO_2Et Et $\xrightarrow[\text{(ii) } H_2-Ni]{\text{(i) hydrol.}}$

Me CO_2H Et $\xrightarrow[\text{(ii) Zn - Hg/HCl}]{\text{(i) } H_2SO_4}$ Me Et $\xrightarrow[(-H_2)]{Pd-C}$ Me Et

Fig. 6.11

An interesting point about the Haworth synthesis is that Birch *et al.* (1946) have shown that heating α-tetralone with a 1 : 1 mixture of NaOH —K.OH at 220° C gives naphthalene (58%). The mechanism proposed is:

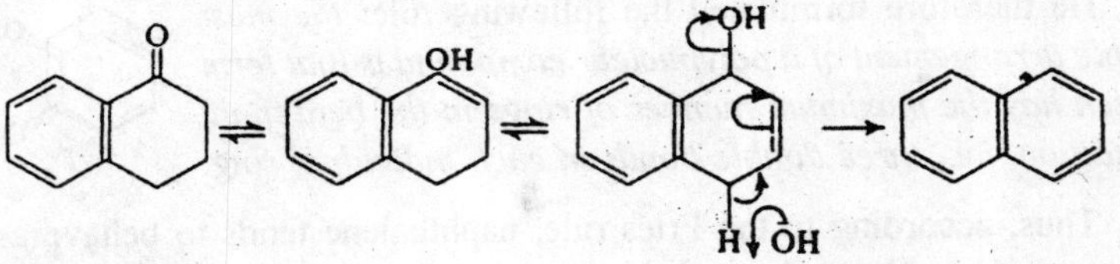

This may prove to be a better way of completing the Haworth synthesis.

Positions of the Double Bonds in Naphthalene

Physico-chemical evidence, *e.g.*, heat of combustion, etc., prints towards naphthalene being a resonance hybrid 'of mainly three resonating structures, (I), (II) and (III):

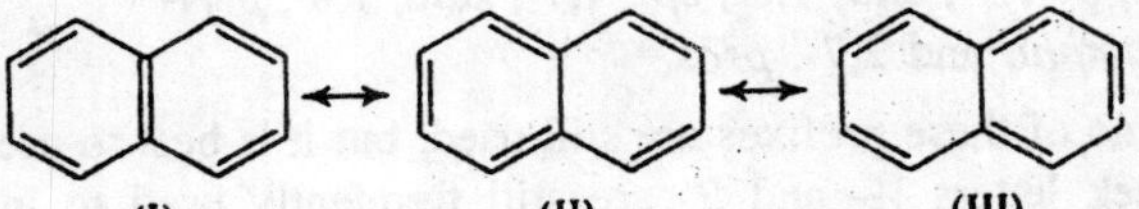

(I) (II) (III)

It should be noted that there are n +1 principal resonating structures for a polynuclear hydrocarbon containing n benzene rings fused together in a *linear* manner.

The double bond character of the various bonds in naphthalene is shown in (IV), and are obtained by taking the average of the three resonating structures (I)-(III) on the assumption that all three contribute equally to the resonance hybrid. Thus, the 1,2-bond would be expected to be shorter than the 2,3-bond, and this has been shown to be the case by X-ray analysis of naphthalene (see V; in Å units).

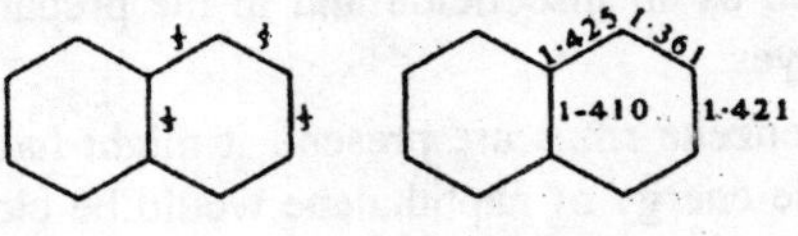

(IV) (V)

Fries rule (1935). Fries compared the possible arrangements of double bonds in polynuclear compounds with benzoquinones. Structures (II) and (III) have arrangements corresponding too-benzoquinone. Since quinones are far more reactive than an aromatic compound, Fries believed that the *stable* form of a polynuclear compound did not contain this quinonoid arrangement.

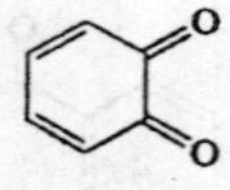

He therefore formulated the following rule: *the most stable arrangement of a polynuclear compound is that form which has the maximum number of rings in the benzenoid condition, i.e., three double bonds in each individual ring.*

Thus, according to the Fries rule, naphthalene tends to behave as structure (1) (with two benzenoid rings) rather than as (II) or (III) (with one benzenoid ring).

ISOMERISM AND NOMENCLATURE OF NAPHTHALENE DERIVATIVES

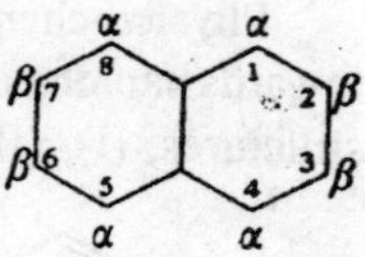

Positions 1, 4, 5 and 8 are identical (a-positions), as are positions 2, 3, 6 and 7 (β-positions). In the old literature, the positions 1,2- were known as o-; 1,3-, m-, 1,4-, p- 1,5-, *ana*; 1,6-, *epi*; 1,7-, *kata*, 1,8-, *peri-* , 2,6-, *amphi*, and 2,7-, *pros*.

Some of these prefixes are still used, but it is best to use numbers; the Greek letters ½- and P- are still frequently used to indicate the position of a single substituent.

Monosubstitution products	$C_{10}H_7X$	Two: 1- and 2-
Disubstitution products	$C_{10}H_6X_2$	10 isomers
	$C_{10}H_6XY$	14 isomers

There are 14 possible isomers for $C_{10}H_5X_3$, 22 for $C_{10}H_4X_4$, 14 for $C_{10}H_3X_5$, 10 for $C_{10}H_2X_6$, 2 for $C_{10}HX_7$ and 1 for $C_{10}X_8$.

Properties of Naphthalene

Naphthalene, m.p. 80°C, has a characteristic odour and is very volatile. It is used as an insecticide and in the preparation of phthalic anhydride and dyes.

Since two benzene rings are present, it might have been expected that the resonance energy of naphthalene would be close to 301.2 (2 × 150.6) $kJmol^{-1}$. Its RE is actually 255.2 $kJmol^{-1}$ and hence naphthalene is less aromatic than benzene (each ring in naphthalene has an RE of 127-6kJ). Consequently, naphthalene will have less aromatic character than benzene and so will be more reactive.

Addition compounds of naphthalene. A number of reduction products of naphthalene can be isolated, the nature of the product depending on the reducing agent used. When reduced with sodium and ethanol, naphthalene gives 1,4-dihydronaphthalene (1,4-dialin), m.p. 25°C :

$e; H^+$

1,4-Dialin is unstable, readily isomerising to 1,2-dialin (m.p. –8°C) when heated with ethanolic sodium ethoxide. 1,2-Dialin is also unstable, fairly readily eliminating hydrogen to form naphthalene.

When reduced with sodium and isopentanol, naphthalene gives 1,2,3,4-*tetrahydronaphthalene* (*tetralin*), b.p. 206-208°C:

$e; H^+$

Tetralin is used as a solvent for varnishes, lacquers, etc. When treated with bromine in the presence of light, tetralin forms the mono- and dibromo-derivatives, substitution occurring in the *alicyclic* ring. When these bromo-derivatives are heated, hydrogen bromide is eliminated, the monobromo-compound forming dialin, and the dibromo-, naphthalene. In the absence of light and in the presence of iron as halogen carrier, tetralin undergoes substitution in the aromatic ring to form 5- and 6-bromo-1,2,3,4-tetrahydronaphthalenes. Catalytic dehydrogenation (Pd—C) converts tetralin into naphthalene.

α-**Tetralone**, b.p. 129-4°C/12 mm, may be prepared by synthesis (iii), or by heating tetralin with air for 50 hours at 70°C and decomposing the peroxide with dilute sodium hydroxide:

$+ O_2 \longrightarrow$ H OOH $\xrightarrow{NaOH}$ O (74–91%)

α-Tetralone may also be synthesised from γ-butyrolactone as follows :

$+$ $CH_2—CH_2$ O CH_2 CO $\xrightarrow{AlCl_3}$ O (91–96%)

When naphthalene is catalytically reduced (under pressure) using nickel, tetralin and their *decahydronaphthalene* (decalin) are obtained:

$$\text{Naphthalene} \xrightarrow[150^\circ C]{H_2-Ni} \text{Tetralin} \xrightarrow{200^\circ C} \text{Decalin}$$

Decalin exists in two geometrical isomeric forms, *cis*- (b.p. 193°C) and trans- (b.p. 185°C). With nickel as catalyst, the main product is the trans-isomer-, with platinum, cis-. The commercial product is a mixture of the two forms, and is used as a solvent for varnishes, lacquers, etc.

Dry chlorine adds on to solid naphthalene to give naphthalene di- and tetrachlorides, $C_{10}H_8Cl_2$ and $C_{10}H_8Cl_4$. In both of these compounds the chlorine atoms are in the same ring (shown by the fact that on oxidation, both form phthalic acid). When naphthalene dichloride is heated at 40°C, hydrogen chloride is eliminated and 1-chloronaphthalene is produced. When naphthalene tetrachloride is treated with alkali, a mixture of dichloronaphthalenes is formed, the 1,3-isomer predominating.

Naphthalene, in dioxan solution, reacts with sodium to form a green-coloured solid. This is an anion free radical, and is produced by the transfer of an electron from the sodium to the lowest vacant MO naphthalene, and is stabilised by resonance:

$$\longleftrightarrow \quad \longleftrightarrow \quad \longleftrightarrow \cdots$$

Sodium naphthalene reacts with water to give 1,4-dihydro-naphthalene.

OXIDATION OF NAPHTHALENE

Naphthalene is oxidised by concentrated sulphuric acid and mercuric sulphate or by air in the presence of vanadium pentoxide, to phthalic anhydride. It is oxidised by acid permanganate to phthalic acid, and by alkaline permanganate to phthalonic acid.

$$\text{Naphthalene} \xrightarrow{O_3} \text{(diozonide, } O_3 \text{)} \xrightarrow[Zn]{H_2O} C_6H_4(CHO)_2 \text{ (CHO, CHO)}$$

Chromic acid oxidises it to 1,4-naphthaquinone. When treated with ozone, naphthalene forms the diozonide and this, on treatment with water, gives phthalaldehyde:

Substitution Products of Naphthalene

Orientation in naphthalene is more complicated than in benzene, due to the presence of two rings in the former. The first group always enters the 1-position in naphthalene except in two cases, when the 2-derivative is the main product; (i) sulphonation at high temperature, and (ii) in the Friedel-Crafts reaction.

The position of substitution may be obtained from a consideration of the nature of the intermediate carbonium ion, the one with the larger number of resonating structures being the more stable. The more important contributions will be by those structures in which one benzene ring is retained (benzene has a large resonance energy). It can be seen that 1-substitution gives rise to four such structures, whereas 2-substitution gives rise to only two. Thus, the former will be preferred (see also the naphthalenesulphonic acids).

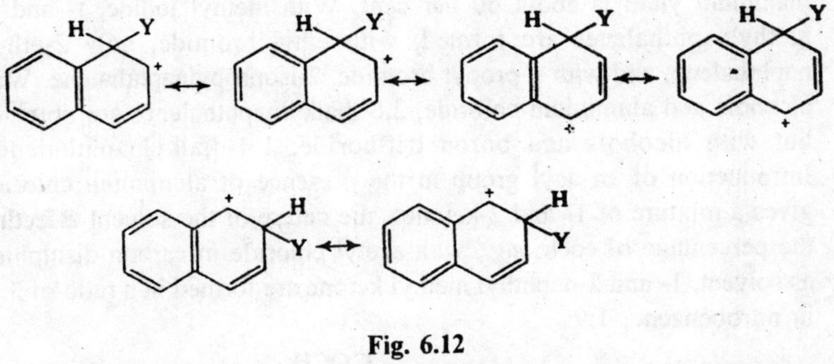

H Y + ↔ H Y + ↔ H Y + ↔ H Y +

+ H Y ↔ + H Y

Fig. 6.12

Introduction of a second substituent can give rise to *homonuclear* (*isonuclear*) substitution (the second substituent entering the *same* ring as the first), or to *heteronuclear* substitution (the second substituent entering the other ring). The following empirical generalisations are useful for predicting the position taken up by the second substituent:

(a) When Cl, Br, OH, CH_3, NHR or $NHCOCH_3$ is in the 1-position, homonuclear substitution takes place mainly in position 4, and to a lesser extent in 2.

(b) When OH, CH_3, NHR or $NHCOCH_3$ is in the 2-position, homonuclear substitution usually takes place in the l-position.

It is worth noting that homonuclear substitution usually occurs when the group already present is o-p-orienting.

(c) When NO_2 or SO_3H is in the 1- or 2-position, heteronuclear substitution occurs in position 5 or 8; if halogen or NH_2 is in the 2-position, heteronuclear substitution also occurs in position 5 or 8.

The infra-red absorption regions of polynuclear hydrocarbons include many of those characteristic of the benzene compounds. Thus, naphthalenes have two bands near 1 600 cm^{-1} and bands in the regions 1520-1 505 and 1400-1 390 cm^{-1}. The C—H deformation (out-of-plane) in the 800 cm^{-1} region may often provide information, *e.g.*, 1-substituted naphthalenes have maxima at 813-784 and 782-760 cm^{-1}, and 2-substituted naphthalenes at 862-835, 830-805 and 758-737 cm^{-1}.

The Friedel-Crafts reaction with naphthalene. Naphthalene is attacked by aluminium chloride when vigorous conditions are used (binaphthyls and compounds with one of the naphthalene rings opened are formed). Hence to carry out the Friedel-Crafts reaction successfully, mild conditions (low temperatures) must be used, and even then the maximum yield is about 60 per cent. With methyl iodide, 1- and 2-methylnaphthalenes are formed; with ethyl bromide, only 2-ethylnaphthalene; and with n-propyl bromide, 2-isopropylnaphthalene. With alcohols and aluminium chloride, 2,6-dialkylnaphthalenes are obtained but with alcohols and boron trifluoride, 1,4-dialkylnaphthalenes. Introduction of an acyl group in the presence of aluminium chloride gives a mixture of 1- and 2-ketones, the nature of the solvent affecting the percentage of each; *e.g.*, with acetyl chloride in carbon disulphide as solvent, 1- and 2-naphthyl methyl ketone are formed in a ratio of 3:1, in nitrobenzene, 1:9:

$$C_{10}H_8 \xrightarrow[AlCl_3]{CH_3COCl} \text{1-}C_{10}H_7COCH_3 + \text{2-}C_{10}H_7COCH_3$$

Chloromethylation of naphthalene using a mixture of para-formaldehyde, hydrochloric acid, glacial acetic acid and phosphoric acid gives mainly the 1-derivative and a small amount of the 1, 5-bischloromethyl-derivative:

$$C_{10}H_8 + CH_2O + HCl \longrightarrow \text{1-}C_{10}H_7CH_2Cl \quad (56\%)$$

Halogen derivatives. Naphthalene is very easily halogenated, *e.g.*, when brominated in boiling carbon tetrachloride solution, naphthalene forms 1-bromonaphthalene (yield 72-75%). Further bromination gives mainly the 1,4-dibromo-derivative, and some 1,2-. Sulphuryl chloride (1 equivalent) in the presence of aluminium chloride at 25°C chlorinates naphthalene to give the 1-chloro-derivative; with 2 equivalents of sulphuryl chloride (at 100-140°C), 1,4-dichloronaphthalene is formed.

2-Halogenonaphthalenes are conveniently prepared from 2-naphthylamine by diazotisation, etc. *e.g.*,

$$C_{10}H_7NH_2 \xrightarrow[HCl]{NaNO_2} C_{10}H_7N_2^+Cl^- \xrightarrow[HCl]{CuCl} C_{10}H_7Cl$$

The halogen atom in naphthalene behaves similarly as in the benzene ring, but is more reactive.

Nitronaphthalenes. Nitric acid attacks naphthalene at room temperature to form l-nitro-naphthalene. This is a yellow solid, m.p. 60°C, which behaves like nitrobenzene, but differs in that it forms l-chloronaphthalene when treated with phosphorus pentachloride (nitrobenzene does not react).

Nitration of naphthalene at high temperature gives a mixture of 1,5- and 1,8-dinitro-naphthalenes. Other dinitro-derivatives are prepared by special means.

2-Nitronaphthalene, m.p. 79°C, may be prepared by heating 2-naphthalenediazonium fluoroborate with sodium nitrite and copper powder.

Naphthalenesulphonic acids. When naphthalene is treated with concentrated sulphuric acid at 40°C, the main product is the l-derivative (m.p. 91°C); at 160°C, the main product is the 2-derivative (m.p. 102°C):

$$\underset{(96\%)}{1\text{-}C_{10}H_7SO_3H} \xrightleftharpoons[40^\circ C]{H_2SO_4} C_{10}H_8 \xrightleftharpoons[160^\circ C]{H_2SO_4} \underset{(85\%)}{2\text{-}C_{10}H_7SO_3H}$$

When heated with concentrated sulphuric acid, the 1-acid is converted into the 2-acid. Thus, the former is the kinetically controlled and the latter the thermodynamically controlled product. The reason for the greater stability of the 2-acid is uncertain; one suggestion is that there is steric repulsion in the 1-acid from the hydrogen atom in the 8-position.

Fig. 6.12 is the energy profile of the sulphonation of naphthalene (N). Since 1-S is the kinetically controlled product, then $k_1 > k_2$, and therefore $E_2 > E_1$. Also, since 2-S is the thermodynamically controlled product, $\Delta H_2 > \Delta H_1$. It therefore follows that the activation energy for the reverse (desulphonation) reaction: 2-S → N, is much greater than that of the reverse reaction: 1-S → N, *i.e.*, E_{-2} (= $E_2 + H_2$) >> E_{-1} (= $E_1 + \Delta H_1$). Consequently, $k_{-1} >> k_{-2}$.

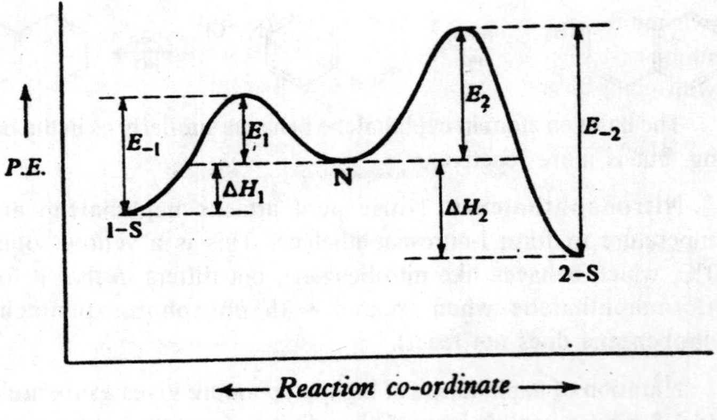

Fig. 6.12

At low temperatures, the equilibrium between the 1- and 2-acids is reached very slowly. Hence, if the reaction is stopped long before equilibrium is reached, 1-S > 2-S since $k_1 > k_2$. If one waited long enough for equilibrium to be reached, then 2-S >> 1-S. This is due to the fact that, since $k_{-1} >> k_{-2}$, 1-S reforms N much faster than does 2-S, and hence 2-S accumulates at the expense of 1-S. Because raising the temperature speeds up a reaction, equilibrium is reached far more quickly at higher temperatures, and therefore to obtain 2-S (the thermodynamically controlled product), the reaction is carried out at some suitably elevated temperature.

1- and 2-Naphthalenesulphonic acids behave similarly to benzenesulphonic acid, but the sulphonic acid group in the former is more easily replaced; *e.g.*, when fused with phosphorus pentachloride, both acids give the corresponding chloronaphthalenes.

2-Naphthalenesulphonic acid is the starting-point of practically all 2-naphthalene derivatives

Sulphonation of the 1-sulphonic acid with concentrated sulphuric acid below 40°C gives 1, 5- (70%) and 1,6-disulphonic acids (25%), at 130°C, the main products are 1,6- and 1,7-disulphonic acids. Sulphonation of the 2-acid at 60°C gives 1,6- (80%) and 1,7-disulphonic acid (20%); above 140°C, the main product is 2,7-disulphonic acid and a small amount of 2,6-. Armstrong and Wynne's rule (1890) for the orientation of naphthalenepolysulphonic acids is useful: two sulpho-groups never occupy positions o,p or *peri* to each other. This rule limits the number of isomers that can be formed by direct sulphonation of naphthalene to two mono-, six di-, three tri- and one tetra-sulphonic acid. The table summarises the products obtained by the sulphonation of naphthalene with concentrated sulphur acid (note the various migrations).

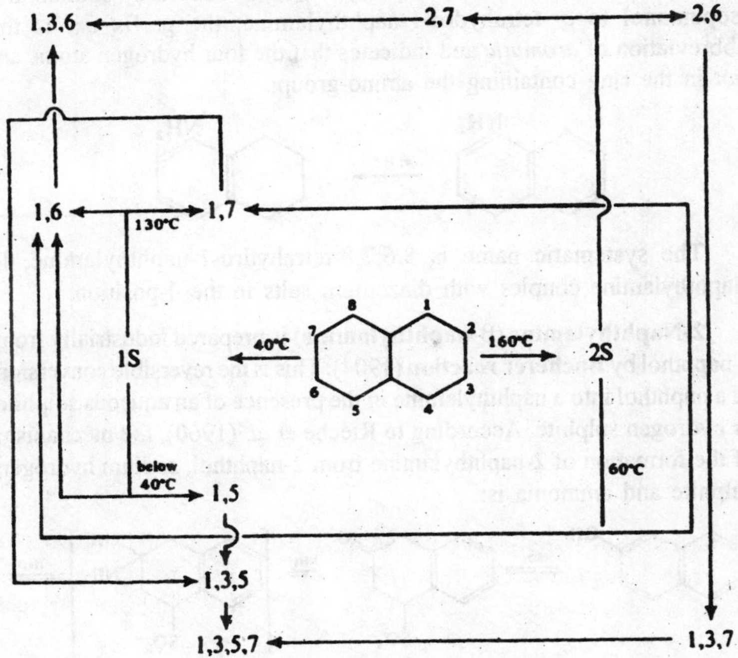

Fig. 6.13

Many of these sulphonic acids are very important dye-intermediates.

Naphthylamines. 1-Naphthylamine (α-naphthylamine) may be prepared by reducing 1-nitro-naphthalene with iron and hydrochloric acid (yield 80-85%). This method is used industrially. 1-Naphthylamine may also be prepared by the Bucherer reaction or by heating 1-naphthol with the double compound of zinc chloride and ammonia (this amination occurs more easily than with phenol):

$$\text{1-naphthol (OH)} + NH_3 \xrightarrow[250°C]{ZnCl_2} \text{1-naphthylamine (NH}_2\text{)} + H_2O$$

1-Naphthylamine, m.p. 50°C, has an unpleasant odour, and turns red on exposure to air. It reduces ammoniacal silver nitrate, and solutions of its salts give a blue precipitate with ferric chloride. Oxidation with boiling chromic acid gives α-naphthaquinone; with permanganate, phthalic acid is obtained. 1-Naphthylamine is reduced by sodium and isopentanol to *ar*-tetrahydro-1-naphthylamine; the prefix *ar-* is the abbreviation of *aromatic* and indicates that the four hydrogen atoms are *not* in the ring containing the amino-group:

$$\text{1-naphthylamine (NH}_2\text{)} \xrightarrow{e;\ H^+} \text{ar-tetrahydro-1-naphthylamine (NH}_2\text{)}$$

The systematic name is 5,6,7,8-tetrahydro-1-naphthylamine. 1-Naphthylamine couples with diazonium salts in the 4-position.

2-Naphthylamine (β-naphthylamine) is prepared industrially from 2-naphthol by **Bucherer reaction** (1904). This is the reversible conversion of a naphthol into a naphthylamine in the presence of an aqueous sulphite or hydrogen sulphite. According to Rieche *et al.* (1960), the mechanism of the formation of 2-naphthylamine from 2-naphthol, sodium hydrogen sulphite and ammonia is:

$$\text{2-naphthol (OH)} \xrightleftharpoons{HSO_3^-} \text{ketone (=O, } SO_3^-\text{)} \xrightleftharpoons{NH_3} \left[\text{(OH, NH}_2\text{, } SO_3^-\text{)}\right] \xrightleftharpoons{-H_2O} \text{imine (=NH, } SO_3^-\text{)} \xrightleftharpoons{H_2O} \text{2-naphthylamine (NH}_2\text{)}$$

Fig. 6.14

The keto-form of the naphthol and the imino-compound produced are stabilised by the addition of the hydrogen sulphite ion (to form the sulphonic acid derivative).

2-Naphthylamine is prepared commercially by heating 2-naphthol with aqueous ammonium hydrogen sulphite at 150° C under pressure (yield: 94-96%).

2-Naphthylamine, m.p. 112°C, is odourless, reduces ammoniacal silver nitrate, but gives *no* coloration with ferric chloride. It is oxidised by permanganate to phthalic acid and reduced by sodium and isopentanol toac-tetrahydro-2-naphthylamine (1,2,3,4-tetrahydro-2-naphthyl-amine): the prefix *ac-* is the abbreviation of *alicyclic* and indicates that the four hydrogen atoms are in the ring containing the amino-group :

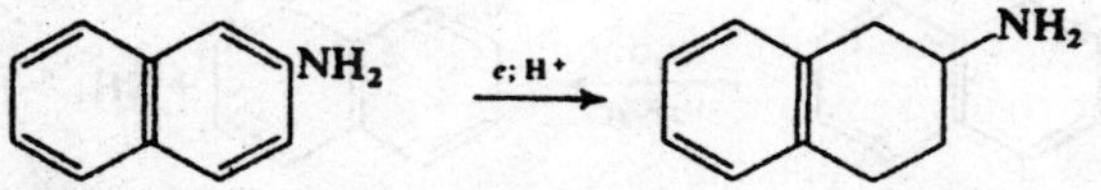

2-Naphthylamine couples with diazonium salts only in the 1-position; if this is occupied, no coupling occurs.

Naphthylaminesulphonic acids. These are very important industrially for making dyes. When heated with excess of concentrated sulphuric acid at 130°C, 1-naphthylamine forms l-naphthylamine-4-sulphonic acid (naphthionic acid). This is used in the preparation of Congo red, and is manufactured by the baking process of naphthylamine hydrogen sulphate (cf. sulphanilic acid). Prolonged action of sulphuric acid at 130°C converts 1 -naphthylamine into 1 -naphthylamine-5-sulphonic acid (*Laurent's acid*). If the heating is prolonged still further, Laurent's acid rearranges to 1-naphthylamine-6-sulphonic acid (*Cleve's acid*).

1-Naphthylamine-8-sulphonic acid (*Scholkopf's acid*) may be prepared by reduction of the corresponding nitro-sulphonic acid. When heated with phosphoryl chloride, l-naphthylamine-8- sulphonic acid forms *naphthsultam*:

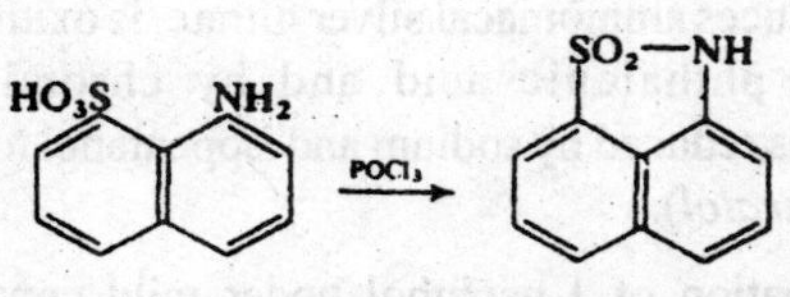

When heated with concentrated sulphuric acid, 2-naphthylamine gives four different sulphonic acid according to the temperature: 2-

naphthylamine-5-sulphonic acid (*Dahl's acid*), 2-naphthylamine-6 sulphonic acid (*Bronner's acid*), 2-naphthylamine-7-sulphonic acid (*F-acid*) and 2-naphthylamine-8 sulphonic acid (*Badische's acid*).

Naphthols. Both 1- and 2-naphthols are present in coal-tar. They are prepared industrially by fusing the corresponding naphthalene-sulphonic acid with sodium hydroxide:

$$C_{10}H_7SO_3Na + 2NaOH \xrightarrow{300\,°C} C_{10}H_7O^-Na^+ + Na_2SO_3 + H_2O$$
(70–80%)

Pure 1-naphthol may be prepared by heating 1-naphthylamine with dilute sulphuric acid a 290°C under pressure (this method is used industrially)

$$\text{1-naphthylamine } (NH_2) \xrightarrow[(H_2SO_4)]{H_2O} \text{1-naphthol } (OH) + NH_3 \quad (95\%)$$

An interesting preparation of 1-naphthol is by the reaction between o-bromofluorobenzene and lithium amalgam in the presence of furan (Wittig *et al.*, 1955). The reaction proceeds via a benzyne intermediate.

$$\text{o-}C_6H_4BrF \xrightarrow{Li} \text{benzyne} + \text{furan (O)} \longrightarrow \text{adduct (O bridge)} \xrightarrow{HCl} \text{1-naphthol (OH)}$$

1-Naphthol (α-naphthol), m.p. 94°C, is readily soluble in alkalis to form naphthoxides, *e.g.*, $C_{10}H_7O^-Na^+$. With ferric chloride it gives a violet-blue precipitate of α-binaphthol (4,4'-bis-l- naphthol):

$$HO\text{–}C_{10}H_6\text{–}C_{10}H_6\text{–}OH$$

1-Naphthol reduces ammoniacal silver nitrate, is oxidised by alkaline permanganate to phthalonic acid and by chromic acid to α-naphthaquinone. It is reduced by sodium and isopentanol to *ar*-tetrahydro-1-naphthol (*ar-*l-*tetralol*).

Direct sulphonation of 1-naphthol under mild conditions gives a mixture of 1-naphthol-2-sulphonic acid (*Schaeffer's acid*) and 1-naphthol-

4-sulphonic acid (*Nevile-Winther's acid*). More vigorous conditions result in the formation of 1-naphthol-2,4-disulphonic acid and finally 1-naphthol-2,4,7- trisulphonic acid.

1-Naphthol-8-sulphonic acid forms an *inner* ester, ***naphthsultone***, when heated:

HO_3S OH $\xrightarrow{-H_2O}$ SO_2—O

Treatment of 1-naphthol with nitrous acid gives mainly the 2-oxime of β-naphthaquinone (I), and a small amount of the 4-oxime of β-naphthaquinone (II) (c.f., nitrosophenol):

OH $\xrightarrow{HNO_2}$ O =NOH (I) + O NOH (II)

1-Naphthol couples with diazonium salts in the 4-position.

2-Naphthol (B-naphthol), m.p. 123°C, resembles l-naphthol in most if its properties, but is more reactive. With ferric chloride it gives a green precipitate of β-*binaphthol* (1,1'-bis-2-naphthol). It reduces ammoniacal silver nitrate, and is oxidised by alkaline permanganate to phthalonic acid. It is reduced by sodium and isopentanol to mainly *ac*-tetrahydro-2-naphthol (*ac-2-tetralol*). With nitrous acid it forms the 1-oxime β-naphthaquinone (nitroso-β-naphthol):

OH OH

OH $\xrightarrow{HNO_2}$ NOH =O (99%)

2-Naphthol couples with diazonium salts only in the adjacent 1-position; if this is occupied, no coupling takes place.

Some 2-naphthol ethers (methyl and ethyl), known as *nerolins*, are used in perfumery. When 2-naphthol is sulphonated, the original product is 2-naphthol-1-sulphonic acid, but this rearranges to 2-naphthol-8-sulphonic acid (*croceic acid*) at low temperature. Croceic acid, at 100°C, rearranges to 2-naphthol-6-sulphonic acid (Schaeffer's β-acid). When sulphonated with larger amounts, of concentrated sulphuric acid, 2-naphthol gives disulphonic acids. At low temperature the main product is 2-naphthol-6,8-disulphonic acid (G-*acid*); at higher temperatures, mainly 2-naphthol-3,6-disulphonic acid (R-*acid*). G- and R-acids are used in the manufacture of dyes.

Woodward (1940) prepared IO-methyldecal-2-one from ar-2-tetralol by means of the Reimer-Tiemann reaction as follows:

$CHCl_3$ / NaOH; $CHCl_2$; H_2, Pd/$BaSO_4$ / KOH/C_2H_5OH; CH_3; OH; H; CrO_3 / CH_3CO_2H; CH_3; 1 2 3 4 5 6 7 8 9 10; O; H

Fig. 6.15

This synthesis is based on the fact that the intermediate product in the Reimer-Tiemann reaction can be isolated when the p-position to the hydroxyl group is occupied by, *e.g.*, a methyl group.

Naphthalenecarboxylic acids. 1-Naphthoic acid (*naphlhalene-1-carhoxylic acid*), m.p. 161°C, may be prepared by the hydrolysis of the corresponding cyanide, or by the oxidation of 1-acetylnaphthalene with sodium hypochlorite. It may also be prepared from l-naphthylmagnesium bromide as follows:

$$1\text{-}C_{10}H_7MgBr + CO_2 \xrightarrow{\text{ether}} 1\text{-}C_{10}H_7CO_2H \ (68\text{-}70\%)$$

2-Naphthoic acid, m.p. 184°C. may be prepared by methods similar to those used for the 1-isomer.

1- and 2-Naphthoic acids both readily eliminate carbon dioxide to form naphthalene when heated with soda-lime. Both undergo most of the usual reactions of a carboxylic acid, but l-naphthoic aci.l shows a steric effect due to the hydrogen atom in position 8: *e.g.*, 2-chloro-l-naphthoic acid is not esterified by the Fischer-Speier method, whereas the isomeric l-chloro-2-naphthoic acid gives the ester under the same conditions.

Naphthalic acid (*naphthalene*-1, 8-*dicarboxylic acid*) may be prepared by oxidising acenaphthene with acid dichromate:

When heated at 180°C. it forms naphthalic anhydride:

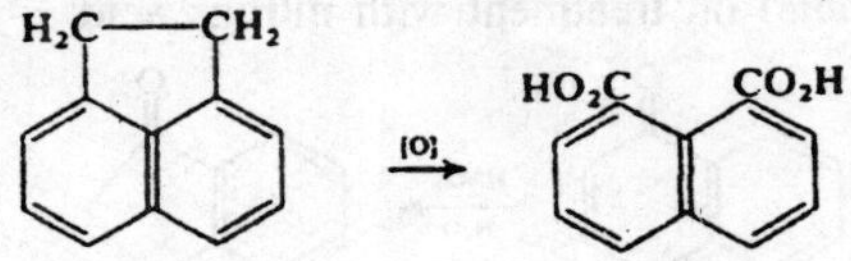

H_2C—CH_2 $\xrightarrow{[O]}$ HO_2C CO_2H

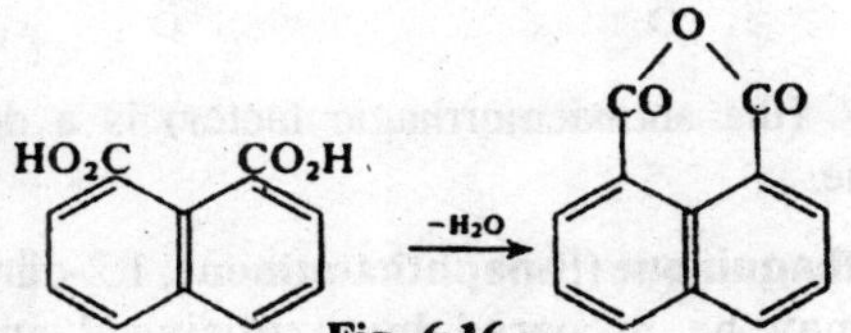

HO_2C CO_2H $\xrightarrow{-H_2O}$ CO O CO

Fig. 6.16

This is in keeping with the fact that all *peri*-(l,8)-substituents interact if possible (cf. sultams and sultones).

Naphthaquinones. Theoretically, six naphthaquinones re possible: 1, 2-, 1,4-, 1,5-, 1,7-, 2,3- and 2,6-. Only three are known, the 1,2-, 1,4- and 2,6-, but it appears that derivatives of 2,3- naphthaquinone have been prepared.

1,4-Naphthaquinone (**α-naphthaquinone,** 1,4-*dihydronaphthalene*-1,4-*dione*) may be prepared by the oxidation of 1,4-diamino-, dihydroxy- or aminohydroxynaphthalene, *e.g.*,

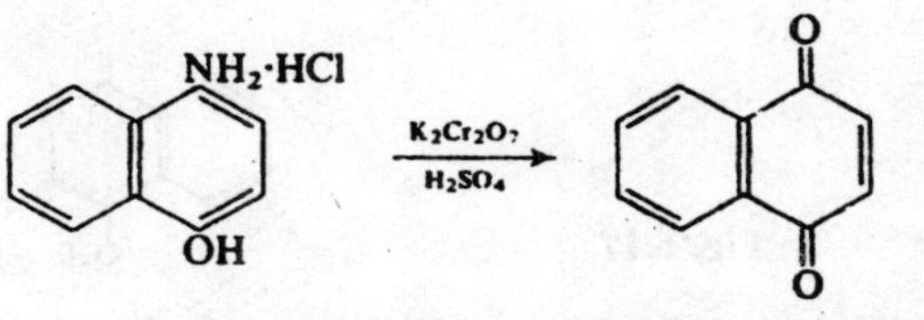

$NH_2 \cdot HCl$, OH $\xrightarrow[H_2SO_4]{K_2Cr_2O_7}$ O, O (78–81%)

It may also be prepared by the direct oxidation of naphthalene with dichromate and sulphuric acid, or chromium trioxide in glacial acetic acid (yield 40%). Catalytic air-oxidation of naphthalene is the commercial method.

1, 4-Naphthaquinone is a volatile yellow solid, m.p. 125°C, with a pronounced odour. It resembles p-benzoquinone in many ways chemically, but it is not reduced by sulphurous acid. It is reduced by metal and acid to 1,4-*dihydroxynaphthalene* (*naphthanlene*-1,4-diol) and oxidised by nitric acid to phthalic acid. It forms a monoxime: this is tautomeric, and in the solid state it exists as the oxime, and in solution this form predominates in equilibrium with the nitroso-phenol form (Having et al., 1955; Hadzi, 1956). It is converted into *indane*-1,3-*dione* (1,3-*dikelohydrindene*) on treatment with nitrous acid:

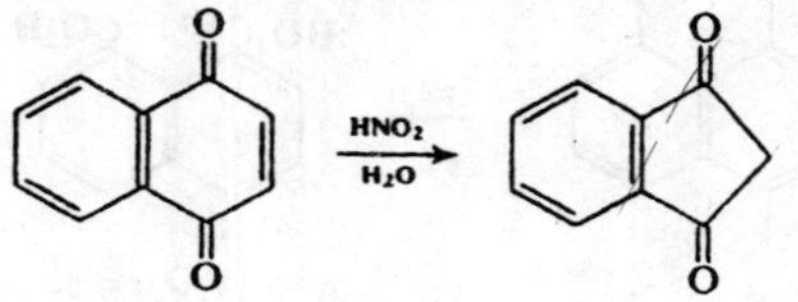

$\xrightarrow[H_2O]{HNO_2}$

Vitamin K (the antihaemorrhagic factor) is a derivative of 1,4-naphthaquinone.

1,2-Naphthaquinone (**β-naphthaquinone**, 1,2-dihydronaphthalene-1, 2-*dione*) may be prepared by oxidising 1-amino-2-naphthol hydrochloride with ferric chloride in hydrochloric acid (93-94%). It is a non-volatile, odourless, red solid which decomposes at 115-120°C. Fission of the quinone ring occurs with simultaneous hydroxylation when 1,2-naphthaquinone is treated with chlorine-water (or with hypochlorous acid); phenylglyceric-o-carboxylic lactone is the final product:

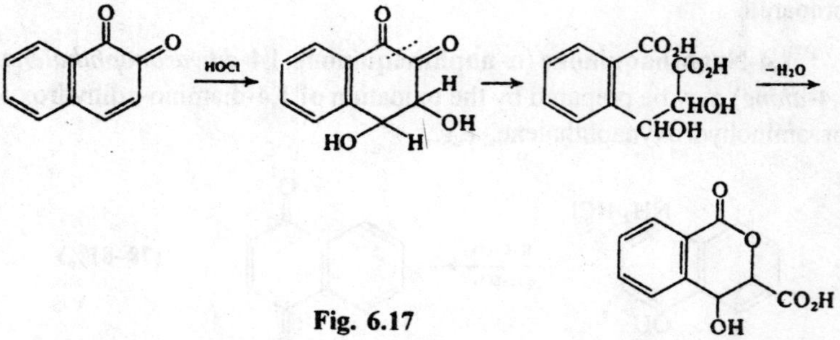

HOCl; H; OH; HO; H; CO_2H; CO_2H; CHOH; CHOH; $-H_2O$; CO_2H; OH

Fig. 6.17

2,6-Naphthaquinone (*amphi*-**naphthaquinone**, 2,6-dihydro-naphthalene-2,6-dione) may be prepared by oxidising 2,6-dihydro-xynaphthalene in benzene solution with 'active' lead dioxide (this may be prepared by decomposing lead tetra-acetate with water: Kuhn *et al.*, 1950):

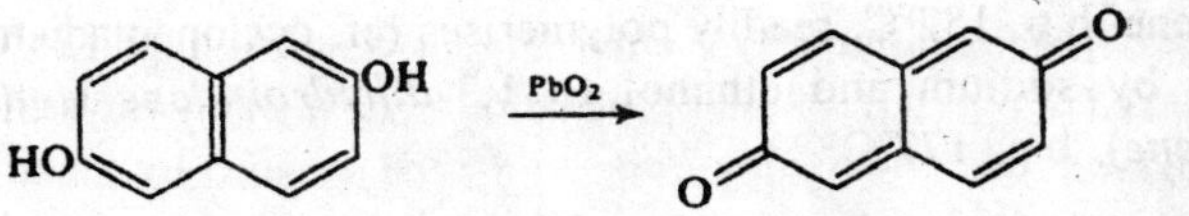

HO, OH —PbO_2→ O, O

It is an orange, non-volatile, odourless solid, m.p. 135°C.

Acenaphthene, $C_{12}H_{10}$, occurs in coal-tar; it may be prepared by passing 1-ethylnaphthalene through a red-hot tube:

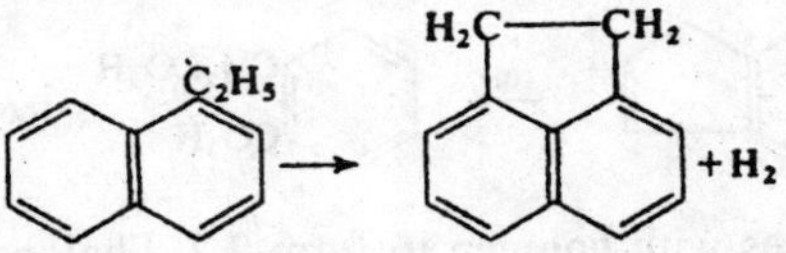

C_2H_5 → H_2C—CH_2 + H_2

It may also be prepared by a Friedel-Crafts reaction using naphthalene and oxalyl chloride and reducing the product, 1,2-acenaphthaquinone, by the Wolf-Kishner method:

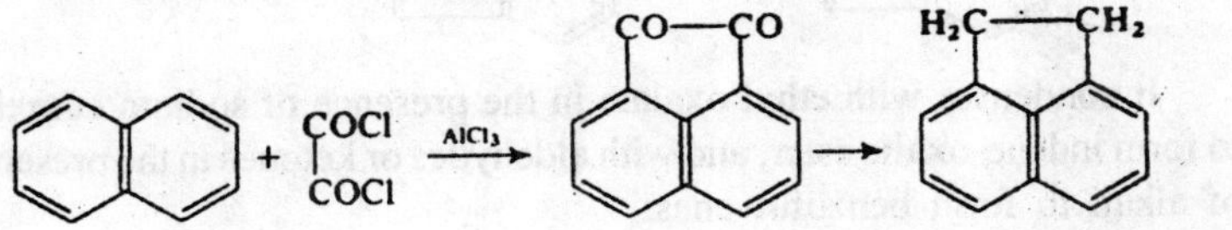

+ COCl—COCl —$AlCl_3$→ CO—CO → H_2C—CH_2

Acenaphthene, m.p. 96°C, on oxidation with dichromate and sulphuric acid, is converted into 1,2-acenaphthaquinone which, by further action of the oxidising agent, gives naphthalic acid.

Substitution in acenaphthene occurs most readily in the 5-position, *e.g.*, bromination, nitration and sulphonation give the corresponding 5-derivatives. When oxidised with lead dioxide or passed through a red-hot tube, acenaphthene forms *acenaphthylene*:

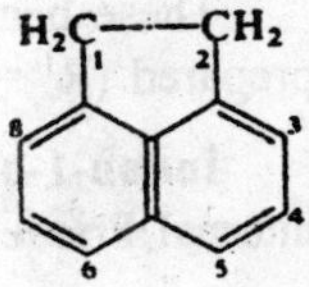

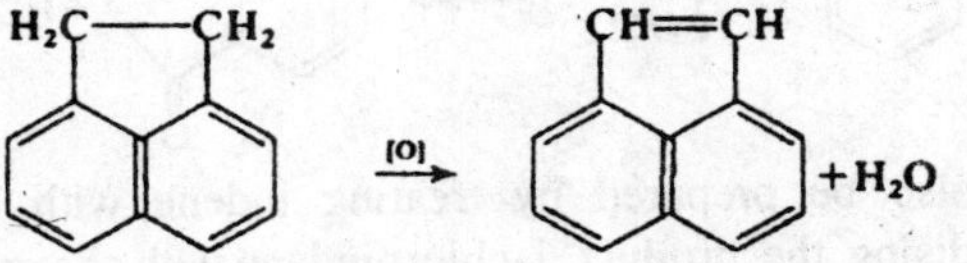

H_2C_1—CH_2 2, 3, 4, 5, 6, 8

H_2C—CH_2 —[O]→ CH=CH + H_2O

Indene (benzocyclopentadiene). C_6H_8, occurs in the coal-tar fraction boiling at 175-185° C. It may be obtained from this fraction by heating with sodium, separating the sodio-indene formed (solid) and steam-distilling it, whereupon indene distils over.

Indene, b.p. 182°C, readily polymerises (cf. cyclopentadiene). It is reduced by sodium and ethanol to 1,3-*dihydroindene* (*indane or hydrindene*), b.p. 177°C.

Oxidation of indene with acid dichromate gives homophthalic acid.

CH_2CO_2H CO_2H (66–77%)

Indene combines with halogen to form 2,3-dihalogenoindane. It is acidic (cf. cyclopentadiene), *e.g.*, it forms a sodio-derivative:

$+Na \longrightarrow$ $Na^+ + \frac{1}{2}H_2$

It condenses with ethyl oxalate in the presence of sodium ethoxide to form indene-oxalic ester, and with aldehydes or ketones in the presence of alkali to form benzofulvenes:

$+ R^1COR^2 \xrightarrow{NaOH}$ CR^1R^2 $+ H_2O$

These benzofulvenes are highly coloured; *benzofulvene* has been prepared ($R^1 = R^2 = H$).

Indan-1-one (α-*hydrindone*). m.p. 42° C. may be synthesised by an internal Friedel-Crafts reaction on β-phenylpropionyl chloride:

CH_2 CH_2 $COCl$ $\xrightarrow{AlCl_3}$ $+ HCl$

It may also be prepared by treating indene with hydrochloric acid and oxidising the product. 1-chloroindane with chromium trioxide in acetic acid:

$\xrightarrow{HCl}$ Cl $\xrightarrow{CrO_3}$ O (50–60%)

Indan-2-one (β-*hydrindone*), m.p. 61°C, may be prepared by heating xylylene-o-dicarboxylic acid

$$C_6H_4(CH_2CO_2H)_2 \longrightarrow \text{indan-2-one}{=}O + CO_2 + H_2O$$

Indane-1,3-dione may be prepared by the action of nitrous acid on 1,4-naphthaquinone

Indane-l,2,3-trione (*ninhydrin*) may be prepared as follows:

$$\text{indane-1,3-dione} \xrightarrow{SeO_2} \text{indane-1,2,3-trione}$$

It has also been synthesised from ethyl phthalate and dimethyl sulphoxide (Becker *et al.*, 1963); note the water of constitution.

$$C_6H_4(CO_2Et)_2 + Me_2SO \xrightarrow[\text{(ii) aq. HCl}]{\text{(i) MeONa}} \text{2-chloro-2-(methylthio)indane-1,3-dione} \xrightarrow[100°C]{H_2O} \text{2,2-dihydroxyindane-1,3-dione} \quad (80\%)$$

Ninhydrin is used as a reagent for amino-acids.

Fluorene (*diphenylenemethane*), $C_{13}H_{10}$, occurs in coal-tar (fraction 270-300°C) and can be separated from the other compounds by means of its sodio-derivative (cf. indene). It may be prepared:

(i) By passing diphenylmethane through a red-hot tube:

$$C_6H_5CH_2C_6H_5 \longrightarrow \text{fluorene} + H_2$$

(fluorene ring positions numbered 1–9)

(ii) By passing o-methylbiphenyl over palladium at 450° C:

$$\text{o-methylbiphenyl} \xrightarrow{Pd} \text{fluorene} + H_2$$

(iii) By heating the calcium salt of diphenic acid, reducing the product, fluorenone with zinc dust and ethanolic sodium hydroxide, and then reducing the fluorenol so produced by the Clemmense method:

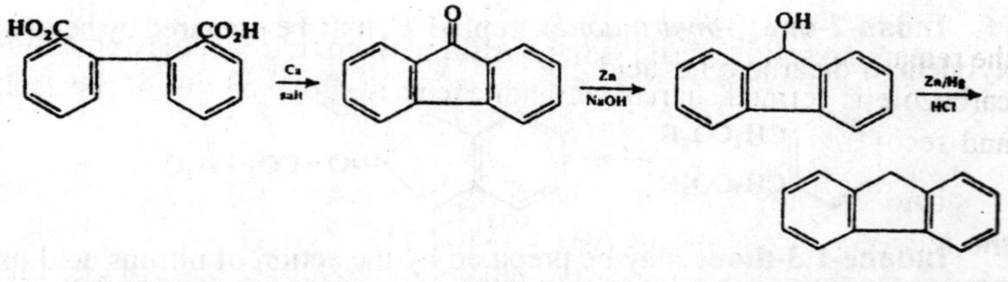

(iv) Fluorene may be prepared by means of the Pschorr synthesis, starting with o-aminodiphenylmethane.

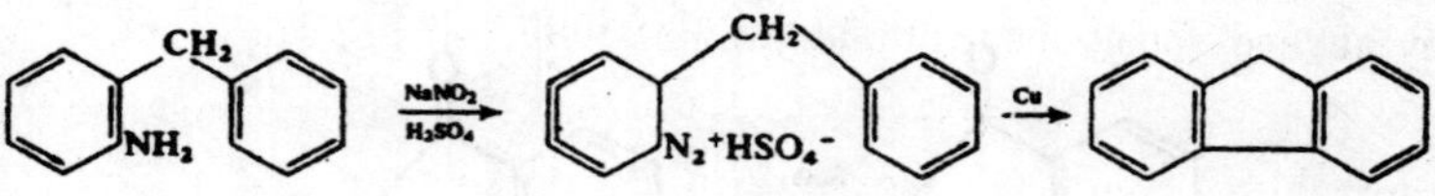

Alternatively, o-aminobenzophenone may be used as the starting material and the product, fluorenone, reduced to fluorene as in (iii).

Methods (iii) and (iv) are useful for preparing substituted fluorenes with the substituents in known positions.

Fluorene, m.p. 116°C, with a blue fluorescence, condenses with aldehydes and ketones in the presence of alkali. When fluorene is halogenated or nitrated, the first substituent enters the 2-position, and the second the 7-position. Fluorene gives 2-acetylfluorene (56-63%) when treated with acetic anhydride and aluminium chloride in carbon disulphide. Fluorene is oxidised by chromium trioxide in glacial acetic acid to *fluorenone* (m.p. 84°C). This is oxidised by permanganate to phthalic acid. When fused with potassium hydroxide, fluorenone is converted into biphenyl-2-carboxylic acid:

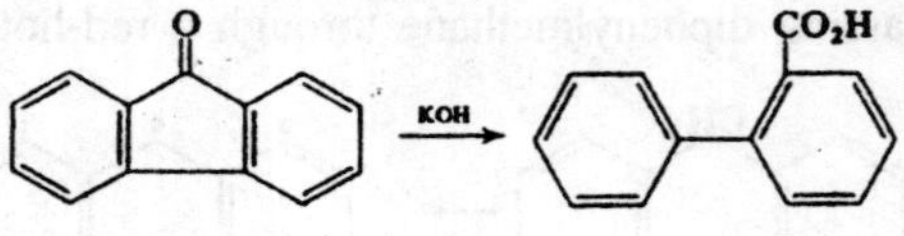

ANTHRACENE

Anthracene, $C_{14}H_{10}$, is obtained from the anthracene oil fraction of coal-tar by cooling the latter and pressing the solid (which crystallises out) free from liquid. The crude anthracene contains phenanthrene and carbazole. The anthracene cake is powdered and washed with 'solvent naphtha' which dissolves the phenanthrene, and the remaining solid is then washed with pyridine which dissolves the carbazole. The anthracene is purified by sublimation. Alternatively, after removal of phenanthrene, the remaining solid is fused with potassium hydroxide, whereby potassio-carbazole is formed; unreacted anthracene is sublimed out of the melt and recovered.

Until recently, there has been very little use for carbazole, and thus the recovery of anthracene was expensive. Since anthracene is mainly used as the starting point of anthra-quinone, a cheaper method of isolating anthracene from coal-tar is to remove the phenanthrene first, and then catalytically oxidise the remaining mixture of anthracene and carbazole by air and vanadium pentoxide at 300-500°C.

Under these conditions, anthracene is oxidised to anthraquinone and carbazole is completely oxidised (to carbon dioxide, etc.).

Synthesis of Anthracene

(i) By a Friedel-Crafts reaction using benzyl chloride; 9,10-dihydroanthracene, which is first formed, readily eliminates two hydrogen atoms under the conditions of the experiment to form anthracene:

CH_2Cl + $ClCH_2$ $\xrightarrow{AlCl_3}$ [ ] $\xrightarrow{-2H}$

Anthracene is also formed by the Friedel-Crafts condensation between benzene and methylene dibromide, or between benzene and acetylene tetrabromide:

2 + $2CH_2Br_2$ $\xrightarrow{AlCl_3}$ [ ] $\xrightarrow{-2H}$

+ Br—CH—Br / Br—CH—Br + $\xrightarrow{AlCl_3}$ + 4HBr

Fig. 6.18

(ii) When phthalic anhydride in benzene solution is treated with aluminium chloride o-benzoylbenzoic acid is formed. This, on heating with concentrated sulphuric acid or P.P.A. at 100°C, forms anthraquinone which, on distillation with zinc dust, gives anthracene:

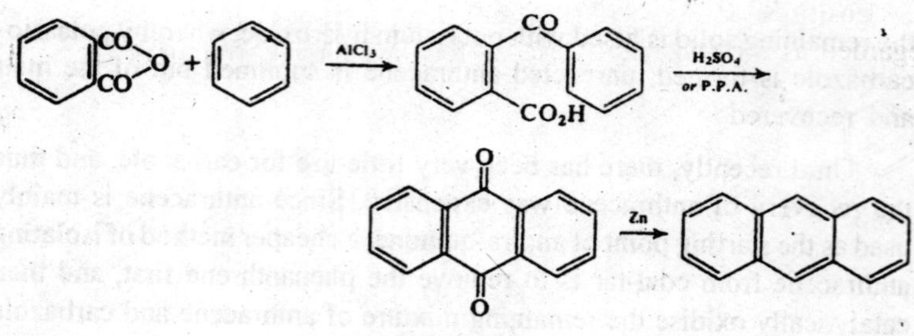

Fig. 6.19

9-Alkylanthracenes may be prepared from o-benzoylbenzoic acid as follows (via anthrone):

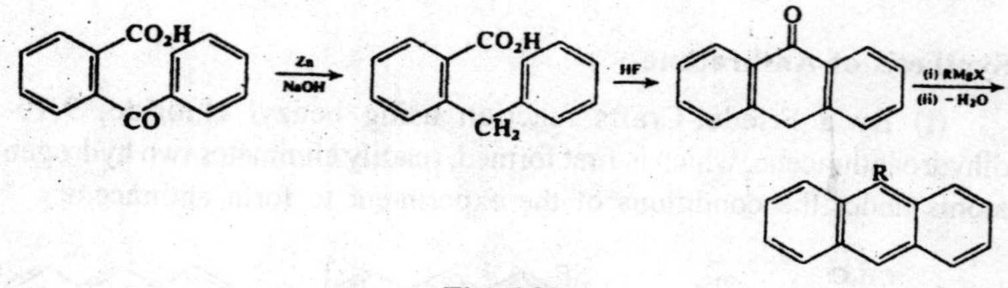

Fig. 6.20

(iii) Anthracene may be prepared by means of the **Elbs reaction** (1884). This is the reaction whereby a polynuclear hydrocarbon is formed by the pyrolysis of a diaryl ketone containing methyl or methylene group *ortho* to the carbonyl group, *e.g.*, o-methylbenzophenone gives anthracene:

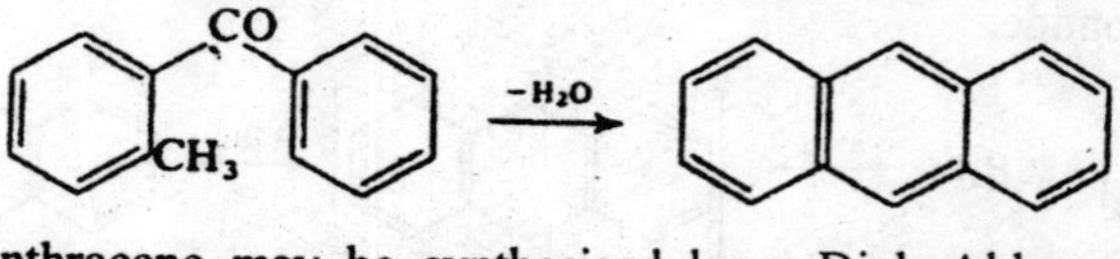

(iv) Anthracene may be synthesised by a Diels-Alder reaction involving 1,4-naphthaquinone and butadiene, followed by oxidation of the product with chromium trioxide in glacial acetic acid and then distillation with zinc dust.

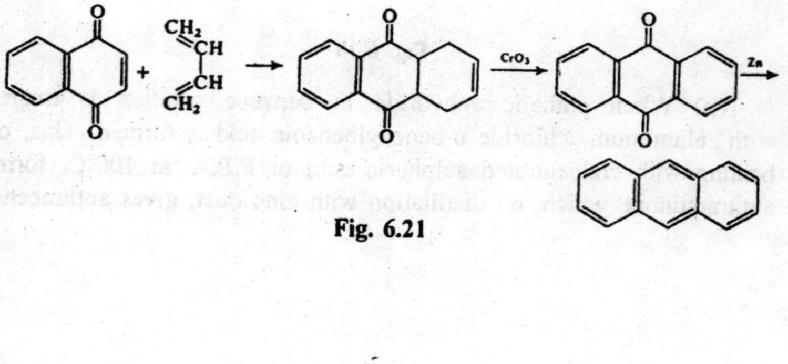

Fig. 6.21

Positions of the double bonds in anthracene. Anthracene is best regarded as a resonance hybrid of four resonating structures:

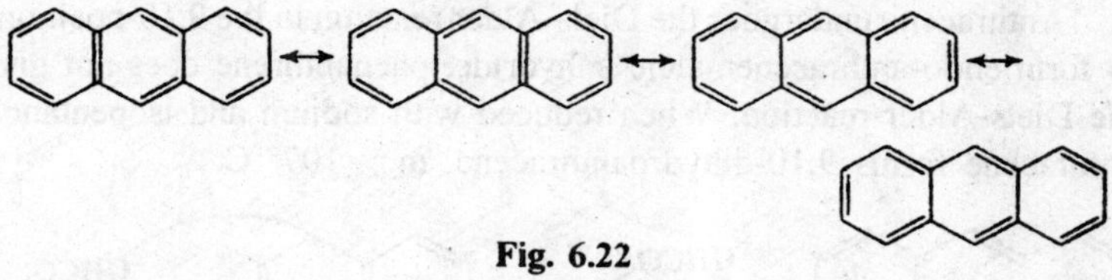

Fig. 6.22

The resonance energy of anthracene is 351.5 kJ mol^{-1}

The double bond character of the various bonds is shown in (I) and bond lengths (in Å) are shown in (II).

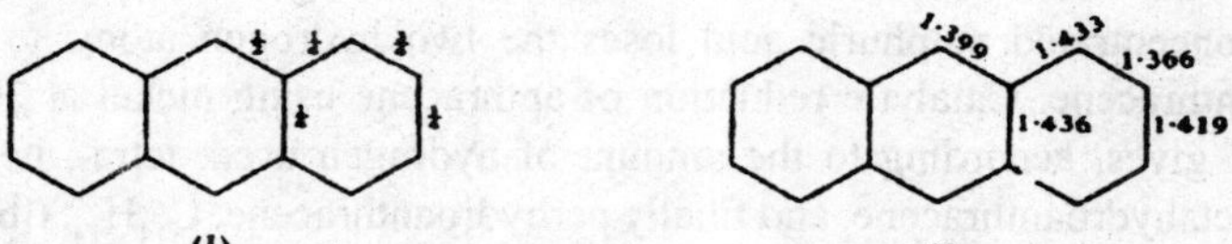

(I) (II)

Fig. 6.23

Isomerism of Anthracene Derivatives

There are three possible mono-substitution products: I-or α-. 2-or β- and 9- or γ- (or *meso*-). There are 15 possible disubstitution products if both substituents are identical; if the substituents are different, the number of isomers is larger.

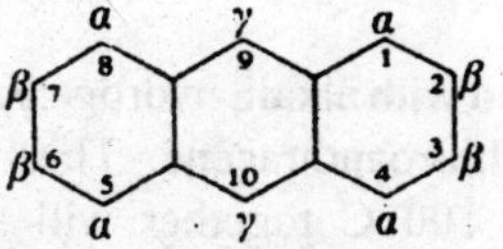

The infra-red absorption regions of the anthracenes are 1640-1 620 and near 1550 cm^{-1} (see also phenanthrenes).

Properties of Anthracene

Anthracene, m.p. 216° C, with a blue fluorescence, is very reactive in the 9, 10-positions. Electro-philic reactions to form σ-complexes at the 1- or 2-position leave a 'naphthalene fragment' with loss of 351.5-255.2 = 96.3 $kJmol^{-1}$, whereas the σ-complex at position 9 leaves two benzene rings intact, and so the loss is 351.5-301.2 = 50.3 kJ mol^{-1}. Hence, for polynuclear aromatic hydrocarbons (with three or more rings), the σ-complex that contains the largest number of *isolated* benzene rings will be the most stable.

Anthracene undergoes the Diels-Alder reaction in the 9,10-positions to form endo-anthracenemaleic anhydride, phenanthrene does not give the Diels-Alder reaction. When reduced with sodium and isopentanol, anthracene forms 9,10-dihydroanthracene. m.p. 107° C.

CHCO, O, CHCO → CHCO, O, CHCO

Fig. 6.24

which is not fluorescent and which, on heating or on treatment with concentrated sulphuric acid loses the two hydrogen atoms to reform anthracene. Catalytic reduction of anthracene using nickel at 200-250° C gives, according to the amount of hydrogen used, tetra-, hexa- and octahydroanthracene, and finally perhydroanthracene, $C_{14}H_{24}$, (the prefix per is often used to denote complete hydrogenation of a ring system).

When chlorine is passed into a cold solution of anthracene in carbon disulphide anthracene dichloride is formed:

$+Cl_2$ → Cl, Cl $\xrightarrow[\text{(heat)}]{-HCl}$ Cl

If this is heated or treated with alkali, hydrogen chloride is eliminated with the formation of 9-chloroanthracene. This is also obtained by chlorinating anthracene at 100 C together will some 9,10-dichloro-anthracene. Bromine reacts similarly, *e.g.*, bromination of anthracene in boiling carbon tetrachloride solution gives 9.10-dibromoanthracene (yield: 83-88%).

The best method of preparing 9-chloro- or 9-bromoanthracene is to heat anthracene with the corresponding cupric halide in carbon tetrachloride solution (yield: 98-99%: Nonhebel. 1963) Sulphuryl chloride, at room temperature, converts anthracene into 9,10-dichloroanthracene.

Anthracene can be chloromethylated in the 9- and 9,10-positions. and can be formylated in the 9-position:

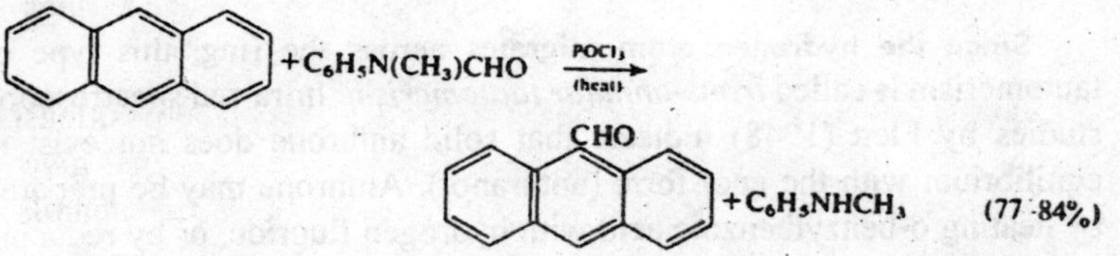

$$\text{Anthracene} + C_6H_5N(CH_3)CHO \xrightarrow[\text{(heat)}]{POCl_3} \text{9-anthraldehyde (CHO)} + C_6H_5NHCH_3 \quad (77\text{-}84\%)$$

Attempts to nitrate anthracene with aqueous nitric acid lead to the formation of anthraquinone by oxidation. If, however, the nitration is carried out in acetic anhydride at 15-20°C, 9-nitroanthracene (m.p. 145°C) and 9,10-dinitroanthracene (m.p. 294°C) can be isolated. Anthracene is readily sulphonated to a mixture of the 1- and 2-sulphonic acids, some disulphonic acids also always being obtained; the 2-position is favoured at high temperature. If the sulphonation of anthracene is carried out in glacial acetic acid, a mixture of about equal amounts of 1- and 2-anthracenesulphonic acid is obtained. The l-sulphonic acid shows no tendency to rearrange to the 2-acid (cf. naphthalenesulphonic acids), With excess of concentrated sulphuric acid anthracene gives disulphonic acids, the 1,8- at low temperatures, and the 2,7- at high temperatures.

According to Gore *et al.* (1966, 1967), the Friedel-Crafts acetylation (CH_3COCl and $AlCl_3$) of anthracene in benzene, nitrobenzene, etc. gives a complex mixture of mono- and di-acetyl derivatives. Furthermore, to obtain pure ketones. It was necessary to use elaborate chromatographic methods. In this way. It was shown that the main product in nitrobenzene as solvent is the 1 -acetyl derivative, and in ethylene dichloride the 9-acetyl derivative.

Hydroxyanthracenes. 1- and 2-*Hydroxyanthracenes* are known as **anthrols**. Each may be obtained from the corresponding anthracenesulphonic acid by alkaline fusion. 1-Anthrol is a yellow solid, m.p. 152°C; 2-anthrol is a brownish solid which decomposes at 200°C.

10-*Hydroxyanthracene,* also known as **anthranol**, is an unstable yellow solid, m.p. 120°C, and when quickly heated, forms **anthrone** (10-*keto*-9,l0-*dihydroanthracene*—the keto-group is numbered last and consequently its enol isomer is known as 10- and not 9-hydroxyanthracene). Anthrone is the stable form and is a colourless solid, m.p. 154°C.

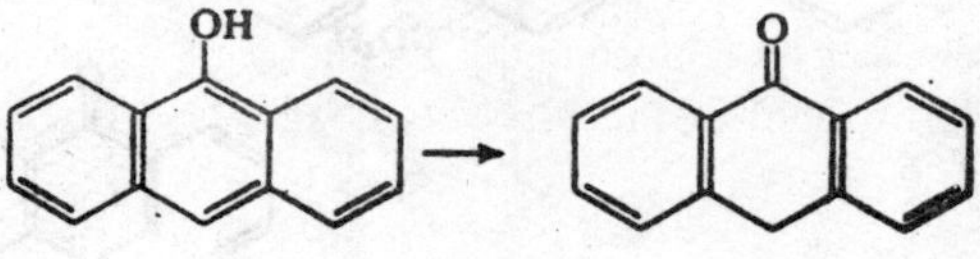

$$\text{Anthranol (OH)} \longrightarrow \text{Anthrone (O)}$$

Since the hydrogen atom migrates across the ring, this type of tautomerism is called *trans-annular tautomerism*. Infra-red spectroscopy studies by Flett (1948) indicate that solid anthrone does not exist in equilibrium with the enol form (anthranol). Anthrone may be prepared by heating o-benzylbenzoic acid with hydrogen fluoride, or by reducing anthraquinone with tin and hydrochloric acid in glacial acetic acid. Anthrone dissolves in warm dilute alkalis, and these solutions, on acidification, precipitate the enol form.

Reduction of anthraquinone with zinc dust and aqueous sodium hydroxide gives *anthracene-9,10-diol* or *anthraquinol* (*anthrahydroquinone*). This is a brown solid, m.p. 180°C; its alkaline solutions (deep red) oxidise in air to give anthraquinone. When anthraquinol in alkaline solution is immediately acidified, it partly tautomerises to *oxanthranol* m.p. 167° C (*trans*-annular tautomerism):

OH OH ⇌ (acid / alkali) O OH

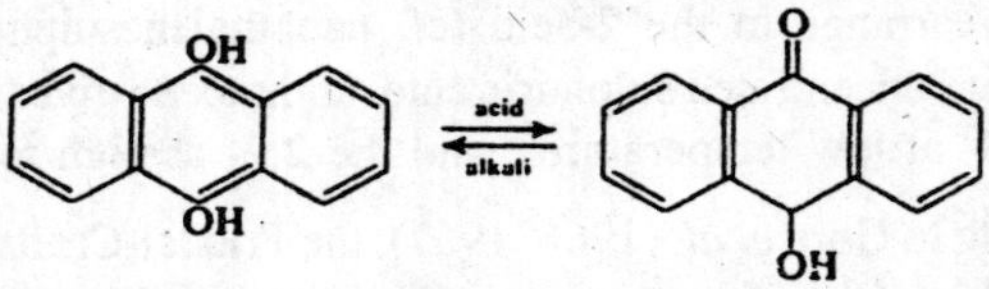

Anthracenecarboxylic acids. 1- and 2-Anthroic acids may be prepared by the hydrolysis of the corresponding cyanides (prepared by fusion of the sodium sulphonate with sodium cyanide). 9-Anthroic acid may be prepared by hydrolysing its acid chloride which may be prepared by heating anthracene with oxalyl chloride at 160 C. 9-Anthroic acid exhibits steric hindrance.

Anthraquinone (9,10-*dihydroanthracene*-9,10-*dione*). There are nine possible isomeric quinones of anthracene, but only three are known: 1,2-, 1,4- and 9,10-. The most important one is the 9,10-compound, and this is referred to simply as anthraquinone. Anthraquinone may be prepared by oxidising anthracene with sodium dichromate and sulphuric acid (yield: 90%). Commercially, it is prepared by the catalytic oxidation of anthracene (containing carbazole;), but the cheapest method is its synthetic preparation:

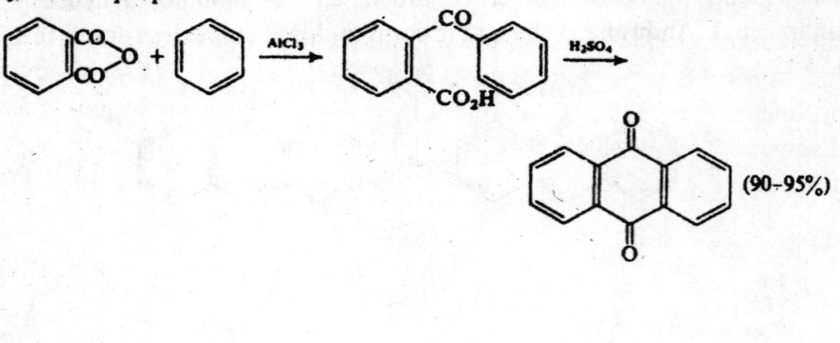

CO, CO, O + $\xrightarrow{AlCl_3}$ CO, CO_2H $\xrightarrow{H_2SO_4}$

O, O (90–95%)

By using chlorobenzene or toluene instead of benzene, chloro- or methylanthraquinone is obtained these are used in the manufacture of dyes.

Anthraquinone is a pale yellow solid which sublimes in needles that melt at 286°C. When distilled with zinc dust, anthraquinone forms anthracene. Anthraquinone is very stable and shows very little resemblance to p-benzoquinone, *e.g.*, it has no smell, is not very volatile, and is not reduced by sulphurous acid. When anthraquinone is reduced, the nature of the reduction product depends on the reducing agent used, *e.g.*, with tin and hydrochloric acid in acetic acid anthrone is formed; using zinc instead of tin, the main product is *bianthryl* (I), with zinc dust and aqueous sodium hydroxide, anthraquinol; and with zinc dust and aqueous ammonium hydroxide, 9,10-dihydroanthrol (II):

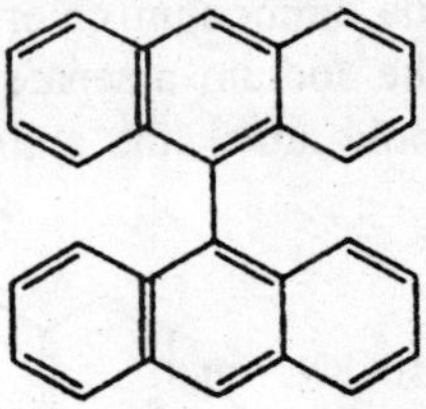

(I)

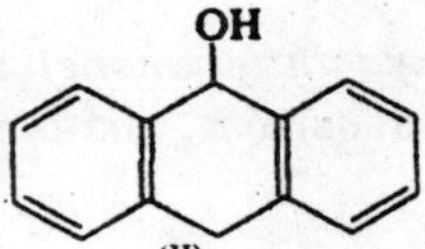

OH

(II)

Fig. 6.25

Nitration of anthraquinone with mixed acid gives 1-nitro-anthraquinone; further nitration gives mainly 1,5- and 1,8-dinitro-anthraquinones, and small amounts of the 1,6- and 1,7-dinitro-compounds. The nitro-group in the 1-position is very reactive, *e.g.*, it is replaced by an amino group when 1-nitroanthraquinone is heated with ammonia.

Anthraquinone is very difficult to sulphonate with concentrated sulphuric acid, but it is readily sulphonated with oleum at 160° C. The first product is the 2-sulphonic acid and a small amount of the I-isomer; prolonged heating gives a mixture of 2;6- and 2,7-anthraquinone-disulphonic acids in about equal amounts. If mercuric sulphate is used as a catalyst, the sulphonation takes an entirely different course. The first product now is the l-sulphonic acid, and then a mixture of the 1,5- and 1,8-disulphonic acids. A possible explanation for this change in orientation is that the reaction now proceeds via a 1-mercurated derivative. The sulphonic acid group in the 1- or 2-position is easily displaced; *e.g.*, when l- or 2-anthraquinonesulphonic acid is treated with chlorine, the corresponding chloroanthraquinone is obtained.

Anthraquinone does not undergo the Friedel-Crafts reaction, and is halogenated with very great difficulty; in fact, monohalogenoanthraquinones cannot be obtained directly.

2-Aminoanthraquinone, m.p. 304 °C, is very important as an intermediate in the preparation of indanthrene dyes. It is prepared industrially by heating the sodium salt of anthraquinone-2-sulphonic acid with a solution of ammonia, ammonium chloride and sodium arsenate under pressure at 200°C. The sodium arsenate oxidises the liberated sulphite which otherwise would attack the amine produced:

$$\text{C}_{14}\text{H}_7\text{O}_2\text{SO}_3\text{Na} + \text{NH}_3 + [\text{O}] \longrightarrow \text{C}_{14}\text{H}_7\text{O}_2\text{NH}_2 + \text{NaHSO}_4$$

Alizarin *(1,2-dihydroxyanthraquinone)* is the most important dihydroxy-derivative of anthraquinone, and is used as a mordant dye.

PHENANTHRENE

Phenanthrene, $C_{14}H_{10}$, is isomeric with anthracene; it is an example of an angular polynuclear hydrocarbon. It occurs in the anthracene oil fraction of coal-tar, and is separated from anthracene by means of solution in solvent naphtha. Phenanthrene is structurally related to certain alkaloids, *e.g.*, morphine, and to the steroids, *e.g.*, cholesterol.

Synthesis of Phenanthrene

Many methods are available, but only two important ones will be described here.

(i) **Pschorr synthesis** (1896). This is carried out by heating o-nitrobenzaldehyde with sodium β-phenylacetate in the presence of acetic anhydride (Perkin's reaction), reducing and diazotising the product, α-phenyl-o-nitrocinnamic acid and treating the diazonium salt with sulphuric acid and copper powder. Phenanthrene-9-carboxylic acid is produced and this, on strong heating forms phenanthrene:

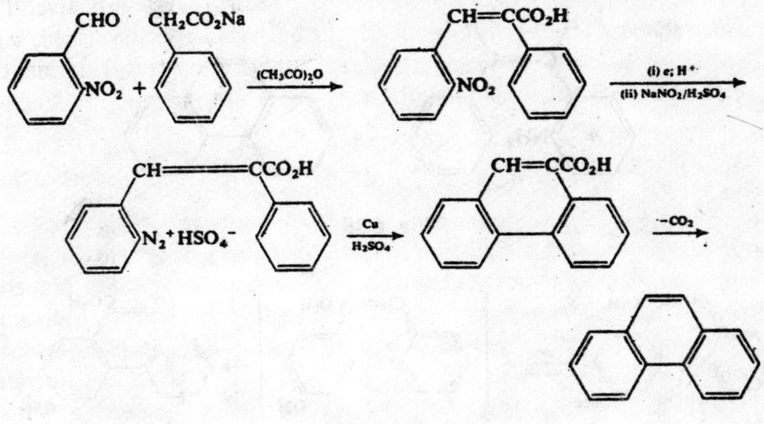

Fig. 6.26

The Pschorr 'synthesis is really an example of intramolecular phenylation, and the decomposition can be effected in several ways. The method described above is the catalysed (copper) decomposition and this is believed to proceed via a free-radical mechanism. Decomposition can also be effected in the absence of copper, in which case the mechanism is believed to be of the nucleophilic unimolecular type. On the other hand, decomposition can be carried out in alkaline solution and in this case it is a free-radical mechanism that operates.

The Pschorr synthesis offers a means of preparing substituted phenanthrenes with the substituents in known positions. In those cases, however, where isomerism in the cyclises product is possible, it is usual to obtain both isomers, *e.g.*,

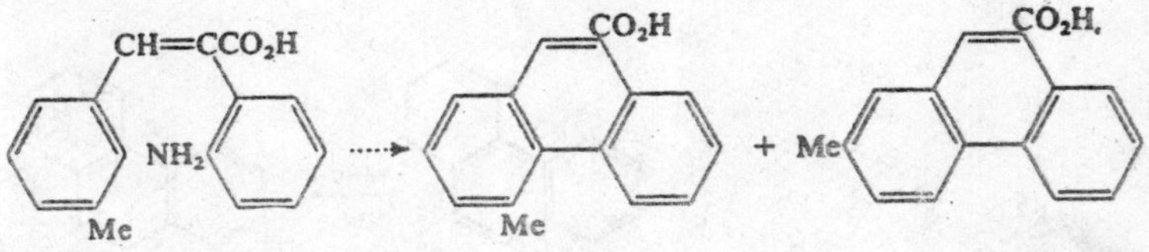

Fig. 6.27

A further point to note is that since ring closure is effected between two rings, these rings must be in the cis position, *e.g.*, cis-o-aminostilbene gives phenanthrene, but the trans isomer does not Kessar *et al.*, (1969) have prepared phenanthrenes from cinnamic acids via what they believe to be benzyne intermediates, *e.g.*,

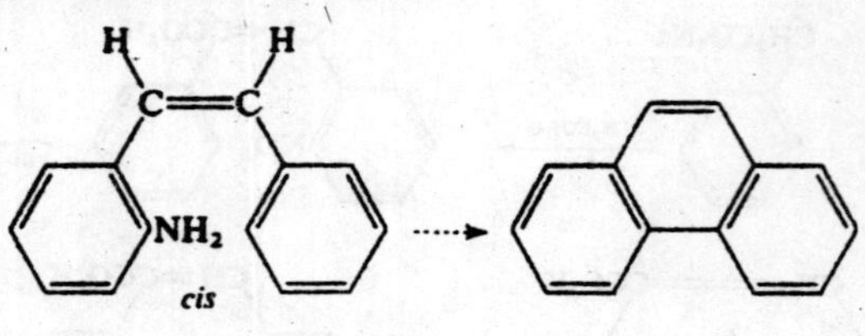

Fig. 6.28

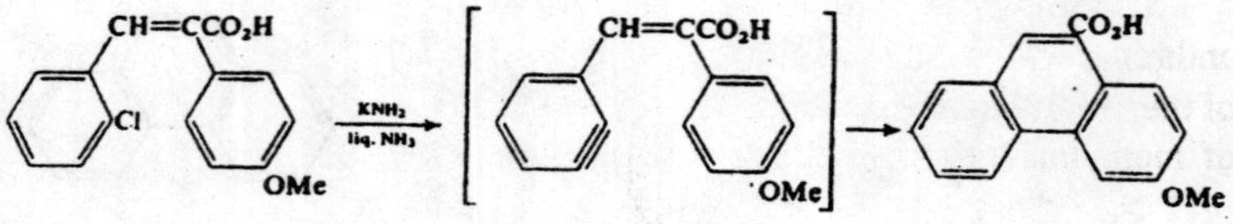

Fig. 6.29

(ii) **Haworth synthesis** (cf. naphthalene):

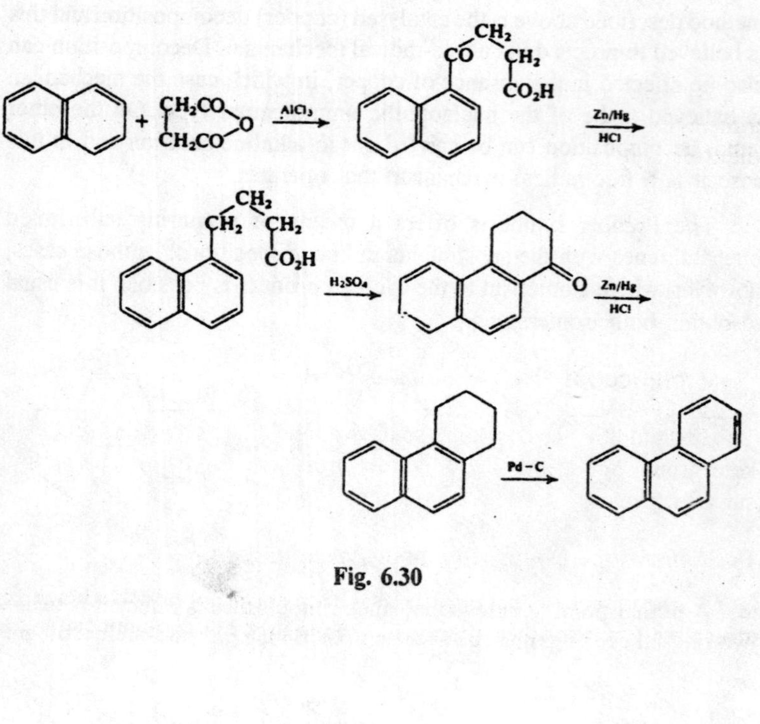

Fig. 6.30

Succinoylation also occurs in the 2-position; the two keto-acids may be readily separated. The 2-compound also gives phenanthrene when treated as above; no anthracene is formed since ring closure occurs only in the 1-position of naphthalene, and not in 3.

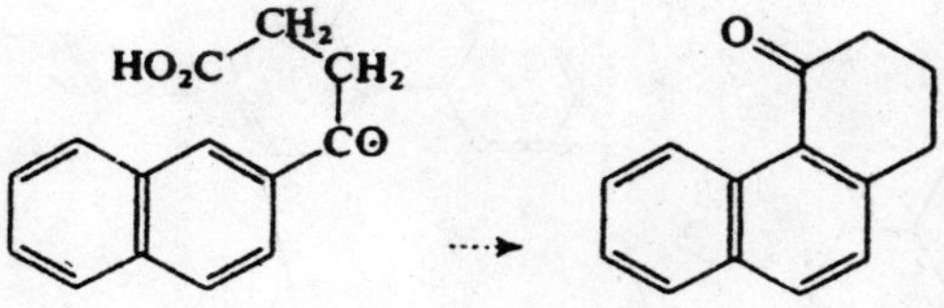

The Haworth synthesis is very useful for preparing alkylphenanthrenes with the alkyl group in known positions; *e.g.*, after ring closure of the 1-derivative, 1-methylphenanthrene may by obtained by the action of methylmagnesium iodide on the ketone, etc.:

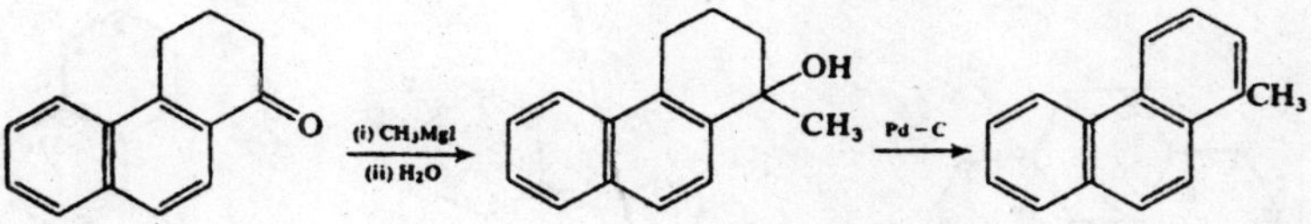

Fig. 6.31

By using methylsuccinic anhydride instead of succinic anhydride, a methyl group can be introduced into the 2- or 3-position, *e.g.*, from the 2-derivative:

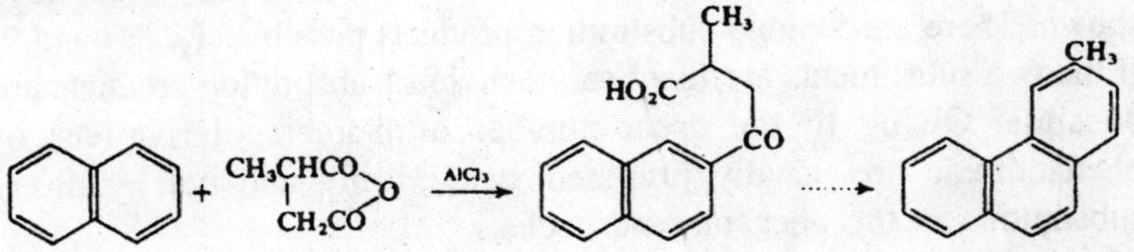

Fig. 6.32

It should be noted that the condensation occurs at the less hindered keto group, *i.e.*, at the one which is further removed from the methyl substituent.

Positions of the Double Bonds in Phenanthrene

Phenanthrene is best represented as a resonance hybrid of 5 resonating structures (its resonance energy is 387.0 $kJmol^{-1}$):

(I)

Fig. 6.33

The double bond character of the various bonds is shown in (1).

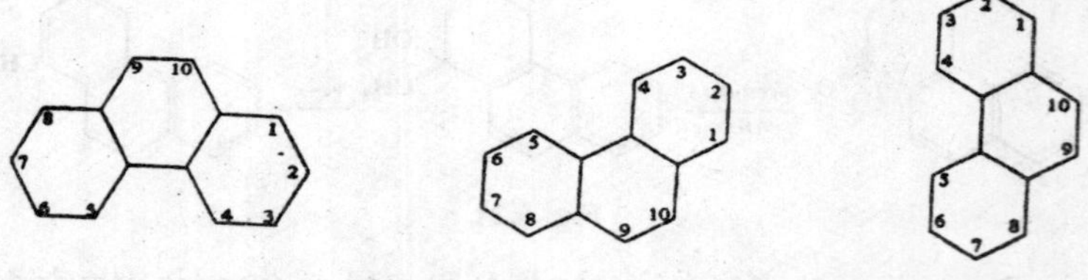

Fig. 6.34

The formula of phenanthrene may be written in the three ways shown. There are 5 mono-substitution products possible: 1,2,3,4 and 9. If the two substituents are identical, then 25 disubstitution products are possible. Owing to the great number of isomers, derivatives of phenanthrene are usually prepared synthetically and not by direct substitution in the phenanthrene nucleus.

Properties of Phenanthrene

Phenanthrene, m.p. 99°C, in benzene shows a blue fluorescence. It is very reactive in the 9,10-positions (cf. anthracene), *e.g.*, phenanthrene is readily catalytically reduced (using copper oxide-chromic oxide) to 9,10-dihydrophenanthrene, and it adds on bromine to form 9,10-phenanthrene dibromide.

These addition reactions occur almost as easily as with a pure ethylenic bond. Ozonolysis (followed by oxidation) or oxidation with peracetic acid gives diphenic acid.

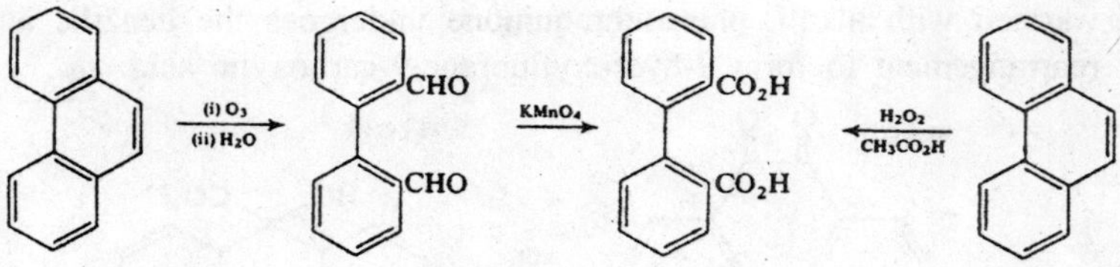

Fig. 6.35

9-Bromophenanthrene (90-94%) is produced by adding bromine to a refluxing solution of phenanthrene in carbon tetrachloride. This is the starting-point of 9-substituted phenanthrenes, *e.g.*, when heated with cuprous cyanide at 260°C, 9-bromophenanthrene forms the corresponding cyano-compound; this may be hydrolysed to phenanthrene-9-carboxylic acid. Phenanthrene undergoes the Friedel-Crafts reaction mainly in the 3-, and to a small extent, in the 2-position. It is chloromethylated in the 9-position. When nitrated, phenanthrene gives a mixture of three mononitro-derivatives, the 3-isomer predominating. Sulphonation of phenanthrene gives a mixture of 1-, 2-, 3- and 9-phenanthrenesulphonic acids, and the ratio of these isomers depends on the temperature.

The infra-red absorption regions of phenanthrenes are 1600 cm^{-1} (two bands) and near 1500 cm^{-1}. Hence, these compounds can be differentiated from anthracenes.

Hydroxyphenanthrenes. Five *phenanthrols* are known: 1-,2-,3-,4- and 9-.

Phenanthraquinone *(9,10-dihydrophenunthrene-9.10-dione)* may be prepared by oxidising phenanthrene with sodium dichromate or chromium trioxide in glacial acetic acid (yield is excellent).

Phenanthraquinone is an orange solid, m.p. 208° C. which is odourless and not steam-volatile. It combines with one or two molecules of hydroxylamine to form phenanthraquinone monoxime and dioxime, respectively, and it is reduced by sulphurous acid to **phenanthrene-9,10-diol** (*phenanthraquinol*). In all of these reactions phenanthraquinone resembles o-benzoquinone. Distillation with zinc dust gives phenanthrene (cf. anthraquinone).

When nitrated, phenanthraquinone gives a mixture of 2- and 4-nitrophenanthraquinones: more vigorous nitration produces a mixture of the 2,7- and 4,5-dinitro-compounds. Oxidation with sodium dichromate and sulphuric acid converts phenanthraquinone into diphenic acid. When warmed with alkali, phenanthraquinone undergoes the benzilic acid rearrangement to form 9-hydroxyfluorene-9-carboxylic acid:

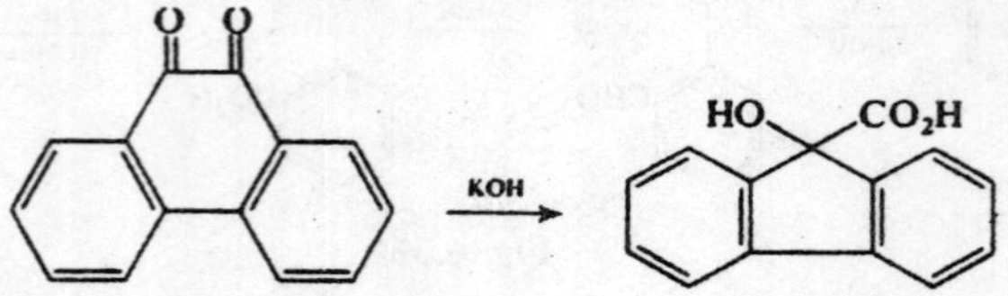

KOH

This, on heating in air, eliminates carbon dioxide and the product, fluorenol, is oxidised to fluorenone:

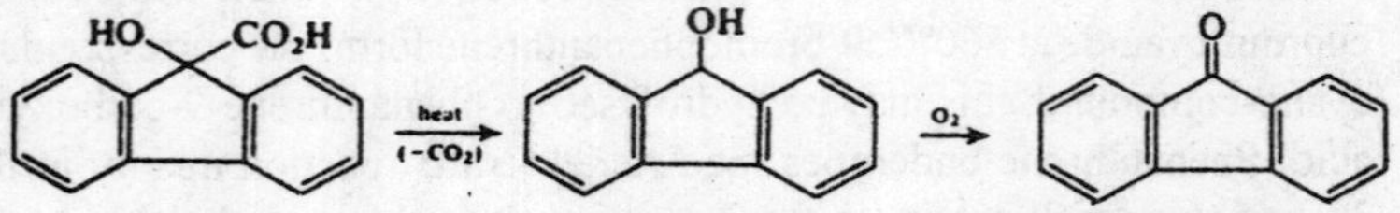

heat (−CO_2) O_2

Phenanthraquinone readily reacts with o-phenylenediamines to form phenazines which, since they are insoluble in many organic solvents, are very useful for characterising o-phenylene diamines.

1, 2-. 1,4- and 3,4-Phenanthraquinones have also been prepared.

Molecular Overcrowding

A molecule becomes overcrowded when at least one pair of *non-bonded* atoms are closer to each other than the sum of their vander Waals radii. Two extremes are now possible. If the atoms were to remain very close to each other and the rest of the molecule remain unaltered, then there would be large steric repulsion. On the other hand if the molecule were deformed so that the overcrowded atoms were separated to the sum of their vander Waals radii, there would be a large strain energy. In practice, a compromise is reached, the steric repulsion and the strain being relieved by bending of substituent bonds and buckling of the molecule as a whole. One example of this is the phenanthrene derivative (I). The phenanthrene nucleus is planar (but see below), and substituents lie in this plane.

If, however, fairly large groups are in positions 4 and 5, there will not be enough room to accommodate both groups in the plane of the nucleus. Such a molecule will not be planar, and hence is asymmetric and consequently should be resolvable. Newman *et al.*, (1940. 1947) have partially resolved (1), and also (in 1955) prepared the optically active forms of (II). It is also interesting to

(I) (II)

Fig. 6.36

note that according to Trotter (1963), the structure of phenanthrene itself is not absolutely plannar in the *crystalline* state and this is probably due to molecular overcrowding.

EXERCISES

1. Arrange the following in order of increasing stability and give your reason: $PhCH_2^+$, Ph_3C^+, Me^+, Ph_2CH^+.
2. Complete the equations :

   (i) $PhCHOHCOPh \xrightarrow[\text{aq. NaOH; heat}]{NaBrO_3}$? (rearr.)

   (ii) $PhNH_2{\cdot}HCl + PH_3COH \xrightarrow{AcOH}$? (not an N-deriv.)
3. Explain the effect of the introduction of one p-NO_2 or p-MeO group in Ph_6C_2 on its dissociation into free radicals.
4. Using PhMgBr, what substrates would lead to Ph_3COH?
5. Arrange Ph_3CH, $(p\text{-}NO_2C_6H_4)_3CH$, and $(3.5\text{-}Me_2\text{-}4\text{-}NO_2\text{—}C_6H_2)_3CH$ in order of increasing acid strength and give your reasons.
6. What are the expected products when the following α-diketons undergo the benzilic acid rearrangement?

   (i) $HO_2CCH_2COCOCH_2CO_2H$;

   (ii) $MeCOCOCO_2Et$;
7. How could you convert naphthalene into 2-nitronaphthalene?
8. Suggest a mechanism for the reaction:

   $2Ph_2CCl_2 + 2NaI \longrightarrow Ph_2C = CPh_2 + 2ICl + 2Cl^-$
9. Synthesise phenanthrene, starting from naphthalene and butyricacid.

10. (i) The u.v. spectrum of a solution of anthracene in H_2SO_4 is similar to that of Ph_2CH^+;

    (ii) The NMR spectrum of anthracene in TFA indicates the presence of a CH_2 group. Offer an explanation.

11. The order of reactivity to electrophilic substitution is fluorene > biphenyl > benzene. Offer an explanation.

12. (i) PhMe $\longrightarrow$ $Ph_2CHCHPh_2$;

    (vi) PhH $\longrightarrow$ Ph_3CCO_2H;

    (ii) phenanthrene $\longrightarrow$ phenanthrene-9-carboxylic acid.

13. Synthesise 2, 9-dimethylanthracene starting with compounds containing one benzene ring.

14. The rate of reaction of benhydryl chloride (Ph_2CHCl) in aqueous EtOH with KF to give benzhydryl fluoride is retarded by the addition of NaCl. Explain.

15. Devise an experiment to show whether the rearrangement of napthanlene-1-sulphponic acid to the 2-sulphonic acid is intramolecular or intermolecular.